Convergence Structures 1984

Mathematical Research · Mathematische Forschung

Wissenschaftliche Beiträge
herausgegeben von der
Akademie der Wissenschaften der DDR
Institut für Mathematik

Band 24

Convergence Structures 1984

Convergence Structures 1984

Proceedings of the Conference on Convergence held in Bechyně (Czechoslovakia) September 24–28, 1984

edited by

Josef Novák
Werner Gähler
Horst Herrlich
Jan Mikusiński

Akademie-Verlag · Berlin 1985

Die Titel dieser Schriftenreihe werden vom Originalmanuskript der Autoren reproduziert.

ISSN 0138 - 3019

Erschienen im Akademie-Verlag Berlin DDR-1086 Berlin Leipziger Str. 3-4

Lizenznummer: 202 · 100/489/85
Printed in the German Democratic Republic
Gesamtherstellung: VEB Kongreß- und Werbedruck, 9273 Oberlungwitz
Lektor: Dr. Reinhard Höppner
Umschlag: Karl Salzbrunn
LSV 1035
Bestellnummer: 763 462 5 (2182/24)
03200

INTRODUCTION

The Conference on Convergence was held in Bechyně, Czechoslovakia, from September 24 to September 28, 1984. The Conference was organized by the Institute of Mathematics of the Czechoslovak Academy of Sciences in cooperation with the Institute of Mathematics of the Slovak Academy of Sciences and the Union of Czechoslovak Mathematicians and Physicists. The Organizing Committee consisted of J. Novák (chairman), V. Koutník (vice-chairman), R. Frič, P. Kratochvíl and A. Rázek (treasurer).

The Conference was attended by 33 specialists from Belgium, Canada, Czechoslovakia, Federal Republic of Germany, German Democratic Republic, Italy, Netherlands, Poland, Switzerland, U.S.A. and Yugoslavia.

The program of the Conference consisted of 28 lectures and short communications. The topics discussed covered the whole field of convergence structures including sequential and filter convergence spaces and groups, completions of convergence structures, categorical and topological aspects of convergence, related questions in set theory as well as applications to analysis, functional analysis and numerical analysis.

These Proceedings contain 28 papers submitted by Conference participants. All papers were read by referees. The Editors would like to thank them for their indispensable services. On recommendation of referees some papers were revised by the authors. However, editorial work on most papers was limited to correcting obvious errors and making minor changes in form. Special thanks of the Editors are due to V. Koutník for his assistance, as executive editor, in editing these Proceedings.

Finally, the Editors wish to express appreciation to Akademie-Verlag, in particular to Dr. R. Höppner, for their assistance in prepering these Proceedings.

J. Novák

CONTENTS

LIST OF PARTICIPANTS

Jiří Adámek, Department of Mathematics, Faculty of Electrical Engineering, Czech Technical University, Suchbátárova 2, 16627 Praha 6, Czechoslovakia

Piotr Antosik, Institute of Mathematics, Polish Academy of Sciences, Wieczorka 8, 40-013 Katowice, Poland

Ronald Beattie, Department of Mathematics, Mount Allison University, Sackville, New Brunswick EOA 3CO, Canada

Herschel L. Bentley, Department of Mathematics, University of Toledo, 2801 W. Bancroft Street, Toledo, OH 43606, U.S.A.

Józef Burzyk, Institute of Mathematics, Polish Academy of Sciences, Wieczorka 8, 40-013 Katowice, Poland

Heinz-Peter Butzmann, Fakultät für Mathematik und Informatik, Universität Mannheim, 6800 Mannheim, Federal Republic of Germany

Eva Copláková, Institute of Mathematics, Slovak Academy of Sciences, Jesenná 5, 04154 Košice, Czechoslovakia

Mario Dolcher, Istituto di Matematica Applicata, Università di Trieste, 34100 Trieste, Italy

Roman Frič, Institute of Mathematics, Slovak Academy of Sciences, Karpatská 5, 04001 Košice, Czechoslovakia

Alfred Frölicher, Section de Mathématiques, Université de Genève, Case postale 240, 1211 Genève 24, Switzerland

Siegfried Gähler, Institut für Mathematik, Akademie der Wissenschaften der DDR, Mohrenstraße 39, 1086 Berlin, German Democratic Republic

Werner Gähler, Institut für Mathematik, Akademie der Wissenschaften der DDR, Mohrenstraße 39, 1086 Berlin, German Democratic Republic

Gabriele H. Greco, Dipartimento di Matematica, Facoltà di Scienze, Università degli Studi, 38050 Povo (Trento), Italy

Matúš Harminc, Institute of Mathematics, Slovak Academy of Sciences, Karpatská 5, 04001 Košice, Czechoslovakia

Klaas P. Hart, Department of Mathematics, Šafárik University, Jesenná 5, 04154 Košice, Czechoslovakia

Horst Herrlich, Fachbereich Mathematik und Informatik, Universität Bremen, Postfach 330440, 2800 Bremen, Federal Republic of Germany

Miroslav Hušek, Institute of Mathematics, Charles University, Sokolovská 83, 18600 Praha 8, Czechoslovakia

Romano Isler, Istituto di Matematica Applicata, Università di Trieste, 34100 Trieste, Italy

Czesław Kliś, Institute of Mathematics, Polish Academy of Sciences, Wieczorka 8, 40-013 Katowice, Poland

Gert Kneis, Institut für Mathematik, Akademie der Wissenschaften der DDR, Mohrenstraße 39, 1086 Berlin, German Democratic Republic

Václav Koutník, Institute of Mathematics, Czechoslovak Academy of Sciences, Žitná 25, 11567 Praha 1, Czechoslovakia

Petr Kratochvíl, Institute of Mathematics, Czechoslovak Academy of Sciences, Žitná 25, 11567 Praha 1, Czechoslovakia

Robert Lowen, Rÿks Universitair Centrum Antwerpen, Dienst Analyse, Groenenborgerlaan 171, 2020 Antwerpen, Belgium

Eva Lowen-Colebunders, Department of Mathematics, Vrÿje Universiteit, Pleinlaan 2, 1050 Brüssel, Belgium

Jan Mikusiński, Institute of Mathematics, Polish Academy of Sciences, Wieczorka 8, 40-013 Katowice, Poland

Ladislav Mišík, Institute of Mathematics, Slovak Academy of Sciences, Obrancov mieru 49, 81473 Bratislava, Czechoslovakia

Ladislav Mišík, Jr., Institute of Mathematics, Slovak Academy of Sciences, Obrancov mieru 49, 81473 Bratislava, Czechoslovakia

Josef Novák, Institute of Mathematics, Czechoslovak Academy of Sciences, Žitná 25, 11567 Praha 1, Czechoslovakia

Endre Pap, Institute of Mathematics, Ilije Djuričica 4, 21000 Novi Sad, Yugoslavia

Steven Purisch, Institute of Mathematics, Czechoslovak Academy of Sciences, Žitná 25, 11567 Praha 1, Czechoslovakia

Caroll V. Riecke, Department of Mathematics and Astronomy, Franklin and Marshall College, P.O. Box 3003, Lancaster, PA 17604, U.S.A.

Peter Vojtáš, Institute of Mathematics, Slovak Academy of Sciences, Karpatská 5, 04001 Košice, Czechoslovakia

Fabio Zanolin, Istituto di Matematica, Università, P. le Europa 1, 34100 Trieste, Italy

LIST OF TALKS

J. Adámek: Topological axioms and co-axioms

P. Antosik: On matrix method in analysis

R. Beattie and H.-P. Butzmann: Convergence spaces determined by their sequences (Presented by R. Beattie)

R. Beattie and H.-P. Butzmann: Some relations between sequential and filter convergence (Presented by H.-P. Butzmann)

H.L. Bentley: Ascoli's theorem for merotropic spaces

J. Burzyk: The example of FLUSHD convergence without Y

R. Frič and P. Vojtáš: Diagonal-type conditions in sequential convergence (Presented by R. Frič)

R. Frič and P. Vojtáš: The space ψ_ω in sequential convergence (Presented by P. Vojtás)

R. Frič and F. Zanolin: Coarse convergence groups (Presented by F. Zanolin)

A. Frölicher: Mackey-convergence in general differentiation theory

S. Gähler: Discrete convergence

W. Gähler: Axioms of convergence structures and functor power series

G.H. Greco: Variational convergences: Γ—limits and K-limits

M. Harminc: Convergence l-groups

K.P. Hart: Converging sequences and examples

H. Herrlich: Sequential structures: categorical aspects

M. Hušek: Uniformly sequential spaces

R. Isler: Some recent results on pseudoradial spaces

G. Kneis: On subspaces of complete and of compact convergence spaces

V. Koutník: Closure and topological sequential convergence

P. Kratochvíl: Multisequences in sequential spaces

E. Lowen-Colebunders: On the function class of Cauchy continuous maps

J. Mikusiński: Approximation by convolution

J. Mikusiński: On the logarithm of the differential operator

J. Novák: Convergence of twofold seuqences

E. Pap: Functional analysis with K-convergence

S. Purisch: The orderability of compact scattered spaces

C. Riecke: Some remarks on d-convergence

TOPOLOGICAL AXIOMS AND COMPLETIONS

Jiří Adámek [1])

Abstract. Topological axioms of Herrlich are used to present the Mac Neille completion and the cartesian closed topological hull of concrete categories. The presentation of mono-topological categories by topological epi-axioms is introduced.

I. Topological Axioms

I.1 We study concrete categories over a given base-category $\mathcal{X}$. An important type of concrete categories, introduced by M. Katětov [K], is defined using set-valued functors $F : \mathcal{X} \to$ Set. The functor-structured category S(F) has

objects : pairs(X, α) where $X \in \mathcal{X}^{obj}$ and $\alpha \subseteq FX$;
morphisms from (X, α) to (Y, β) : $\mathcal{X}$-maps $f : X \to Y$ with $Ff(\alpha) \subseteq \beta$.

The forgetful functor from S(F) to $\mathcal{X}$ sends (X, α) to X and f to f.

These categories are extremely well-behaved, in fact, they are topological, i.e.

(a) initially (= finally) complete which means that every source of structured maps $f_i : X \to |K_i|$ with $K_i \in \mathcal{K}^{obj}$ has an initial $\mathcal{K}$-object;
(b) small-fibred which means that for each $X \in \mathcal{X}^{obj}$ the collection $\{K \in \mathcal{K}^{obj} \mid |K| = X\}$ is small.

Example. Let H_{N^*} denote the functor from Set to Set assigning to each set X the set X^{N^*} of all functions from $N^* = N \cup \{\infty\}$ to X (which we denote by $\varphi \to x$ for $\varphi : N \to X$ and $x \in X$). To each map $f : X \to Y$, the functor H_{N^*} assigns the map $X^{N^*} \to Y^{N^*}$ given by

$$(\varphi \to x) \mapsto (f . \varphi \to f(x)) .$$

The category $S(H_{N^*})$ can be considered as the category of "sequential spaces without axioms" : an object on a set X is given by specifying a set of sequences converging to points in X , and a morphism is a map preserving this convergence.

I.2 H. Herrlich introduced in $[H_4]$ the concept of a topological axiom in S(F) : this is an identity-carried morphism $id : (X, \alpha) \to (X, \alpha')$ ($\alpha \subseteq \alpha' \subseteq FX$), which we denote as follows

[1]) Faculty of Electrical Engineering ČVUT
Suchbátarova 2, 166 27 Praha 6 , Czechoslovakia

$$\alpha \vdash \alpha'[X] .$$

An object (Z, γ) of $S(F)$ is said to <u>satisfy</u> this axiom if each morphism $f : (X, \alpha) \to (Z, \gamma)$ is a morphism $f : (X, \alpha') \to (Z, \gamma)$.

<u>Example</u>. Consider the following axiom in $S(H_N*)$:

(c) $$\emptyset \vdash \{\dot{0} \to 0\} \quad [\{0\}] ,$$

where dot denotes the constant sequence. An object (Z, γ) satisfies (c) iff for each map $f : \{0\} \to Z$ we have morphism

$$f : (\{0\}, \{\dot{0} \to 0\}) \to (Z, \gamma) ,$$

i.e., $\dot{f(0)} \to f(0)$ in γ. In other words, if for each point $z(= f(0))$ we have $\dot{z} \to z$ in γ. Another axiom :

(SUB) $$\{id_N \to \infty\} \quad \vdash \quad \{\rho \to \infty \mid \rho : N \to N \text{ strictly monotonic}\} \ [N^*].$$

Morphisms from $(N^*, \{id_N \to \infty\})$ into an object (Z, γ) correspond naturally to elements $\varphi \to x$ of γ. Hence, an object (Z, γ) satisfies (SUB) iff for each $\varphi \to x$ in γ we have $\varphi . \rho \to x$ for all strictly monotonic $\rho : N \to N$. In other words, if each subsequence of a convergent sequence converges.

We see that the category Seq of sequential spaces is the full subcategory of $S(H_N*)$ given by the satisfaction of (c) and (SUB).

I.3 We are interested in presenting a given concrete category $\mathcal{K}$ by axioms in some functor-structured category. By a <u>presentation</u> of $\mathcal{K}$ is understood a <u>realization</u> (= full and concrete embedding) $E : \mathcal{K} \to S(F)$ together with a collection of topological axioms satisfied by precisely those $S(F)$-objects which lie in $E(\mathcal{K})$. For example the axioms (c) and (SUB) form a presentation of Seq in $S(H_N*)$.

A realization $E : \mathcal{K} \to S(F)$ of a topological category $\mathcal{K}$ defines a presentation of $\mathcal{K}$ iff it is initiality preserving, see $[H_4]$. The following result, originally proved in $[H_4]$, will be a consequence of Theorem II.4 below.

<u>Theorem</u> (Herrlich). A concrete category has a presentation iff it is topological.

I.4 An important type of concrete categories over Set are those in which each constant map is a morphism. These are called <u>c-categories</u>. For example, $S(F)$ is never a c-category, but imposing the following topological axiom

(c) $$\emptyset \vdash F\{0\} \qquad [\{0\}] ,$$

we obtain a full subcategory $S_c(F)$ which is a c-category.

As an example, let H_2 denote the hom-functor of a two-point set (i.e., $H_2X = X^2 = X \times X$ and $H_2f = f \times f$). Then $S(H_2)$ is the category of graphs and compatible maps, whereas $S_c(H_2)$, presented by the following axiom

(c) $\emptyset \vdash \{(0,0)\} \quad [\{0\}]$,

is the category of reflexive graphs.

I.5 We use below a smallness condition on a concrete category slightly stronger than fibre-smallness. Two structured maps $X \xrightarrow{a} |A|$ and $X \xrightarrow{a} |A'|$ in a concrete category are said to be <u>equivalent</u> if for each object K and each map $f : |K| \to X$ the following holds : $a . f : K \to A$ is a morphism iff $a' . f : K \to A'$ is a morphism. A concrete category is <u>strongly fibre-small</u> if each object X has a small set of representatives w.r.t. the equivalence of structured maps. (For example, any fibre-small concrete category over Set which has epi--mono factorization of morphisms is strongly fibre-small, see $[A_1]$.) The following has been proved in $[AHS_2]$:

<u>Proposition</u>. A concrete category has a realization in some functor-structured category iff it is strongly fibre-small.

II. Topological $\mathcal{E}$-axioms

II.1 Topological categories are actually rather scarce, and they have a more frequent, close relative, which is "nearly" as good : the mono-topological categories. These are the fibre-small categories in which each mono-source has an initial lift. For example, Hausdorff spaces and posets form mono-topological categories which are not topological. By a natural generalization of the concept of axiom we are able to present all mono-topological categories.

In this section, the base category $\mathcal{X}$ is assumed to have a factorization structure consisting of a class $\mathcal{E}$ of epis and a class $\mathcal{M}$ of sources such that

(a) each source factors as an $\mathcal{E}$-epi followed by an $\mathcal{M}$-source;

(b) diagonal fill-in : given a map $e : X \to Y$ in $\mathcal{E}$, a source $\{m_i : A \to B_i\}_{i \in I}$ in $\mathcal{M}$ and commuting squares

$$\begin{array}{ccc} X & \xrightarrow{e} & Y \\ {\scriptstyle f}\downarrow & & \downarrow{\scriptstyle g_i} \\ A & \xrightarrow[m_i]{} & B_i \end{array}$$

there exists $d : Y \to A$ with $f = d . e$;

(c) both $\mathcal{E}$ and $\mathcal{M}$ are closed under composition with isomorphisms;

(d) $\mathcal{X}$ is $\mathcal{E}$-cowell-powered.

The basic example is $\mathcal{X}$ = Set, $\mathcal{E}$ = epis and $\mathcal{M}$ = points-separating sources.

A concrete category $\mathcal{K}$ is said to be <u>$\mathcal{M}$-topological</u> if each structured $\mathcal{M}$-source has an initial object, and if $\mathcal{K}$ is fibre-small, see [H_1].

II.2 <u>Definition.</u> A <u>topological $\mathcal{E}$-axiom</u> in S(F) is a morphism $e : A \longrightarrow A'$ with $e \in \mathcal{E}$. An object K is said to satisfy the $\mathcal{E}$-axiom if K is injective with respect to e, i.e., any morphism from A to K factors through e.

Presentation is defined analogously as for topological axioms.

II.3 <u>Examples</u>. (i) The category of Hausdorff sequential spaces (i.e., those in which limits are unique) is presented in $S(H_{N^*})$ by the above axioms (c) and (SUB) together with the following epi-axiom (Haus). Put

$$N^{\#} = N \cup \{\infty_1, \infty_2\}$$

and let $e : N^{\#} \longrightarrow N^*$ be the map forgetting the index of ∞, then

(Haus) $\quad (N^{\#}, \{id \longrightarrow \infty_1,\ id \longrightarrow \infty_2\}) \xrightarrow{e} (N^*, \{id \rightarrow \infty\})$.

An object (X, α) of $S(H_{N^*})$ satisfies (Haus) iff each morphism from $(N^{\#}, \{id \longrightarrow \infty_1,\ id \longrightarrow \infty_2\})$ into (X, α) [i.e., each pair $\varphi \rightarrow x_1$ and $\varphi \longrightarrow x_2$ in α] factors through e [i.e., fulfils $x_1 = x_2$].

(ii) The category of posets is presented in $S_c(H_2)$, the category of reflexive graphs, by a topological axiom (T) (the transitivity) and a topological epi-axiom (A) (the antisymmetry) as follows :

(T)

(A)

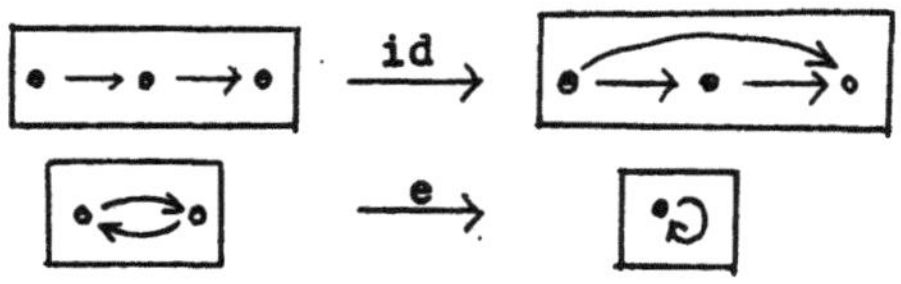

II.4 <u>Theorem</u>. A concrete category is $\mathcal{M}$-topological iff it can be presented by topological $\mathcal{E}$-axioms in a functor-structured category.

Proof. Let $\mathcal{K}$ be an $\mathcal{M}$-topological category. As proved in [AHS_2], $\mathcal{K}$ has a realization $R : \mathcal{K} \rightarrow S(F)$ such that $R(\mathcal{K})$ is $\mathcal{E}$-reflective in S(F). The collection of all reflections $L \rightarrow L^*$ for $L \in S(F)$ defines a presentation of $\mathcal{K}$ by topological $\mathcal{E}$-axioms : any object RK with $K \in \mathcal{K}$ satisfies each of these $\mathcal{E}$-axioms and, conversely, whenever L satisfies the axiom $L \rightarrow L^*$, then L is isomorphic with $L \in R(\mathcal{K})$. It follows that $L \in R(\mathcal{K})$, since given $L^* = RK^*$, $K^* \in \mathcal{K}$, we have an initial object K in $\mathcal{K}$ for the isomorphism $|L| \longrightarrow |L^*| = |K^*|$, and then $L = RK$.

Conversely, let $\mathcal{K}$ be presented in S(F) by topological $\mathcal{E}$-axioms. We assume, for convenience, that $\mathcal{K}$ is the full subcategory of S(F) consisting of all objects which satisfy each of the given $\mathcal{E}$-axioms. Then $\mathcal{K}$ is $\mathcal{M}$-topological because it is closed under initial lifts of $\mathcal{M}$-sources in S(F) : let $m_i : L \to L_i$ $(i \in I)$ be an initial $\mathcal{M}$-source with $L_i \in \mathcal{K}$ for all $i \in I$. To show that $L \in \mathcal{K}$, it is sufficient to verify that any topological $\mathcal{E}$-axiom $e : A \to A'$ satisfied by each L_i is also satisfied by L . Given a morphism $f : A \to L$, for each $i \in I$ we have f_i such that the following square

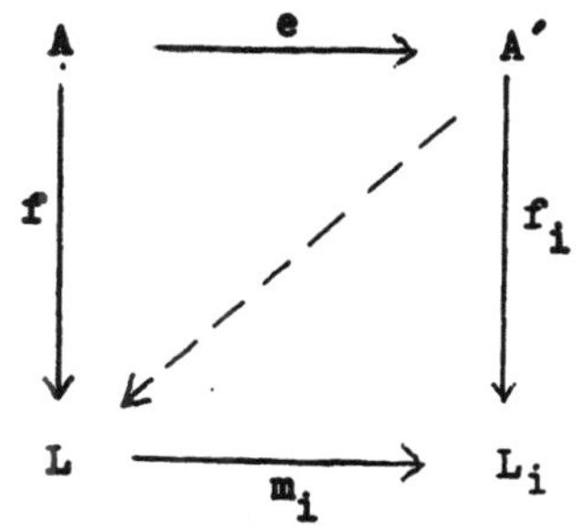

commutes. By the diagonal fill-in, f factors through e . Consequently, L satisfies the $\mathcal{E}$-axiom $e : A \to A'$ and hence, $L \in \mathcal{K}$. Thus, $\mathcal{K}$ has initial objects for all $\mathcal{M}$-sources, which was to be proved.

II.5 <u>Examples</u>. (i) If we choose $\mathcal{E}$ = all isomorphisms and $\mathcal{M}$ = all sources, then we obtain Herrlich's presentation theorem : a concrete category is topological (= $\mathcal{M}$-topological) iff it can be presented by topological axioms.

(ii) The category of Pgr partial groupoids is mono-topological. It can be realized in $S(H_3)$, the category of ternary relations, by assigning to each partial groupoid (X, +) the set $\alpha = \{(x, y, z) \in X^3 \mid x + y = z\}$. Then Pgr is presented in $S(H_3)$ by the following epi-axiom (G) . Put $A = \{x, y, z_1, z_2\}$ and $A' = \{x, y, z\}$, and let $e : A \to A'$ by the map forgetting the index of z . Then we have

$$\text{(G)} \qquad (A, \{(x, y, z_1), (x, y, z_2)\}) \xrightarrow{e} (A', \{(x, y, z)\}) .$$

III. Mac Neille Completion

III.1 The least topological extension of a concrete category $\mathcal{K}$ is called the Mac Neille completion of $\mathcal{K}$. We are going to show how Mac Neille completion is presented by topological axioms.

A concrete category $\mathcal{K}$ has a realization in a topological category iff $\mathcal{K}$ is strongly fibre-small (see [AHS_1]), in which case there exists a smallest such realization (i.e., a topological realization $\mathcal{K} \hookrightarrow \mathcal{K}^*$ such that any topological realization of $\mathcal{K}$ can be extended to a realization of $\mathcal{K}^*$). Then $\mathcal{K}^*$ is called a Mac Neille completion of $\mathcal{K}$. The intrinsic characterization of $\mathcal{K}^*$ is the following : it is a topological extension of $\mathcal{K}$ such that $\mathcal{K}$ is

(a) finally dense, i.e., each object of $\mathcal{K}^*$ is final for some sink whose domains are $\mathcal{K}$-objects, and
(b) initially dense, i.e., each object of $\mathcal{K}^*$ is initial for some source whose co-domains are $\mathcal{K}$-objects.

III.2 Example. The category Seq_h of all Hausdorff sequential spaces (II.3) has Mac Neille completion Seq. It is clear that Seq_h is finally dense - in fact, the space N^* alone is finally dense in Seq. Further, Seq_h is initially dense because each space K is initial with respect to its Hausdorff reflection.

III.3 Theorem. Any strongly fibre-small category has initiality preserving realization in some S(F). The Mac Neille completion of $\mathcal{K}$ is then presented in S(F) by the collection of all $\mathcal{K}$-axioms, i.e., topological axioms satisfied by each $\mathcal{K}$-object.

Remark. For each c-category $\mathcal{K}$ we have an analogous result with $S_c(F)$ in place of S(F).

Furthermore, there is a useful generalization of the theorem above with respect to an arbitrary initially dense subcategory $\mathcal{H}$ of $\mathcal{K}$: the Mac Neille completion of $\mathcal{K}$ is presented in S(F) by the collection of all $\mathcal{H}$-axioms. We prove this stronger form.

Proof. Let $\mathcal{K}^*$ be the Mac Neille completion of $\mathcal{K}$. This is a topological category and hence, it has following initiality-preserving realization $E : \mathcal{K}^* \to S(F)$. F is the structure functor of $\mathcal{K}^*$: for each $X \in \mathcal{X}$, the set FX consists of precisely all $\mathcal{K}^*$-objects with the underlying object X, and for each morphism $f : X \to X$, the map Ff sends each object K to the f-final object Ff(K). The embedding E is defined by $EK = (|K|, \alpha_K)$ where α_K consists of those L for which $id_X : L \to K$ is a morphism. A detailed proof that E is a well-defined, full, concrete embedding has been given in [AHS_2]. It is easy to see that moreover, the realization E is initiality-preserving:

In fact, let $(K \xrightarrow{f_i} K_i)_{i \in I}$ be an initial source in $\mathcal{K}^*$. For each object (X, α) of $S(F)$ and each map $h : X \to |K|$ such that $f_i \cdot h : (X, \alpha) \to EK_i$ are morphisms we are to show that $h : (X, \alpha) \to K$ is a morphism. Given $L \in \alpha$ then $F(f_i \cdot h)$ sends L to an object $\leq K_i$ and therefore, $f_i \cdot h : L \to K_i$ are morphisms in $\mathcal{K}^*(i \in I)$. Consequently, $h : L \to K$ is a morphism in $\mathcal{K}^*$, and we conclude that $Fh(\alpha) \subseteq \alpha_K$.

Thus, we have an initiality-preserving realization $E : \mathcal{K}^* \to S(F)$ and hence, $E(\mathcal{K}^*)$ is presented in $S(F)$ by a collection of topological axioms. Consequently, the collection $\mathcal{A}(\mathcal{K}^*)$ of all $\mathcal{K}^*$-axioms presents $\mathcal{K}^*$. It remains to prove that this collection is equal to the collection $\mathcal{A}(\mathcal{H})$ of all $\mathcal{H}$-axioms, assuming that $\mathcal{H}$ is initially dense in $\mathcal{K}$. The inclusion $\mathcal{A}(\mathcal{K}^*) \subseteq \mathcal{A}(\mathcal{H})$ follows from $\mathcal{H} \subseteq \mathcal{K}^*$. For the converse, let $\alpha \vdash \alpha'[X]$ be an $\mathcal{H}$-axiom. We show that it is satisfied by any object EK^* with $K^* \in \mathcal{K}^*$. First we observe that $\mathcal{H}$ is initially dense in $\mathcal{K}^*$ (because the embedding $\mathcal{K} \hookrightarrow \mathcal{K}^*$ is both initiality-preserving, and initially dense). Let $h_i : K^* \to H_i$ $(i \in I)$ be an initial source, $H_i \in \mathcal{H}$. Given a morphism $f : (X, \alpha) \to EK^*$, for each $i \in I$ we have a morphism $h_i \cdot f : (X, \alpha) \to EH_i$, and since $H_i \in \mathcal{H}$, also a morphism $h_i \cdot f : (X, \alpha') \to EH_i$. By the initiality, $f : (X, \alpha') \to EK^*$ is a morphism, proving that $\alpha \vdash \alpha'$ is a $\mathcal{K}^*$-axiom.

III.4 <u>Example</u>. Let $\mathcal{K} = \mathbf{Pos}$ be the category of posets. It is a full subcategory of the category of directed graphs, which is $S(H_2)$ for the hom-functor H_2 (given by $X \mapsto X^2$ and $f \mapsto f^2 = f \times f$). Since $\mathcal{K}$ is a c-category, we shall rather use the category $S_c(H_2)$ of reflexive graphs.

The two-point chain $\underset{\sim}{2} = \{0, 1\}$ forms a singleton, initially dense subcategory of Pos because any poset can be embedded into a power of $\underset{\sim}{2}$. On the other hand, $\underset{\sim}{2}$ is finally dense in all of $S_c(H_2)$ because the structure of a reflexive graph is determined by compatible maps from $\underset{\sim}{2}$.

The Mac Neille completion of Pos is thus presented by all $\underset{\sim}{2}$-axioms in $S_c(H_2)$. A topological axiom $\alpha \vdash \alpha'[X]$ is satisfied by $\underset{\sim}{2}$ iff for each subset $Y \subseteq X$ [i.e., each map $f : X \to \underset{\sim}{2}$, where $Y = f^{-1}(1)$] such that α has the following property :

$$\text{given } y \in Y \text{ and } (y, z) \in \alpha, \text{ then } z \in Y$$

[i.e., $f : (X, \alpha) \to (\underset{\sim}{2}, \leq)$ is compatible], α' also has the corresponding property. This is the case iff each α'-arrow is a "composite" of α-arrows, more precisely, if α and α' have the same transitive hull,. All of these topological axioms are easily seen to be equivalent to the following one :

(T) 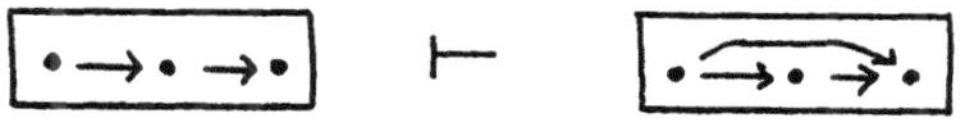

A graph G satisfies the axiom (T) (and all $\mathcal{Z}$ -axioms) iff G is transitive. The resulting subcategory is that of preordered sets, i.e., reflexive and transitive relations.

IV. Cartesian Closed Topological Hulls

IV.1 A topological c-category is called <u>cartesian closed</u> provided that it has "well-behaved" function spaces. That is, for each pair of objects K and K', the set $\hom(K, K')$ of all morphisms from K to K' can beequipped with the structure of an object $[K, K']$ having the following universal property :

for each object A, the morphisms $a : A \to [K, K']$ are precisely those maps for which $\hat{a} : A \times K \to K'$, given by $\hat{a}(x, k) = (a(x))(k)$, is a morphism.

Function spaces with the above property have a large collection of further pleasant properties, a list of which can be found in $[H_2]$.

Unfortunately, the category Top of topological spaces is not cartesian closed. This has led to a search of cartesian closed topological c-categories (shortly CCT-categories) extending Top with finite products preserved. Antoine [A] proved the existence of the least such extension. For general c-categories, this procedure has been studied by Herrlich and Nel [HN]. Given a finitely productive c-category $\mathcal{K}$, they proved that the existence of a finitely productive CCT-realization $\mathcal{L}$ of $\mathcal{K}$ implies that there is the smallest such $\mathcal{L}$, called the <u>CCT-hull</u> of $\mathcal{K}$. Internally, the CCT-hull is characterized as a CCT-category $\mathcal{K}^{\#}$ in which $\mathcal{K}$ has a realization such that

(a) $\mathcal{K}$ is finally dense, and

(b) $[\mathcal{K}, \mathcal{K}]$ is initially dense, where $[\mathcal{K}, \mathcal{K}]$ is the full subcategory of $\mathcal{K}^{\#}$ of all $[K, K']$ with $K, K' \in \mathcal{K}$.

IV.2 <u>Examples</u>. (i) The category Seq of sequential spaces is CCT : for arbitrary spaces K and K', the pointwise convergence defines the apropriate structure of $[K, K']$. Thus, Seq is the CCT hull of Seq_h .

(ii) The category $S_c(F)$ is CCT for a number of set-functors, including all hom-functors, the power-set functor, any product or composite of such functors, etc. Nevertheless, it is not always CCT . A necessary and sufficient condition has been presented in $[AK_2]$. Here, we shall use a simpler sufficient condition. Recall that a commuting square of maps

$$\begin{array}{ccc} T & \xrightarrow{\;f'\;} & Y \\ {\scriptstyle g'}\downarrow & & \downarrow{\scriptstyle g} \\ X & \xrightarrow[\;f\;]{} & Z \end{array}$$

is a <u>weak pullback</u> provided that for each pair $(x, y) \in X \times Y$ with $f(x) = g(y)$ there exists $t \in T$ with $x = g'(t)$ and $y = f'(t)$. A set functor is said to <u>weakly preserve pullbacks</u> if it maps any such square to a weak pullback again. If so, $S_c(F)$ is a CCT-category.

IV.3 Which topological c-categories have a CCT hull ? This question has been answered in [AK_2] by the following smallness condition. Two structured maps $X \xrightarrow{a} A$ and $X \xrightarrow{a} A'$ are said to be productively equivalent if for each object K the maps $X \times |K| \xrightarrow{a \times 1} |A \times K|$ and $X \times |K| \xrightarrow{a \times 1} |A' \times K|$ are equivalent. A topological c-category has a CCT hull iff it is <u>strictly fibre-small</u>, i.e., for each set X there exists a (small) set of representatives w.r.t. productive equivalence of structured maps from X.

Whereas all the usual categories used by topologists turn out to be strictly fibre-small, there exists a set functor F such that $S_c(F)$ fails to be strictly fibre-small, see [AK_2].

IV.4 <u>Theorem</u>. Every strictly fibre-small c-category has an initiality-preserving realization in $S_c(F)$ for some set functor F weakly preserving pullbacks. The CCT hull of $\mathcal{K}$ is then presented in $S_c(F)$ by the collection of all productive $\mathcal{K}$-axioms, i.e., topological axioms $(X, \alpha) \xrightarrow{id} (X, \alpha')$ such that the products $(X, \alpha) \times K \xrightarrow{id} (X, \alpha') \times K$ are $\mathcal{K}$-axioms for all $K \in \mathcal{K}$.

Proof. We use the following universally topological hull $\mathcal{K}^*$ of $\mathcal{K}$, the existence of which is equivalent to strict fibre-smallness, see [A_2]. A topological category is said to be universally topological if the formation of pullbacks along any morphism $f : A \to B$ preserves finality, i.e., from any final sink $(B_i \to B)$ we get, by pointwise pullbacks, a final sink $(A_i \to A)$. The least universally topological category in which $\mathcal{K}$ has a finally dense realization is called the universally topological hull of $\mathcal{K}$. By any easy modification, for each strictly fibre-small c-category we have a universally topological c-hull, i.e., the least universally topological c-category in which $\mathcal{K}$ has a finally dense realization. We denote this hull by $\mathcal{K}^*$.

The advantage of universally topological categories is that obviously the structure functor F (see the proof of III.3 above) weakly preserves pullbacks and hence, $S_c(F)$ is universally topological. (Whereas for a CCT-category, the structure functor need not render a cartesian closed category $S_c(F)$!). We have an initiality-preserving realization $E : \mathcal{K}^* \to S_c(F)$ as in the proof of III.3 above. Since $\mathcal{K}^*$ is CCT, it contains $\mathcal{K}^\#$, the CCT hull. Moreover, since $\mathcal{K}$ is finally dense in $\mathcal{K}^*$, so is $\mathcal{K}^\#$ and hence, the restriction of E gives an initiality - preserving realization $E : \mathcal{K}^\# \to S_c(F)$. Thus $\mathcal{K}^\#$ is presented in $S_c(F)$ by the collection $\mathcal{A}(\mathcal{K}^\#)$

of all $\mathcal{K}^{\#}$-axioms. We are going to prove that $\mathcal{A}(\mathcal{K}^{\#})$ is equal to the collection of all productive $\mathcal{K}$-axioms, denoted by $\mathcal{A}_p(\mathcal{K})$.

We first observe that $\mathcal{K}^{\#}$ is closed in $S_c(F)$ under power-objects, i.e., $E[K,K'] = [EK,EK']$ for all K and K' in $\mathcal{K}^{\#}$. In fact, since $\mathcal{K}^{\#}$ (being finally dense) is clearly closed under power objects in $\mathcal{K}^{*}$, it is sufficient to show that $\mathcal{K}^{*}$ is closed under them in $S_c(F)$. To this end, let $a: (X,\alpha) \to E[K,K']$ be a morphism in $S_c(F)$. We are going to show that then $a: (X,\alpha) \times EK \longrightarrow EK'$ is a morphism in $S_c(F)$; since the converse is obvious, this will establish the desired universal property of $E[K,K']$. Given an object $A \in \alpha$, the final object $Fa(A)$ lies in $\alpha_{[K,K']}$ and thus, $id: Fa(A) \to [K,K']$ is a morphism in $\mathcal{K}^{\#}$. Therefore, $\widehat{id} = eval: Fa(A) \times K \to K'$ is a morphism and hence, $\hat{a}: A \times K \to K'$ is a morphism in $\mathcal{K}^{\#}$. Consequently, $a: (X,\alpha) \times EK \longrightarrow EK'$ is a morphism in $S_c(F)$.

(i) $\mathcal{A}(\mathcal{K}^{\#}) \subseteq \mathcal{A}_p(\mathcal{K})$. For $K \in \mathcal{K}$ and $\mathcal{K}^{\#}$-axiom $\alpha \vdash \alpha'[X]$ we prove that $(X,\alpha) \times EK \xrightarrow{id} (X,\alpha') \times EK$ is a $\mathcal{K}$-axiom, i.e., is satisfied by any EH. In fact, each morphism from $(X,\alpha) \times EK$ into EH has the form $\hat{f}$ for some morphism $f: (X,\alpha) \to [EK, EH]$ (because $S_c(F)$ is cartesian closed, E is full and it preserves power objects and products). Since $(X,\alpha) \xrightarrow{id} (X,\alpha')$ is a $\mathcal{K}^{\#}$-axiom, $\hat{f}: (X,\alpha') \to [EK,EH]$ is also a morphism and hence, $f: (X,\alpha') \times EK \to EH$ is a morphism.

(ii) $\mathcal{A}_p(\mathcal{K}) \subseteq \mathcal{A}(\mathcal{K}^{\#})$. Let $(X,\alpha) \xrightarrow{id} (X,\alpha')$ be a productive $\mathcal{K}$-axiom, then we prove that any object $K^{\#}$ of $\mathcal{K}^{\#}$ satisfies it. Since $[\mathcal{K},\mathcal{K}]$ is initially dense in $\mathcal{K}^{\#}$, we have an initial source $h_i: K^{\#} \to [K_i, H_i]$ for $i \in I$, where all K_i and H_i are $\mathcal{K}$-objects. Given a morphism $f: (X,\alpha) \to EK^{\#}$, it is sufficient to prove that $h_i \cdot f: (X,\alpha') \to E[K_i, H_i]$ are morphisms for all $i \in I$ (since E preserves initiality). Equivalently, that $\widehat{h_i \cdot f}: (X,\alpha') \times EK_i \to EH_i$ are morphisms. In fact, $(X,\alpha) \times EK_i \xrightarrow{id} (X,\alpha') \times EK_i$ is a $\mathcal{K}$-axiom and $H_i \in \mathcal{K}$. Since $\widehat{h_i \cdot f}: (X,\alpha) \times EK_i \to EH_i$ is a morphism (corresponding to $h_i \cdot f$, the composition of two morphisms), $h_i \cdot f: (X,\alpha') \times EK_i \to EH_i$ is a morphism too.

IV.5 <u>Main example</u>. The example that has initiated the study of CCT hulls (in the work of Antoine) is **Top**, the category of topological spaces. Following the preceding theorem, we are looking for a set functor F weakly preserving pullbacks and such that **Top** is finally dense in the category $S_c(F)$. Then we try to characterize the CCT hull by topological axioms.

Since topology is determined by filter-convergence, we can try the filter-functor. But ultrafilter convergence is also sufficient, and thus we take the smaller ultrafilter functor:

$$\mathcal{U}X = \{(\mathfrak{U}, a) \mid \mathfrak{U} \text{ is an ultrafilter on } X, \text{ and } a \in X\},$$

and for each map f, the map

$$\mathcal{U}f : \mathcal{U}X \to \mathcal{U}Y$$

sends $(\mathfrak{U}, a)$ to $(f[\mathfrak{U}], f(a))$, where

$$f[\mathfrak{U}] = \{B \subseteq Y \mid f^{-1}(B) \in \mathfrak{U}\} .$$

(The elements of $\mathcal{U}X$ are usually denoted by $\mathfrak{U} \to a$.)

It is easy to verify that $\mathcal{U}$ weakly preserves pullbacks.

Observations. (i) $S_c(\mathcal{U})$ is the category of pseudotopological spaces.

In fact, each pseudotopological space is given by the filter-convergence structure such that a filter converges to $a \in X$ iff each ultrafilter containing it converges to a. Hence, a pseudotopological space on a set X is completely described by the set $\alpha \subseteq \mathcal{U}X$ of all $\mathfrak{U} \to a$, where $\mathfrak{U}$ is an ultrafilter converging to a. The axioms for a pseudotopological space reduce to the single one, when expressed by ultrafilters, viz.,

(c) $\dot{a} \to a$ for each $a \in X$

where $\dot{a} = \{A \subseteq X \mid a \in A\}$. In this sense, pseudotopological spaces are precisely the objects of $S_c(\mathcal{U})$. And continuous maps are precisely those preserving ultrafilter convergence - and these are just the morphisms in $S_c(\mathcal{U})$.

(ii) The category Top is a full subcategory of $S_c(\mathcal{U})$, presented by the following topological axioms (for each set X and each $\alpha \subseteq \mathcal{U}X$) :

(TOP_1) $\alpha \vdash \{\mathfrak{U} \to a \mid \exists \mathfrak{B}_i \xrightarrow{\alpha} a,\ i \in I, \text{ with } \bigcap_{i \in I} \mathfrak{B}_i \subseteq \mathfrak{U}\}$ $[X]$

(TOP_2) $\alpha \vdash \{\mathfrak{U} \to a \mid \exists \mathfrak{B}_i \xrightarrow{\alpha} b_i,\ i \in I, \exists\ \mathcal{C} \to a \text{ with } \coprod_{\mathcal{C}} \mathfrak{B}_i \subseteq \mathfrak{U}\}$ $[X]$

where $\coprod_{\mathcal{C}} \mathfrak{B}_i$ is the filter defined iff each $C \in \mathcal{C}$ contains b_i for some $i \in I$, and then its base is formed by all $\bigcup_{b_i \in C} B_i$, where $C \in \mathcal{C}$ is arbitrary and for each i with $b_i \in C$, $B_i \in \mathfrak{B}_i$ is arbitrary.

In fact, a pseudotopological space satisfies (T_1) iff it is a pretopological space, i.e., each point a has the smallest filter converging to a. A pretopological space satisfies (T_2) iff the closure operator is idempotent.

Theorem (Bourdaud [Bo_1]). The CCT hull of Top is presented in the category $S_c(\mathcal{U})$ of pseudotopological spaces by the following topological axioms (for all sets X and all $\alpha \subseteq \mathcal{U}X$) :

(LIM) $\alpha \vdash \{\mathfrak{U} \to a \mid \exists \mathfrak{B} \xrightarrow{\alpha} a,\ \lim \mathfrak{B} \in \mathfrak{U}\}$ $[X]$,

where $\lim \mathfrak{B} = \{b \in X \mid \mathfrak{B} \to b\}$,

(STARR) $\alpha \vdash \{\mathfrak{U} \to a \mid \exists \mathfrak{B} \xrightarrow{\alpha} a,\ B^* \in \mathfrak{U} \text{ for all } B \in \mathfrak{B}\}$ $[X]$, where $B^* = \{b \in X \mid \dot{b} \to x$ for some $x \in B\}$.

IV.6 **A survey of CCT hulls.** For some of the basic topological categories, an explicit description of the CCT hull is known. The proofs of most of the examples below are quite involved, and a general technique has been described by [A R S], where all the full proofs are also given. The examples 1 - 5 live in the category $S_c(\mathcal{U})$ of pseudotopological spaces, and we mention what topological axioms present the CCT hull. The remaining examples could also be presented in some $S_c(F)$, but since the presentation is clumsy, we omit it.

A SURVEY OF CARTESIAN CLOSED TOPOLOGICAL HULLS

Category; [reference]	Name of the CCT hull	Axioms in $S_c(\mathcal{U})$
1. Pretopological sp.; Bourdaud [Bo$_2$]	Pseudotopological sp.	none
2. Topological sp.; Antoine [A], Machado [M], Bourdaud [Bo$_1$]	Antoine sp.	(LIM) & (STARR)
3. Completely regular topological sp.; Bourdaud [Bo$_2$]	c-embedded sp.; see [Bi], [KMRS], [Sc].	(LIM^{CR}) & $(STARR^{CR})$
4. 0-dimensional topological sp.; Adámek, Reiterman and Strecker [ARS]	0-embedded topological sp.	(LIM^0) & $(STARR^0)$
5. Metrizable topological sp.; Adámek Reiterman and Strecker [ARS]	functionally sequential topological sp.	(TOP_1) & (TOP_2) & (SEQ)
6. Uniform sp.; Adámek and Reiterman [AR]	Bornological uniform sp.	- - - - -
7. Metric sp. and contractions; Adámek and Reiterman [AR]	Bornological metric sp.	- - - - -
8. Compact Hausdorff topological sp.; Nel [N], see also [ARS]	Compactly generated uniform sp.	- - - - -

(LIM^{CR}) denotes the variant of the axiom (LIM) above using the completely regular modification α^{CR} of α (in which $\mathcal{U} \to a$ iff $f[\mathcal{U}] \to f(a)$ for each continuous real function f on (X, α)):

$$(LIM^{CR}) \quad \alpha \vdash \{\mathcal{U} \to a \mid \exists\, \mathcal{B} \xrightarrow{\alpha^{CR}} a \;,\; \lim \mathcal{B} \in \mathcal{U}\}.$$

Analogously, (LIM^0) denotes the variant of (LIM) using the 0-dimensional modification α^0 of α.

(STARR^{CR}) denotes the variant of the axiom (STARR) above using B^{*CR} in place of B^*, where $B^{*CR} = \{b \in X \mid \dot{b} \xrightarrow{\alpha^{CR}} x$ for some $x \in B\}$. Analogously (STARR^0) denotes the variant with B^* replaced by $B^{*0} = \{b \in X \mid \dot{b} \xrightarrow{\alpha^0} x$ for some $x \in B\}$.

(Seq) is the following axiom which claims that the space (X, α) is initial w.r.t. the set $C_{seq}(X, \alpha)$ of all sequentially continuous real functions on (X, α) :

$$\text{(SEQ)} \quad \alpha \vdash \{\mathcal{U} \to a \mid f[\mathcal{U}] \to f(a) \text{ for all } f \in C_{seq}(X, \alpha)\} \; [X].$$

A bornological uniform space (see [AR]) is a triple $(X, \alpha, \mathcal{B})$ where α is a uniformity on X and $\mathcal{B}$ is a bornology on X such that α is $\mathcal{B}$-final (i.e., a cover is uniform iff it is uniform on each subspace $B \in \mathcal{B}$) and the following holds :

(*) a set $A \subseteq X$ is in $\mathcal{B}$ iff for each uniform cover $\mathcal{V} \in \alpha$ there is $B \in \mathcal{B}$ with $A \subseteq st_{\mathcal{V}}(B) = \bigcup \{V \in \mathcal{V} \mid V \cap B \neq \emptyset\}$.

A bornological metric space is defined analogously, with a modification of (*). Morphisms of these structures are the (uniformly) continuous maps preserving bounded sets.

References

[A_1] Adámek J. : Theory of mathematical structures, Reidel Publish. Comp., Dordrecht 1983.

[A_2] Adámek J. : Classification of concrete categories, Proc. of Category Theory Conference, Oberwolfach 1983, Tech. Report of Fernuniversität Hagen.

[AHS_1] Adámek J., Herrlich H., Strecker G.E. : Least and largest initial completions, Comment. Math. Univ. Carolinae 20 (1979), 43-77.

[AHS_2] Adámek J., Herrlich H., Strecker G.E. : The structure of initial completions, Cah. Topo. Géom. Differ. 20 (1979), 333-352.

[AK_1] Adámek J., Koubek V. : Cartesian closed functor-structured categories, Comment. Math. Univ. Carolinae 21 (1980), 573-589.

[AK_2] Adámek J., Koubek V. : Cartesian closed topological completions, Topology and Appl. 11 (1980), 1-16.

[AR] Adámek J., Reiterman J. : Cartesian closed hull of the category of uniform spaces, to appear in Topology and Appl.

[AS] Adámek J., Strecker G.E. : Construction of cartesian closed topological hulls, Comment. Math. Univ. Carolinae 22 (1981), 235-254.

[A] Antoine P. : Étude élémentaire des catégories d'ensenbles structures I., II., Bull. Soc. Math. Belg. 18 (1966), 142-164 and 387-414.

[Bi] Binz E. : Kompakte Limesräume und limitierte Funktionenalgebren, Comm. Math. Helv. 43 (1968), 195-203.

[Bo_1] Bourdaud G. : Espaces d'Antoine et semi-espaces d'Antoine, Cahiers Topo. Géom. Differentielle 16 (1975), 107-134.

[Bo_2] Bourdaud G. : Some cartesian closed topological categories of convergence spaces, Lecture Notes in Mathematics 540, Springer--Verlag 1976, 93-108.

[H_1] Herrlich H. : Topological functors, Topology and Appl. 4 (1974), 125-142.

[H_2] Herrlich H. : Cartesian closed topological categories, Math. Coll. Univ. Cape Town 9 (1974), 1-16.

[H_3] Herrlich H. : Initial completions, Math. Z. 150 (1976), 101-110.

[H_4] Herrlich H. : Universal topology, Proc. Categorical Conf. University of Toledo 1983.

[HN] Herrlich H., Nel L.D. : Cartesian closed topological hulls, Proc. Amer. Math. Soc. 62 (1977), 215-222.

[K] Katětov M. : Allgemeine Stetigkeitstrukturen, Proc. Inter. Conf. Math. Stockholm 1962, 473-479.

[KMRS] Kent D.C., McKennon K., Richardson G.D., Schroder M. : Continuous convergence in C(X), Pacific J. Math. 52 (1974), 457-465.

[M] Machado A. : Espaces d'Antoine et pseudo-topologies, Cahiers Topo. Géom. Differentielle 14 (1973), 309-327.

[N] Nel L.D. : Cartesian closed coreflective hulls, Questiones Math. 2 (1977), 269-283.

[Sc] Schroder M. : Solid convergence spaces, Bull. Austral. Math. Soc. 8 (1973), 443-459.

[ARS] Adámek J., Reiterman J., Strecker G.E.: Realizations of cartesian closed topological hulls, submitted.

ON K, M AND KM-SEQUENCES AND UNIFORM CONVERGENCE

Piotr Antosik [1/]

1. INTRODUCTION.

We are concerned with conditions under which pointwise convergence of sequences of mappings implies their uniform convergence on subsets of the domain. The conditions are expressed in terms of K or M or KM-properties of sequences. As simple consequences of the theorems one can get: classical Banach theorems on uniform boundedness and equicontinuity, Nikodym theorems on uniform countability and uniform boundedness, Orlicz-Pettis theorem, theorems on joint continuity and others classical results. The method of proving theorems is simple and leads to various generalizations. It is based on a lemma on matrices proved in the next section.

2. A LEMMA ON MATRICES.

Throughout this paper N denotes the set of all positive integers. By a matrix (double sequence) we mean a function $\{x_{ij}\}$ defined on $N \times N$. The sequence $\{x_{ii}\}$ is called the diagonal of the matrix. The lemma is concerned with conditions under which the diagonal $\{x_{ii}\}$ of $\{x_{ij}\}$ converges to zero. We limit ourselves to matrices with entires in a quasi-normed group. We recall that a functional $|\cdot|$ on an abelian group X is a quasi-norm if $|0|=0$, $|-x|=|x|$ and $|x+y|=|x|+|y|$. We assume the usual definitions of the convergence of sequences and series in a quasi-normed group.

<u>Lemma</u>. Assume that X is a quasi-normed group and $x_{ij} \in X$ for $ij \in N$. If for each increasing sequence of positive integers m_i there is a subsequence $\{n_i\}$ of $\{m_i\}$ such that

(i) $$x_{n_i n_j} \to 0 \quad \text{as} \quad i \to \infty \text{ for } j \in N$$

and

(ii) $$\sum_{j=1}^{\infty} x_{n_i n_j} \to 0,$$

then

$$x_{ii} \to 0.$$

1/ Institute of Mathematics, Polish Academy of Sciences, Wieczorka 8, 40-013 Katowice, Poland

Proof. Assume that the Lemma does not hold. Then there exists an $\varepsilon > 0$ and an increasing sequence of positive integers m_i such that

$$|x_{m_i m_i}| > \varepsilon \tag{1}$$

for $i \in N$. Let $\{n_i\}$ be a subsequence of $\{m_i\}$ so that the conditions (i) and (ii) hold. We have $|x_{n_i n_j}| \to 0$ as $i \to \infty$ for $j \in N$ and $|x_{n_i n_j}| \to 0$ as $j \to \infty$ for $i \in N$. We put $p_1 = n_1$ and we find a member p_2 of the sequence $\{n_i\}$ such that $p_1 < p_2$ and

$$|x_{p_i p_j}| < 2^{-i-j}$$

for $i,j = 1,2$ and $i \neq j$. By induction, we select a subsequence $\{p_i\}$ of $\{n_i\}$ such that

$$|x_{p_i p_j}| < 2^{-i-j} \tag{2}$$

for $i,j \in N$ and $i \neq j$. By the assumption, there is a subsequence $\{q_i\}$ of $\{p_i\}$ such that

$$\sum_{j=1}^{\infty} x_{q_i q_j} \to 0 \tag{3}$$

From properties of the quasi-norm it follows that

$$|x_{q_i q_i}| \leqq \sum_{j=1}^{i-1} |x_{q_i q_j}| + \sum_{j=i+1}^{\infty} |x_{q_i q_j}| + \left|\sum_{j=1}^{\infty} x_{q_i q_j}\right|$$

Hence, by (2), we get

$$|x_{q_i q_i}| < 2^{-i} + \left|\sum_{j=1}^{\infty} x_{q_i q_j}\right|$$

Consequently, by (3), $|x_{q_i q_i}| \to 0$ which leads to contradiction with (1). This contradiction completes the proof of the Lemma.

It can be shown that the assumption (i) can be droped in the Lemma when X is a locally convex space. It is unknown whether or not (i) can be droped in the general case.

As a simple consequence of the Lemma we have the following

Corollary 1. Assume that X is a linear space equiped with a quasi-norm $|\cdot|$ such that the multiplication by scalars is separately continuous, i.e.,

(a) $\alpha_n \to 0$ and $x \in X$ implies $\alpha_n x \to 0$

and

(b) $x_n \to 0$ and $\alpha \in R$ implies $\alpha x_n \to 0$.

Then we have

$\alpha_n x_n \to 0.$

Proof. To prove the corollary consider the matrix $\{\alpha_j x_i\}$ and note that the matrix satisfies the conditions of the Lemma since the field of scalars is complete. Consequently, $\alpha_n x_n \to 0$.

A linear space X equiped with a functional $|\cdot|$ such that (a) and (b) hold is called a quasi-normed linear space. By Corollary 1, the multiplication in a quasi-normed linear space is jointly continuous.

We end this section with recalling basic definitions used in the next sections. We shall refer to the following properties of convergences:

F. Limits of sequences are limits of their subsequences;

L. Algebraic operations are continuous;

U. If each subsequence of a given sequence has a subsequence converging to a given element, then the sequence converges to the element;

S. Constant sequence converge to their elements.

A set X equipped with an FS-convergence is a convergence space. A subset A of a convergence space X is compact in X if each sequence of elements in A has a subsequence converging to an element in X. Finite subsets of X are compact in X. Sets of members of convergent sequences are compact in X.

By a convergence group we mean an abelian group X endoved with an FLS-convergence. Quasi-normed groups are convergence groups satisfying, additionally, U.

A linear space X equiped with an FLS-convergence is a convergence linear space. In this case we assume that the multiplication by scalars is jointly continuous. Quasi-normed linear spaces are convergence linear spaces. A subset A of a convergence linear space X is bounded if $\alpha_n x_n \to 0$ whenever $x_n \in A$ and $\alpha_n \to 0$. Finite subsets of X are bounded. Sets of members of convergent sequences are bounded.

3. K-SEQUENCES.

Assume that X is a convergence group. A sequence of elements x_n in X is said to be a K-sequence if for each subsequence $\{u_n\}$ of $\{x_n\}$, there are a subsequence $\{v_n\}$ of $\{u_n\}$ and an element v in X such that

$$v_1 + v_2 + \dots = v.$$

If $x_n = 0$ for $n \in N$, then $\{x_n\}$ is a K-sequence in X. Sequences converging to zero in a Banach space, or more generally, in a complete matric group are K-sequences, respectively. If Q is the group of rational numbers with the usual convergence, then $\{10^{-2^n}\}$ is not a K-sequence in Q.

Suppose that X is a convergence space, Y is a quasi-normed group and $C(X,Y)$ is the set of all continuous mappings from X to Y, i.e., $f \in C(X,Y)$ iff $f(x_n) \to f(x)$ whenever $x_n \to x$. Obviously, $C(X,Y)$ is an abelian group. We assume that $C(X,Y)$ is equippedwith the pointwise convergence, i.e., $f_n \to f$ iff $f_n(x) \to f(x)$ for each $x \in X$.

We have the following theorem.

Theorem 1. K-sequences in $C(X,Y)$ converge to zero uniformly on compact sets in X.

Proof. Assume that $\{f_n\}$ is a K-sequence in $C(X,Y)$ and A is a compact set in X. Let $\{x_n\}$ be a sequence of elements in A. We assert that $f_n(x_n) \to 0$. In fact, let $\{g_n(u_n)\}$ be a subsequence of $\{f_n(x_n)\}$ and let $\{h_n(v_n)\}$ be a subsequence of $\{g_n(u_n)\}$ such that $v_n \to v$ for same $v \in X$. Then the matrix $\{h_j(v_i) - h_j(v)\}$ satisfies the assumptions of the Lemma. Consequently $h_i(v_i) - h_i(v) \to 0$, or equivalently, $h_i(v_i) \to 0$. Hence, by U, $f_n(x_n) \to 0$ which proves the theorem.

From Theorem 1 it follows that subseries convergent series in $C(X,Y)$ with respect to the pointwise convergence are subseries convergent with respect to the uniform convergence on compact sets in X.

A subset A of a convergence group X is said to be K-compact in X if for each sequence of elements x_n in A there are a subsequence $\{u_n\}$ of $\{x_n\}$ and $u \in X$ such that $\{u_n - u\}$ is a K-sequence in X. Finite subsets of X are K-compact in X. Sets of members of K-sequences are K-compact in X. We recall that a subset F of $C(X,Y)$ is equicontinuous if $f_n(x_n) - f_n(x) \to 0$ whenever $x_n \to x$. Finite subsets of $C(X,Y)$ are equicontinuous.

Corollary 1. K-compact sets in $C(X,Y)$ are equicontinuous.

In what follows $A(X,Y)$ denotes the set of all additive and continuous mappings from a convergence group X to a quasi-normed group Y. We assume that $A(X,Y)$ is endowed with the pointwise convergence.

Theorem 2. The pointwise convergence in $A(X,Y)$ and the uniform convergence on K-compact sets in X are equivalent.

Proof. Obviously, the uniform convergence on K-compact sets in X is the pointwise convergence. To prove the converse, assume that $f_n \in A(X,Y)$ for $n \in N$, $f_n \to 0$, A is a K-compact subset of X and $x_n \in A$ for $n \in N$. We assert that $f_n(x_n) \to 0$. In fact, let $\{g_n(u_n)\}$ be a subsequence of $\{f_n(x_n)\}$ and let $\{h_n(v_n)\}$ be a subsequence of $\{g_n(u_n)\}$ such that $\{v_n - v\}$ is a K-sequence in X for some $v \in X$. It is easy to verify that the matrix $\{h_i(v_j - v)\}$ satisfies the conditions of the Lemma. Therefore, $h_i(v_i - v) \to 0$, or equivalently, $h_i(v_i) \to 0$ since $h_i(v) \to 0$. Hence, by U, $f_n(x_n) \to 0$, or equivalently, $\{f_n\}$ converges to zero uniformly on A, which proves the theorem.

A convergence group X is said to be a K-convergence group if each sequence converging to zero is a K-sequence in X. Banach spaces, complete quasi-normed groups are K-convergence groups. By Theorem 2, the pointwise convergence in $A(X,Y)$ and the uniform convergence on compact subsets of X are equivalent if X is a K-convergence group.

<u>Corollary 2</u>. If X is a K-convergence group, then compact sets in $A(X,Y)$ are equicontinuous.

4. M-SEQUENCES.

A sequence of elements x_n in a convergence group X is said to be M sequence if for each subsequence $\{u_n\}$ of $\{x_n\}$ there are a subsequence $\{v_n\}$ of $\{u_n\}$ and a sequence of positive integers m_n such that $m_n \to \infty$ and $m_n v_n \to 0$. Sequences converging to zero in quasi-normed groups are M-sequences. If X is a convergence linear space $\{x_n\}$ is a bounded sequence in X and $\alpha_n \to 0$, then $\alpha_n x_n$ is an M-sequence.

M-sequences in X may not be K-sequences in X. In fact, the sequence $\{10^{-2^n}\}$ is an M-sequence in Q but it is not a K-sequence in Q. Also K-sequences may not be M-sequences. In fact, let m_o be the subspace of l^∞ which consists of those sequences $\{t_n\}$ with finite range. Define a norm on m_o, by means of the formula

$$||\{t_j\}|| = \sum_{j=1}^{\infty} |j^{-2} t_j|$$

We assume that $x_n \to x$ in m_o iff $||x_n'-x|| \to 0$ and

$$\sup_n \max\{|t_{jn}| : j \in N\} < \infty$$

with $x_n = t_{jn}$. Then the sequence of unit vectors e_j is a K-sequence in m_o but not M-sequence.

A convergence group X is said to be M-convergence group if each sequence converging to zero in X is an M-sequence. Quasi-normed groups are M-convergence groups. If Y is a quasi-normed linear space, then, by Theorem 2, pointwise bounded subsets of $A(X,Y)$ are uniformly bounded on K-compact sets in X. Hence, we get the corollary.

<u>Corollary 3.</u> If X is a K and M-convergence group, then pointwise bounded subsets of $A(X,Y)$ are equicontinuous.

When X is a complete metric linear space, Corollary 3 reduces to the Banach-Steinhaus theorem.

A subset A of a convergence linear space X is said to be K-bounded if $\alpha_n x_n$ is a K-sequence in X whenever $x_n \in A$ for $n \in N$ and $\alpha_n \to 0$.

Bounded subsets of K-convergence linear spaces are K-bounded.

Assume that X is a convergence linear space and Y is a quasi-normed linear space. By $L(X,Y)$ we denote the set of all linear and continuous mappings from X to Y. Obviously, $L(X,Y)$ is a linear space. We assume that $L(X,Y)$ is equipped with the pointwise convergence.

Theorem 3. M-sequences in $L(X,Y)$ converge to zero uniformly on K-bounded subsets of X.

Proof. Assume that $\{f_n\}$ is an M-sequence in $L(X,Y)$, A is a K-bounded subset of X and $x_n \in A$ for $n \in N$. To show that $f_n(x_n) \to 0$ we take a subsequence $\{g_n(u_n)\}$ of $\{f_n(x_n)\}$. Let $\{h_n(v_n)\}$ be a subsequence of $\{g_n(u_n)\}$ and let $\{m_n\}$ be a scalar sequence such that $m_n \to \infty$ and $m_n h_n \to 0$. We have

$$h_n(v_n) = m_n h_n(m_n^{-1} v_n)$$

We note that $\{m_n^{-1} v_n\}$ is a K-sequence in X. This implies that the matrix $\{m_i h_i(m_j^{-1} v_j)\}$ satisfies the assumptions of the Lemma. Consequently, $m_i h_i(m_i^{-1} v_i) \to 0$, or equivalently, $h_n(v_n) \to 0$. Hence, by the Urysohn condition, $f_n(x_n) \to 0$, or equivalently, $\{f_n\}$ converges to zero uniformly on A which proves the theorem.

Corollary 4. Pointwise bounded subsets of $L(X,Y)$ are uniformly bounded on K-bounded subsets of X.

5. KM-SEQUENCES.

Along with K-sequences and M-sequences, we deal with KM-sequences. A sequence of elements x_n in a convergence group X is called a KM-sequence if for each subsequence $\{u_n\}$ of $\{x_n\}$ there are a subsequence $\{v_n\}$ of $\{u_n\}$ and a sequence of integers m_n such that $m_n \to \infty$ and $\{m_n v_n\}$ is a K-sequence in X. If X is an M and K-convergence group, then sequences converging to zero are KM-sequences in X. KM-sequences are M-sequences. Assume that $X=(0,1)$ is two element group with the addition such that $0+0=0$, $0+1=1+0=1$ and $1+1=0$ and assume that $x_n \to x$ iff $x_n = x$ for sufficiently large n. Then the sequence $\{x_n\}$ with $x_n = 1$ for $n \in N$ is a KM-sequence in X but not K-sequence. The sequence of unit vectors e_j in m_o (see sect. 4) is a K-sequence in m_o but it is not a KM-sequence in m_o.

Justification for introducing KM-sequences is the following theorem.

Assume that X is a set, Y is a quasi-normed linear space and $B(X,Y)$ is the set of all bounded on X mappings from X to Y. We assume that $B(X,Y)$ is equipped with the pointwise convergence.

Theorem 4. KM-sequences in $B(X,Y)$ converge to zero uniformly on X.

Proof. Assume that $\{f_n\}$ is a KM-sequence in $B(X,Y)$ and $x_n \in X$ for $n \in N$. To show that $f_n(x_n) \to 0$ we take a subsequence $\{g_n(u_n)\}$ of $\{f_n(x_n)\}$ Let $\{h_n(v_n)\}$ be a subsequence of $\{g_n(u_n)\}$ and let m_n be a sequence of integers such that $m_n \to \infty$ and $\{m_n h_n\}$ is a K-sequence in $B(X,Y)$. Then we have

$$h_n(v_n) = m_n^{-1}[m_n h_n(v_n)]$$

for $n \in N$. Now its easy to see that the matrix

$$m_i^{-1}[m_j h_j(v_i)]$$

satisfies the conditions of the Lemma. Hence we get $h_i(v_i) \to 0$. Consequently, by the Urysohn property, $f_n(x_n) \to 0$, or equivalently, $\{f_n\}$ converges to zero uniformly on X.

6. THEOREMS ON SET FUNCTIONS

Applying the Lemma we prove Nikodym type theorems for set functions. Assume that Σ is a σ-ring of sets, Y is a quasi-normed group and $ca(\Sigma,Y)$ is the set of all countably additive set functions from Σ to Y.

Theorem 5. Setwise convergence in $ca(\Sigma,Y)$ is equivalent to the uniform convergence on pairwise disjoint subfamilies of Σ

Proof. Obviously, the uniform convergence on pairwise disjoint subfamilies of Σ implies the setwise convergence. To prove the converse assume that $\mu_n \in ca(\Sigma,Y)$ for $n \in A$, $\mu_n \to 0$ setwise, A is a pairwise disjoint subfamily of Σ and $E_n \in A$ for $n \in N$. Then the matrix $\{\mu_i(E_j)\}$ satisfies the conditions of the Lemma. Consequently, $\mu_i(E_i) \to 0$ which proves the theorem.

Corollary 5. If Y is a quasi-normed linear space, then setwise bounded subsets of $ca(\Sigma,Y)$ are uniformly bounded on disjoint subfamilies of Σ.

REFERENCES:

[1] Antosik, P., A diagonal theorem for nonnegative matrices and equicontinuous sequences of mappings, Bull. Acad. Polon. Sci. Sér. Math. Astronom. Phys., I. 24, 1976, 855-853.

PSEUDO-RADIAL SPACES AND AN OTHER GENERALIZATION OF SEQUENTIAL SPACES

A.V.Arhangel'skiĭ, R.Isler, G.Tironi [1)]

1. Introduction and basic definitions.

Pseudo-radial or chain-net spaces were first introduced by Herrlich [7] in 1967. The same class of topological spaces was then considered by Meyer, Mrówka, Rajagopalan, ... [12], [14], and systematically examined by Arhangel'skiĭ [1], [2]. Some questions presented there stimulated the publication of other papers [9], [15], [6], [8], [16].

In order to avoid trivial exceptions we will suppose that all topological spaces are (T_1).

We shortly recall the basic equivalent definitions of a pseudo-radial or chain-net space:

Definition A. A topological space X is called pseudo-radial or chain-net if, for every non closed subset A there are a point $x \in \bar{A} \setminus A$ and a family $\mathcal{P}$ of subsets of X such that $\mathcal{P}$ is linearly ordered by inclusion and:

(i) $P \cap A \neq \emptyset$ for every P in $\mathcal{P}$

(ii) for every neighbourhood U of x there is P in $\mathcal{P}$ such that $P \subset U$

(iii) $\bigcap \mathcal{P} = \{x\}$.

Definition B. X is a chain-net or pseudo-radial space if for every non closed subset A of X there are a point $x \in \bar{A} \setminus A$ and a λ-sequence $(x_\alpha)_{\alpha<\lambda}$ in A converging to x.

The equivalence of the two definitions is easy to prove if we observe that from every family $\mathcal{P}$ linearly ordered by inclusion we can choose a well-ordered subfamily.

Moreover there is a third definition (originally a theorem of Arhangel'skiĭ [1]) equivalent to definition B.

Definition C. X is a pseudo-radial space if and only if for every non closed set A there exist a point $x \in \bar{A} \setminus A$ and a subset B of A of regular cardinality, such that for any neighbourhood U of x, $|B \setminus U| < |B|$. (see [1]).

Radial or Fréchet chain-net spaces are topological spaces such that every point in $\bar{A} \setminus A$ satisfies the properties of the three equivalent definitions.

We now give the following

Definition 1.1 A subset B of a topological space X is said to be topologically directed

1) Address of the authors: A.V.Arhangel'skiĭ, Chair of Geometry, Moscow State Univ. USSR
R.Isler, G.Tironi, Institute of Applied Mathematics, University of Trieste, Italy.

(in X) if $|B|$ is a regular cardinal number and there exists a point x such that for every neighbourhood U of x, $|B\setminus U|<|B|$. In this case we also say that every neighbourhood of x contains "almost all" points of B. Note that the above condition for x is rather stronger than the other one to be a complete accumulation point. We can call such point x as a "strong accumulation point" for B.

If X is a Hausdorff space, then the point x is unique. In case the point x is unique it will be called the <u>end</u> of B.

The following definition introduces a new cardinal invariant, which seems rather interesting.

<u>Definition 1.2</u>. Let X be a topological space and x a point of it. We call $q\chi(x,X)$: <u>quasi-character of X at x</u> the least cardinal number τ such that, for any $A\subset X$ such that $x\in \bar{A}\setminus A$ there is a family γ of subsets of A such that $|\gamma|\le\tau$, $x\notin\bar{B}$ for any $B\in\gamma$, but $x\in\overline{\bigcup\gamma}$.

We call <u>quasi character of X</u> the cardinal number $q\chi(X) = \sup\{q\chi(x,X):x\in X\}$.

2. Almost-radial spaces.

Pseudo-radial spaces are a generalization of sequential spaces. However converging sequences in sequential spaces have better properties than converging transfinite sequences or topologically directed sets in pseudo-radial spaces.

Consider the property given in the following definition:

<u>Definition 2.1</u>. Let X be a topological space and let be given a subset A and a point x of X. We call the point x a <u>target point</u> for A if for any B, subset of A, the following assertions are equivalent

(i) $x\in\bar{B}$

(ii) $|B| = |A|$.

This property will be denoted by $A \overset{t}{\hookrightarrow} x$ or x(tar)A or by saying that (A,x) is an <u>arrow</u>.

The following are straightforward consequences of our definition 2.1.

<u>Fact 2.1</u>. If $|A|>1$, then x(tar)A implies $x\in\bar{A}\setminus A$.

<u>Fact 2.2</u>. If X is a T_2 space, $A\subset X$ and x,y in X are target points for A, then $x=y$.

In fact observe that if x(tar)A then U contains "almost all" points of A for every neighbourhood U of x, (i.e. A is topologically directed towards x).

<u>Fact 2.3</u>. If A is infinite and x(tar)A in the topological space X, then x is a strong accumulation point of A. Moreover we can observe that not only every neighbourhood U of x contains almost all points of A, but also every subset B of A such that $|B|<|A|$ doesn't have x as accumulation point.

Note that if the space is Hausdorff and x(tar)A, then x is the only complete accumulation point of A.

We recall the following

Definition 2.2. Given a topological space X, a set $A \subset X$ and a point $x \in \bar{A}$, we call primitive tightness of x with respect to A the least infinite cardinality of a set $B \subset A$, such that $x \in \bar{B}$ [5],[10], and we denote it with pt(x,A). Then

Fact 2.4. If x(tar)A in the topological space X, then pt(x,A) $=|A|$.

Definition 2.3. A topological space X is called almost-radial if for every non-closed subset A of X there exist a subset B of A and a point $x \in \bar{A} \setminus A$ such that $|B|$ is a regular cardinal number and x(tar)B.

It is easy to see that definition 2.3 is equivalent to the following

Definition 2.3'. A topological space is called almost-radial if for every non-closed subset A of X there exist a λ-sequence $(x_\alpha)_{\alpha<\lambda}$ of points in A and a point $x \in \bar{A} \setminus A$ such that:

(i) $(x_\alpha)_{\alpha<\lambda}$ converges to x

(ii) for every $\beta<\lambda$, if we call $S_\beta = \{x_\alpha : \alpha<\beta\}$ the initial segment of lenght β, $x \notin \overline{S_\beta}$.

In fact it is enough to observe that λ can be chosen initial and regular and the (x_α) all distinct. Obviously every almost-radial space is pseudo-radial.

Proposition 2.5. Every radial space X is almost-radial.

Proof. Let A be a non-closed set in X, $x \in \bar{A} \setminus A$ and $\tau =$ pt(x,A). Take $M \subset A$ such that $|M| = \tau$ and $x \in \bar{M}$. Since X is radial there is (see [2]) a subset B of M of regular cardinality and topologically directed towards x, i.e. such that $|B \setminus U| < |B|$ for any neighbourhood U of x. Now if $P \subset B$ is such that $x \notin P$, then $|P| < |B|$.

Conversely, if for $P \subset B (\subset A)$, we have $|P| < |B|$, then $|P| < |M| =$ pt(x,A) and $x \notin \bar{P}$. It follows that x(tar)B. Then every radial space is an almost-radial one.

Something more can be said (and in fact was proved above): in a radial space X for every $x \in \bar{A} \setminus A$, there exists $B \subset A$ such that x(tar)B.

Proposition 2.6. Every sequential space is almost-radial.

Theorem 2.7. For every almost-radial space X we have $q\chi(X) = t(X)$.

Proof. It is enough to show that $t(X) \leq \tau = q\chi(X)$. Let A be a non-closed subset of X. Since X is almost-radial there are $x \in \bar{A} \setminus A$ and a set $B \subset A$ such that x(tar)B. As $q\chi(X) = \tau$ there is a family $\gamma \subset \operatorname{Exp} B$ such that $|\gamma| \leq \tau$, $|P| < |B|$ for any P in γ, but $x \in \overline{\bigcup \gamma}$. This fact implies that $|\bigcup \gamma| = |B|$, and from the regularity of $|B|$ and $|P| < |B|$ for any P in γ, it follows $|\gamma| = |B|$. Thus $\text{pt}(x,A) \leq |B| \leq \tau$. Since this holds for any non-closed subset A of X, we conclude that the weak tightness, and hence the tightness is not greater than τ.

Theorem 2.8. Let X be almost-radial with countable tightness. Then X is sequential.

Proof. Let A be an infinite non-closed subset of X. There are $x \in \bar{A} \setminus A$ and $B \subset A$, of regular cardinality, such that x(tar)B. Since $x \in \overline{B} \setminus B$ and $t(X) \leq \aleph_0$, there is also a countable subset C of B such that $x \in \overline{C} \setminus C$. But, since the space is almost-radial, then $|C| = |B| = \aleph_0$. This means that from B a sequence can be extracted which converges to x.

So, for T_1 spaces, a topological space is sequential if and only if it is almost-radial and $t(X) \leq \aleph_0$.

We recall that for pseudo-radial space $\sigma_c(X)$ is the least cardinal number (initial ordinal number) such that for every non-closed set A there is a sequence in A of length $\lambda \leq \sigma_c(X)$ converging to a point $x \in \overline{A} \setminus A$.

More in general we have the following

<u>Theorem 2.9</u>. Let X be almost-radial. Then $t(X) = \sigma_c(X)$.

<u>Proof</u>. It is enough to prove that $\sigma_c(X) \leq t(X)$ (the converse is well known [6]). Put $\tau = t(X)$ and let $A \neq \overline{A}$. There exist a λ-sequence $(x_\alpha)_{\alpha<\lambda}$ and $x \in \overline{A} \setminus A$ such that (x_α) converges to x, x_α all distincts and $x_\alpha \in A$ $\forall \alpha$. Put $B = \bigcup_\alpha \{x_\alpha\}$. Because $x \in \overline{B} \setminus B$, there exists B' ($\subset B$) such that $x \in \overline{B'}$, $|B'| \leq t(X)$. But $|B'| = |B|$ X being almost radial and then the thesis follows.

From the known inequalities for pseudo-radial spaces $t(X) \leq bt(X) \leq div(X) \leq \sigma_c(X)$ (if bt(X) exists), it follows that for almost radial spaces the four invariants coincide. So every pseudo-radial space for which $t(X) < \sigma_c(X)$ cannot be almost-radial. In particular in the well known example (see [9]) of $X = R \cup \{p\}$ with the topology: on the real line R the natural topology, a neighbourhood basis of p are the complements of closed countable sets in R, we have $t(X) = \aleph_0$, $\sigma_c(X) = \aleph_1$.

We recall that [6] the closure character kc(X) of a topological space X is the least cardinal number τ such that C is closed if and only if $C \cap K$ is closed for every closed subset K with $|K| \leq \tau$.

In [5] we proved the following <u>theorem</u>: "For any topological space X the following conditions are equivalent

(i) $kc(X) \leq 2^{t(X)}$

(ii) for any $A \subset X$ with more then one point $|\overline{A}| \leq |A|^{t(X)}$.

From the above result and theorem 2.9 we obtain, for almost-radial spaces, that if bt(X) exists (in particular if X is T_2), $kc(X) \leq 2^{t(X)}$, being well known that $|\overline{A}| \leq |A|^{bt(X)}$.

Hence $|\overline{A}| \leq |A|^{t(X)}$ and $|X| \leq d(X)^{t(X)}$.

References.

[1] A.V.Arhangel'skiĭ: Structure and classification of topological spaces and cardinal invariants, Russian Math. Surveys 33 (1977), 33-95.

[2] A.V.Arhangel'skiĭ: Some properties of radial spaces, Mat. Zametki (Russian) 27 (1980), 95-104.

[3] A.V.Arhangel'skiĭ: Bicompacta that satisfy the Suslin condition hereditarily. Tightness and free sequences, Soviet Math. Dokl. 12 (1971), 1253-1257.

[4] A.V.Arhangel'skiĭ: The frequency spectrum of a topological space and the product operation, Trans. Moscow Math. Soc. 2 (1981), 163-200.

[5] A.V.Arhangel'skiĭ, R.Isler, G.Tironi: On pseudo-radial spaces. To appear.

[6] Z.Frolík, R.Isler, G.Tironi: Some results on chain-net and sequential spaces, Colloquia Math. Soc. J.Bolyai 41. Topology and applications, Eger(Hungary), 1983.

[7] H.Herrlich: Quotienten geordneter Räume und Folgenkonvergenz, Fund. Math. 61 (1967), 79-81.

[8] R.Isler, G.Tironi: Alcuni recenti risultati sugli spazi pseudo-radiali. To appear in the proceedings of the II Italian Meeting on Topology (Taormina), 1984.

[9] I.Jané, P.R.Meyer, P.Simon, R.G.Wilson: On tightness in chain-net spaces, CMUC, 22 (1981), 809-817.

[10] I.Juhász: Cardinal functions in topology: ten years later, Math. Centre Tracts 123, Amsterdam (1980).

[11] I.Juhász, K.Kunen, M.E.Rudin: Two more hereditarily separable non-Lindelöf spaces, Can. Math. J. 28 (1976), 998-1005.

[12] K.Malliha Devi, P.R.Meyer, M.Rajagopalan: When does countable compactness imply sequential compactness?, General Topology and Appl. 6 (1967), 279-289.

[13] P.R.Meyer, R.G.Wilson: Cardinal functions on topological spaces satisfying order conditions, Coll. Math. Soc. J.Bolyai 23, Topology 1978, 839-847.

[14] S.Mrówka, M.Rajagopalan, T.Soundararajan: A characterization of compact scattered spaces through chain-limits, TOPO 72, L.N.M. Vol. 378, Springer, Berlin 1974, 288-297.

[15] G.Tironi: Some remarks on radial and pseudo-radial spaces, Proc. Fifth Prague Topol. Symp. 1981, Helderman, Berlin 1982, 608-612.

[16] G.Tironi: Spazi chain-net e spazi sequenziali. To appear in the proceedings of the II Italian Meeting on Topology, (Taormina) 1984.

STRONGLY FIRST COUNTABLE CONVERGENCE SPACES

R. Beattie[1)] and H.-P. Butzmann[2)]

1. INTRODUCTION

First countable topological spaces are completely determined by their sequences: if two first countable topologies on a set have the same convergent sequences, they must coincide. Consequently closure, completeness and continuity may all be described by sequential behaviour. This is not the case with first countable convergence spaces. While adherence and completeness are sequential, continuity is not. There exist distinct first countable convergence structures on the same set having the same converging sequences. As a result, sequential continuity and continuity are different notions.

In this paper we introduce a notion of countability for convergence spaces strong enough to guarantee that such spaces are completely determined by their convergent sequences. We examine permanence properties of such spaces and illustrate with a number of examples. Finally we show that a convergence space satisfying this stronger countability condition has the Urysohn property if and only if it is countably Choquet.

Throughout, convergence spaces will be abbreviated to cs and, unless otherwise specified, will mean filter convergence spaces. All limits and colimits will be taken in the category <u>CS</u> of convergence spaces and continuous mappings. If (x_n) is a sequence, $F(x_n)$ will denote the Fréchet filter generated by (x_n).

A topological space X is called first countable if the neighbourhood filter of each $p \in X$ has a countable base. This is generalised to convergence spaces by relaxing the condition of neighbourhood filter:

<u>Definition 1.1</u> A convergence space X is called first countable if for every filter $F \longrightarrow p$ in X, there is a coarser filter $G \subset F$ such that $G \longrightarrow p$ and G has a countable basis.

[1)] Dept. of Mathematics, Mount Allison Univ.
Sackville, N.B., EOA 3CO, Canada

[2)] Fakultät für Mathematik und Informatik
Schloß, D-6800 Mannheim

In first countable cs adherence and sequential adherence agree, as was shown by W. Gähler ([8], 3.16.3). Likewise, he proved that in a uniform cs derived from a uniformity with a countable basis - so in particular in every first countable convergence vector space - completeness and sequential completeness agree ([8], 4.12.8). However sequential continuity and continuity are different for first countable cs as the following example shows:

Example 1.2 Let X be any first countable topological space which is not locally compact. Let X_s be the sequential modification in the sense of Frič (cf. e.g. [6]) : A filter F converges in X_s to a point $p \in X$ iff there is a sequence (x_n) in X converging in X to p such that $F \supset F(x_n)$.
Then both X and X_s are first countable and they both have the same convergent sequences. But clearly $X \neq X_s$ since X_s is locally compact and so the identity mapping $id : X \longrightarrow X_s$ is sequentially continuous but not continuous.

2. STRONG FIRST COUNTABILITY

We generalise the notion of first countability in topological spaces to convergence spaces by viewing the countable basis of the neighbourhood filter as a countable local basis:

Defintion 2.1 A cs X is strongly first countable if, at each point $p \in X$, there is a countable local basis of X , i.e. there is a collection $\mathcal{B}_p$ of countably many subsets of X with the property that, for any filter $F \longrightarrow p$ in X , there is a coarser filter $G \subset F$ such that $G \longrightarrow p$ in X and G contains a basis consisting of elements of $\mathcal{B}_p$.

Before we discuss strongly first countable cs in more detail we give some examples:

Examples

i) A pretopological cs is strongly first countable if and only if it is first countable. So in particular every metrisable topological space is strongly first countable.

ii) Every convergence space satisfying the second axiom of countability, and so in particular the continuous dual L_cF of a separable Fréchet space F , is strongly first countable.

iii) The prototype of a strongly first countable cs is the web space $F_{\mathcal{W}}$ associated with a convergence vector space F carrying a vector web $\mathcal{W}$. (See [1] and [2] for more details.)

As was noticed above, "second countable" implies "strongly first countable" and this, in turn, implies "first countable". Each of the converse statements is false. For example, a Fréchet space is strongly first countable but not second countable unless it is separable. An inductive limit of uncountably many Fréchet spaces is first countable but in general not strongly first countable.

As the above examples show, the class of strongly first countable cs is very large. As well, as the following theorem shows, this class enjoys excellent permanence properties.

Theorem 2.2 Let X and $\{X_n : n \in \mathbb{N}\}$ be strongly first countable cs.

i) Any subspace of X is strongly first countable.

ii) Any quotient of X is strongly first countable.

iii) The sum $\sum_{n\in\mathbb{N}} X_n$ is strongly first countable.

iv) The product $\prod_{n\in\mathbb{N}} X_n$ is strongly first countable.

v) If the X_n's form an inductive system, then their inductive limit is strongly first countable.

vi) If the X_n's form a projective system, then their projective limit is strongly first countable.

Proof i), ii) and iii) can easily be seen; v) follows from ii) and iii) while vi) follows from i) and iv). So we have to show iv):

Let $p = (p_n) \in \prod_{n\in\mathbb{N}} X_n$ and choose for each $n \in \mathbb{N}$ a local basis $\mathcal{B}_p$ of X_n at p_n. Then the collection

$$\{B_1 \times B_2 \times \ldots \times B_k \times \prod_{n>k} X_n : k \in \mathbb{N} \text{ and } B_n \in \mathcal{B}_n \text{ for all } n \leq k\}$$

is a countable local basis for $\prod_{n\in\mathbb{N}} X_n$ at p.

Before we state our next result we notice for easier reference:

Remark 2.3 Let X be a cs, $p \in X$ and $\mathcal{B}_p$ a local basis of X at p. If F is a filter on X which does not converge to p then

$$\tilde{\mathcal{B}}_p = \{B \in \mathcal{B}_p : B \notin F\}$$

is a local basis of X at p with $\tilde{\mathcal{B}}_p \cap F = \emptyset$.

Theorem 2.3 Let X be a strongly first countable cs, $p \in X$ and F a filter on X with a countable basis. Then $F \longrightarrow p$ if and only if any sequence (x_n) in X with $F(x_n) \supset F$ converges to p.

Proof The "only if" direction is trivial and so we prove the "if" direction:

Assume that F does not converge to p, then by Remark 2.3 there is a local basis $\mathcal{B} = \{B_n : n \in \mathbb{N}\}$ of X at p with $\mathcal{B} \cap F = \emptyset$.
Choose a basis $\{F_n : n \in \mathbb{N}\}$ of F with $F_{n+1} \subset F_n$ for all $n \in \mathbb{N}$.
Then $F_m \not\subset B_n$ for all m,n and so there are elements $x_{m,n} \in F_m \setminus B_n$ for all $m,n \in N$. List the collection $\{x_{m,n} : m \geq n\}$ to obtain a sequence $(x_i)_{i \in \mathbb{N}}$. We show $F(x_n) \supset F$:
For all $m \in \mathbb{N}$ the set $\{(\mu,\nu) : \nu \leq \mu \leq m\}$ is finite and so there is some $i_o \in \mathbb{N}$ such that

$$x_i = x_{\mu,\nu} \quad \text{with} \quad \mu \geq m \quad \text{for all} \quad i \geq i_o .$$

Hence
$$x_i \in F_\mu \subset F_m \qquad \text{for all} \quad i \geq i_o .$$

Thus $F(x_i)$ converges to p by assumption so that there is some $r_o \in \mathbb{N}$ with $B_{r_o} \in F(x_i)$. But then

$$\{x_i : i \geq i_1\} \subset B_{r_o}$$

for some $i_1 \in \mathbb{N}$. However

$$x_{\nu,r_o} \notin B_{r_o} \qquad \text{for all} \quad \nu \geq r_o ,$$

and so there are infinitely many $i \in \mathbb{N}$ with $x_i \notin B_{r_o}$
This contradiction completes our proof.

Theorem 2.4 Let X be a first countable cs, Y a strongly first countable cs and $f : X \longrightarrow Y$ a sequentially continuous mapping. Then f is continuous.

Proof Let $F \longrightarrow p$ in X and assume, without loss of generality, that F has a countable basis $\{F_n : n \in \mathbb{N}\}$ with $F_{n+1} \subset F_n$ for all n.
Then $f(F)$ has the countable basis $\{f(F_n) : n \in \mathbb{N}\}$. Let (y_n) be a sequence in Y such that $F(y_n) \supset f(F)$. Then for each $n \in \mathbb{N}$ there is some $k(n) \in \mathbb{N}$ such that

$$f(F_n) \supset \{y_k : k \geq k(n)\}$$

W.l.o.g. again, we may assume $k(n+1) > k(n)$ for all $n \in \mathbb{N}$.

For each $k(n) \leq i < k(n+1)$, choose $x_i \in F_n$ with $f(x_i) = y_i$. Then since $\{x_i : i \geq k(n)\} \subset F_n$ for all $n \in \mathbb{N}$, we have $F(x_i) \supset F$. Thus $(x_k)_{k \geq k(1)}$ converges to p in X and so $(f(x_k)_{k \geq k(1)})$ converges to $f(p)$ in Y . But then (y_k) converges also to $f(p)$. By Theorem 2.3 it follows that $f(F) \longrightarrow f(p)$ and so f is continuous.

Corollary 2.5 Let X_1 and X_2 be two strongly first countable cs on the same set having the same convergent sequences. Then X_1 and X_2 coincide.

Theorem 2.4 and Corollary 2.5 show that for strongly first countable cs, sequential and filter behaviour are very close. However, an examination of the proof of Theorem 2.4 reveals that what is actually required is not strong first countability but rather the property stated in Theorem 2.3 that a filter with a countable basis converges to a point if every finer sequence does. We study cs with this property in much more detail in [3] and [5].

3. THE URYSOHN PROPERTY IN STRONGLY FIRST COUNTABLE CS

Since strongly first countable cs are completely determined by their convergent sequences it seems to be worthwile to examine it in the framework of sequential convergence in the sense of Novak (cf. e.g. [6]). The sequential convergence associated with a cs X satisfies for all elements $p \in X$ the axioms:

($L1$) the constant sequence (p) converges to p

($L2$) if (x_n) converges to p , then every subsequence of (x_n) converges to p .

One of the interesting properties of sequential convergence is the Urysohn property:

($L3$) a sequence (x_n) converges to p if every subsequence of (x_n) contains a subsequence which converges to p .

We say that a cs has the Urysohn property if its associated sequential convergence satisfies the axiom ($L3$), and are now able to characterise strongly first countable cs having the Urysohn property in well-known terms:

<u>Definition 3.1</u> A cs X is called solid or a Choquet space if a filter F on X converges to p whenever every ultrafilter $G \supset F$ converges to p. Similarly, X is called countably Choquet when this property holds for every F with a countable basis.

<u>Theorem 3.2</u> Let X be a Hausdorff strongly first countable cs. Then X has the Urysohn property if and only if X is countably Choquet.

<u>Proof</u> Assume that X has the Urysohn property and let F be a filter with a countable basis which does not converge to $p \in X$. We show the existence of an ultrafilter $G \supset F$ which does not converge to p:

By Theorem 2.3 there is a sequence (x_n) in X with $F(x_n) \supset F$ which does not converge to p, and by the Urysohn property we get a subsequence (x_{n_k}) without subsequence converging to p. Since X is first countable we know by 3.16.2 in [8] that p is not a point of adherence of (x_{n_k}). Take any ultrafilter $G \supset F(x_{n_k})$, then $G \supset F$ and $G \not\to p$, as desired.

Assume now that X is countably Choquet and let (x_n) be a sequence in X which does not converge to $p \in X$. Then we know that there is an ultrafilter $G \supset F(x_n)$ with $G \not\to p$ and by Remark 2.3 there is a local basis $B = \{B_n : n \in \mathbb{N}\}$ of X at p with $B \cap G = \emptyset$. And so $X \setminus B_n \in G$ and hence

$$X \setminus (B_1 \cup \ldots \cup B_n) \in G \qquad \text{for all } n \in \mathbb{N} .$$

Since $\{x_n : n \in \mathbb{N}\} \cap (X \setminus B_1) \neq \emptyset$, we can choose $x_{n_1} \in X \setminus B_1$. Similarly, since $\{x_n : n > n_1\} \cap [X \setminus (B_1 \cup B_2)] \neq \emptyset$, we can choose an element $x_{n_2} \in X \setminus (B_1 \cup B_2)$. Continuing in this way, we obtain a subsequence (x_{n_k}) such that $x_{n_k} \notin B_1 \cup \ldots \cup B_k$ for all $k \in \mathbb{N}$. Thus, for all $r \geq k$ we have $x_{n_r} \notin B_1 \cup \ldots \cup B_k$. It follows that (x_{n_k}) is a subsequence of (x_n) no subsequence of which converges to p.

We finish this paper by showing that the conditions in Theorem 3.2 are independent from each other for first countable cs. Consider the following examples:

Example 3.3 We define a sequential convergence L on $\mathbb{R}$ as follows: A sequence (x_n) of reals converges to $p \in \mathbb{R}$ with respect to L if and only if

$$\sum_{n=1}^{\infty} |x_n - p| < \infty .$$

The Frič convergence space X associated with L is a Choquet space since every sequential cs is c-embedded (cf. [7], Prop. 4). However X does not have the Urysohn property: every subsequence of $(1/n)$ has a subsequence which converges to 0 , but $(1/n)$ does not converge to 0

Example 3.4 Let (r_n) be a sequence in $\mathbb{R}$ with the property

$$(r_n + \mathbb{Q}) \cap (r_k + \mathbb{Q}) = \emptyset \qquad \text{for all } n \neq k$$

Let $A_n = r_n + \mathbb{Q}$ for all $n \in \mathbb{N}$ and F_n the filter generated by

$$\{(\mathbb{R} \setminus A_n) \cap \ldots \cap (\mathbb{R} \setminus A_{n+k}) \cap U : r \in \mathbb{N} \cup \{0\} , U \in N(0)\} ,$$

where $N(p)$ denotes the neighbourhood filter of a point $p \in \mathbb{R}$ in the natural topology.

We define a convergence structure μ on $\mathbb{R}$ as follows:

$F \longrightarrow p$ in $(\mathbb{R},\mu)$ if and only if $F \supset N(p)$ $p \neq 0$

$F \longrightarrow 0$ in $(\mathbb{R},\mu)$ if and only if $F \supset F_k \cap F(x_n)$, for some $k \in \mathbb{N}$ and some sequence (x_n) converging to 0 in the natural topology.

It is clear that μ is a first countable convergence structure on $\mathbb{R}$ which is finer than the natural topology but which has precisely the same convergent sequences. Thus $(\mathbb{R},\mu)$ has the Urysohn property. On the other hand, $(\mathbb{R},\mu)$ is not countably Choquet since $N(0)$ does not converge to 0 in $(\mathbb{R},\mu)$, whereas every finer ultrafilter $G \supset N(0)$ does.

REFERENCES

[1] Beattie, R., Convergence spaces with webs, Math. Nachr. 116 (1984) 159-164.

[2] Beattie, R., The closed graph theorem in convergence spaces with webs, to appear in the Proc. Intern. School on Convergence Structures and Appl. II, Schwerin 1983, Abh. Akad. Wiss. DDR (in print).

[3] Beattie, R. and Butzmann, H.-P., Convergence spaces determined by their sequences, to appear.

[4] Binz, E., Continuous convergence on C(X), Lecture Notes in Mathematics 469, Springer-Verlag, Berlin-Heidelberg-New York.

[5] Butzmann, H.-P. and Beattie, R., Some relations between sequential and filter convergence, these Proceedings.

[6] Frič, R., On convergence spaces and groups, Math. Slovaca 27 (1977) 323-330.

[7] Frič, R. and Kent, D.C., on c-embedded sequential convergence spaces, in Convergence Structures and Applications to Analysis, Abh. Akad. Wiss. DDR, Abt. Math.-Naturwiss.-Technik, 1979, Nr. N4, Akademie Verlag, Berlin 1980, 33-36.

[8] Gähler, W., Grundstrukturen der Analysis Band I, Berlin Basel 1977.

[9] Schroder, M., Solid convergence spaces, Bull. Austral. Math. Soc. 8 (1973), 443-459.

ASCOLI'S THEOREM FOR A CLASS OF MEROTOPIC SPACES

H.L. Bentley[1] and H. Herrlich[2]

ABSTRACT: We prove two new theorems of Ascoli type - i.e. theorems which characterize totally bounded subspaces of certain function spaces in terms of uniform equicontinuity and pointwise total boundedness. The setting for these results is a certain cartesian closed category of merotopic spaces: the coreflective hull of the contigual spaces.

* * * *

John W. Gray has presented a category theoretic study of Ascoli theorems [2], and Oswald Wyler asks for more examples of Ascoli theorems to clarify the status of a satisfactory general theory [8]. Our objective here is to present two rather general Ascoli theorems for a certain class of merotopic spaces, a comprehensive category introduced by M. Katětov containing as nicely embedded full subcategories both the category Top of all symmetric topological spaces and the category Unif of all uniform spaces [4], [3]. We won't reproduce all the definitions and basic results here; rather, we refer the reader to the works of M. Katětov and H. Herrlich cited above.

We are concerned here initially with the category Cont of all contigual merotopic spaces; it is interesting largely because Cont ∩ Top is the category of all compact symmetric topological spaces and because Cont ∩ Unif is the category of all precompact uniform spaces (= the category of all proximity spaces). Cont is a bireflective subcategory of the category Grill of all grill determined merotopic spaces (= Katětov's filter merotopic spaces); besides Katětov, W.A. Robertson also studied the category Grill [7].

Even more generally, we will be concerned with the coreflective hull of Cont in Mer, which for simplicity we denote by $\underline{A}$, whose objects were characterized internally by H.L. Bentley and H. Herrlich [1] as follows:

---

1) Department of Mathematics, University of Toledo, Toledo, Ohio 43606 USA

2) Fachbereich Mathematik und Informatik, Universität Bremen, Postfach 330440, 28 Bremen 33, Fed. Rep. Germany

Proposition 1. $\underline{A}$ = Grill $\cap$ LTB, i.e. a merotopic space X is an object of $\underline{A}$ iff X is grill determined and is locally totally bounded. More simply, X is an object of $\underline{A}$ iff for every micromeric collection $\mathcal{G}$ in X there exists a Cauchy filter $\mathcal{F}$ in X which corefines $\mathcal{G}$ and which contains some totally bounded member.

The reader is reminded that a merotopic space X is totally bounded (Katětov says full bounded) iff each uniform cover of X is refined by a finite cover of X (not necessarily uniform), equivalently, iff every ultrafilter on X is micromeric. TB, the category of totally bounded spaces, is bireflective in Mer. A merotopic space is locally totally bounded iff X has a uniform cover each member of which is totally bounded. The subcategory LTB of locally totally bounded spaces is bicoreflective in Mer.

Proposition 2. A merotopic space is totally bounded iff its Grill-coreflection is.

Proof:

Let GX → X be the Grill coreflection of X. Since total boundedness is preserved by uniformly continuous surjections, GX being totally bounded implies X is. Assume X is totally bounded and let $\mathcal{U}$ be an ultrafilter on GX. Then $\mathcal{U}$ is near on X.
Since $\mathcal{U}$ is a grill, $\mathcal{U}$ is near on GX.

Corollary 3. If X is totally bounded, then its Grill-coreflection is the same as its $\underline{A}$-coreflection.

Corollary 4. A merotopic space X is totally bounded iff its $\underline{A}$-coreflection is.

Proposition 5. If $(f_i\colon X \to X_i)_{i\in I}$ is an initial source in $\underline{A}$ with each X_i totally bounded, then X is totally bounded.

Proof:

Let $(f_i\colon Z \to X_i)_{i\in I}$ be the corresponding initial source in Grill. In order to show Z is totally bounded, let $\mathcal{U}$ be an ultrafilter on Z. Then each $f_i\mathcal{U}$ generates an ultrafilter on X_i which is therefore micromeric in X_i. It follows that $\mathcal{U}$ is micromeric in Z and so Z is totally bounded. Since Z is therefore an $\underline{A}$-space, Z = X.

L.D. Nel studied the category $\underline{A}$ as an example in his theory of cartesian closed coreflective hulls [6], and he proved that $\underline{A}$ is cartesian closed. Let Mer(X,Y) denote the set of all uniformly continuous maps f: X → Y. We make Mer(X,Y) into a merotopic space, denoted Mer[X,Y], by defining a cover $\mathcal{D}$ of Mer(X,Y) to be uniform iff for some uniform cover $\mathcal{B}$ of Y, $C(\mathcal{B})$ refines $\mathcal{D}$, where $C(\mathcal{B})$ is the collection of all subsets of Mer(X,Y) of the form

$$S(B,\mathcal{A},\mathcal{B}) = \{h \mid \forall A \in \mathcal{A}\ h[A] \subset B(A)\}$$

where $\mathcal{A}$ is a uniform cover of X and B: $\mathcal{A} \to \mathcal{B}$ is any map.

The $\underline{A}$-power, denoted $\underline{A}[X,Y]$, for X contigual, is the $\underline{A}$-coreflection of Mer[X,Y]. For a general X in $\underline{A}$, we take a final epi sink $(K_i \to X)_{i \in I}$ in $\underline{A}$ with each K_i contigual and we decree the mono source

$$(\underline{A}[X,Y] \longrightarrow \underline{A}[K_i,Y])_{i \in I}$$

to be initial in $\underline{A}$.

For our Ascoli theorems, we need the concept of a uniformly equicontinuous set of maps. Let X and Y be merotopic spaces and let F ⊂ Mer(X,Y). Then we say that F is uniformly equicontinuous if for every uniform cover $\mathcal{B}$ of Y,

$$\bigwedge_{f \in F} f^{-1}\mathcal{B} = \{ \bigcap_{f \in F} f^{-1}[B_f] \mid B_f \in \mathcal{B} \}$$

is a uniform cover of X. Note that

$$\mathcal{A} = \bigwedge_{f \in F} f^{-1}\mathcal{B}$$

refines each $f^{-1}\mathcal{B}$.

<u>Lemma 6</u>. Let X and Y be contigual and let F be a uniformly equicontinuous subspace of Mer[X,Y]. Then F is totally bounded.

Proof:

Let $\mathcal{U}$ be an ultrafilter on F; we must show that $\mathcal{U}$ is micromeric on Mer[X,Y]. Let $\mathcal{B}$ be a uniform cover of Y; without loss of generality we may assume $\mathcal{B}$ is finite. Since F is uniformly equicontinuous and since X is contigual, there exists a finite uniform cover $\mathcal{A}$ of X such that for every $f \in F$, $\mathcal{A}$ refines $f^{-1}\mathcal{B}$. For each $A \in \mathcal{A}$ and each $B \in \mathcal{B}$, let

$$L^A_B = \{ f \in F \mid f[A] \subset B \}$$

For each $A \in \mathcal{A}$, we have

$$\bigcup \{ L^A_B \mid B \in \mathcal{B} \} = F \in \mathcal{U}$$

so there exists $B(A) \in \mathcal{B}$ with $L^A_{B(A)} \in \mathcal{U}$. The rule $A \longmapsto B(A)$ gives a map $B\colon \mathcal{A} \to \mathcal{B}$ and we have

$$F \cap S(B,\mathcal{A},\mathcal{B}) = \cap \{L^A_{B(A)} \mid A \in \mathcal{A} \} \in \mathcal{U}.$$

Thus $S(B,\mathcal{A},\mathcal{B}) \in \mathcal{C}(\mathcal{B})$ and $S(B,\mathcal{A},\mathcal{B})$ meets every member of $\mathcal{U}$. It follows that $\mathcal{U}$ is micromeric.

Theorem 7. Let Y be contigual and let X be any $\underline{A}$-space. Then for any subspace F of the $\underline{A}$-power $\underline{A}[X,Y]$, the following statements are equivalent:

(1) F is uniformly equicontinuous.

(2) F is totally bounded.

Proof:

(1) ⇒ (2): Consider first the case where X is contigual. Let G be the subspace of Mer[X,Y] with the same underlying set as F. Then G is uniformly equicontinuous so by Lemma 6, G is totally bounded. F is the $\underline{A}$-coreflection of G so from Proposition 2 and Corollary 3 it follows that F is totally bounded. Consider next any $\underline{A}$-space X and let $(K_i \to X)_{i \in I}$ be a final epi sink in $\underline{A}$ with each K_i contigual. Then the source

$$(g_i\colon \underline{A}[X,Y] \longrightarrow \underline{A}[K_i,Y])_{i \in I}$$

is an initial mono source in $\underline{A}$. Clearly, for each $i \in I$, $g_i[F]$ is uniformly equicontinuous so by the above, $g_i[F]$ is totally bounded. By restriction we have that

$$(g_i \mid F: F \longrightarrow g_i[F])_{i\in I}$$

is an initial mono source in $\underline{A}$. By Proposition 5, F is totally bounded.

(2) ⇒ (1): Consider first the case where X is contigual. Let $\mathcal{B}$ be a uniform cover of Y. Since the $\underline{A}$-coreflection $\underline{A}[X,Y] \to \mathrm{Mer}[X,Y]$ is uniformly continuous, $C(\mathcal{B})$ is a uniform cover of $\underline{A}[X,Y]$. Since F is totally bounded, there exists a finite subset $\mathcal{H}$ of $C(\mathcal{B})$ with $F \subset \cup\, \mathcal{H}$. For each $H\in\mathcal{H}$ there exists a uniform cover $\mathcal{A}_H$ of X and there exists a map $B_H: \mathcal{A}_H \to \mathcal{B}$ with $H = S(B_H, \mathcal{A}_H, \mathcal{B})$. Then

$$\bigwedge_{H\in\mathcal{H}} \mathcal{A}_H$$

is a uniform cover of X which refines

$$\bigwedge_{f\in F} f^{-1}\mathcal{B}$$

and it follows that F is uniformly equicontinuous. Consider next any $\underline{A}$-space X and let $(f_i: K_i \to X)_{i\in I}$ be a final epi sink with each K_i contigual. Then the source

$$(g_i: \underline{A}[X,Y] \longrightarrow \underline{A}[K_i,Y])_{i\in I}$$

is an initial mono source in $\underline{A}$. Since uniform surjections preserve total boundedness and since F is totally bounded, we have that for each $i\in I$, $g_i[F]$ is totally bounded. By the above, since K_i is contigual, it follows that each $g_i[F]$ is uniformly equicontinuous. To show F is uniformly equicontinuous, let $\mathcal{B}$ be a uniform cover of Y. Then for each $i\in I$,

$$\mathcal{A}_i = \bigwedge_{f\in g_i[F]} f^{-1}\mathcal{B} = \bigwedge_{f\in F} f_i^{-1}f^{-1}\mathcal{B}$$

is a uniform cover of K_i. Therefore

$$\mathcal{A} = \cup\ \{f_i\mathcal{A}_i \mid i\in I\}$$

is a uniform cover of X (because the sink $(f_i: K_i \to X)_{i\in I}$ is final in Mer). Clearly $\mathcal{A}$ refines

$$\bigwedge_{f\in F} f^{-1}\mathcal{B}$$

and it follows that F is uniformly equicontinuous.

The preceding theorem was for the case where Y is contigual. In the next theorem, we don't need Y to be contigual; rather we require Y to be uniform.

<u>Theorem 8</u>. Let X and Y be $\underline{A}$-spaces with Y uniform and let F be a subspace of $\underline{A}[X,Y]$. Then F is totally bounded iff

(1) F is uniformly equicontinuous.

and

(2) For each $x \in X$ $F(x) = \{f(x) \mid f \in F\}$ is totally bounded as a subspace of Y.

<u>Proof</u>:

$\Rightarrow$: Assume F is totally bounded. The proof that (2) $\Rightarrow$ (1) in Theorem 7 works here also to show that F is uniformly equicontinuous. The evaluation map

$$ev: X \times A[X,Y] \longrightarrow Y$$

from the $\underline{A}$-product is uniformly continuous. Restricting, we get the uniformly continuous map $X \times F \to Y$. For each $x \in X$, we have the uniformly continuous map $F \to X \times F$, $f \longmapsto (x,f)$. Composing these, we get a map $F \to Y$ with image $F(x)$. Since F is totally bounded, so is $F(x)$.

$\Leftarrow$: Assume that conditions (1) and (2) hold and assume first that X is contigual. Let G be the subspace of the $\underline{A}$-product

$$\prod_{x \in X} F(x)$$

with the same underlying set as F. We shall show that the restriction

$$g = ev \mid (X \times G): X \times G \to Y$$

of the evaluation map is uniformly continuous. Let $\mathcal{B}$ be a uniform cover of Y and let $\mathcal{H}$ be a uniform cover of Y which starrefines $\mathcal{B}$. Let $\mathcal{A}$ be a finite uniform cover of X which refines

$$\bigwedge_{f \in F} f^{-1}\mathcal{H}.$$

For each nonempty $A \in \mathcal{A}$, select $a_A \in A$ and let

$$T = \{a_A \mid \Phi \neq A \in \mathcal{A}\}.$$

For each $x \in X$, let

$$\pi_x : \prod_{x \in X} F(x) \longrightarrow F(x)$$

be the canonical projection map and let

$$\mathcal{D} = \{ A \times \bigcap_{x \in T} \pi_x^{-1}[H_x] \mid A \in \mathcal{A} \text{ and each } H_x \in \mathcal{H} \}.$$

Then $\mathcal{D}$ is a uniform cover of $X \times \prod_{x \in X} F(x)$.

Since, as is easily shown, $\mathcal{D} \wedge \{X \times G\}$ refines $g^{-1}\mathcal{B}$, it follows that g is uniformly continuous. Therefore its transpose

$$\hat{g}: G \longrightarrow \underline{A}[X,Y]$$

is also uniformly continuous. Since G is totally bounded, so is $F = \hat{g}[G]$. For the general case, where X is an $\underline{A}$-space which may not be totally bounded, the proof of the corresponding generalization for (1) ⇒ (2) of Theorem 7 works here also.

REFERENCES

[1] H.L. Bentley and H. Herrlich, The coreflective hull of the contigual spaces in the category of merotopic spaces, In: Lecture Notes in Mathematics 915, Springer-Verlag, Berlin (1982) 16-26.

[2] John W. Gray, Ascoli's theorem for topological categories, In: Lecture Notes in Mathematics 915, Springer-Verlag, Berlin (1982) 86-104.

[3] H. Herrlich, Topological structures, In: Topological structures, Math. Centre Tracts 52, Amsterdam (1974) 59-122.

[4] M. Katětov, Allgemeine Stetigkeitsstrukturen, Proc. Intern. Congr. Mathematicians, (Stockholm 1962) 473-479.

[5] M. Katětov, Spaces defined by a family of centered systems, Russian Math. Surveys 31:5 (1976) 111-113, from Uspekhi Mat. Nauk 31:5 (1976) 95-107.

[6] L.D. Nel, Cartesian closed coreflective hulls, Quaest. Math. 2 (1977) 269-283.

[7] W.A. Robertson, Convergence as a Nearness Concept, Thesis, Carleton University (1975).

[8] Oswald Wyler, Ascoli theorems for cartesian closed categories, In: Sigma Series in Math., Heldermann-Verlag, to appear.

ACKNOWLEDGEMENT

The first author hereby acknowledges financial support by a FRAF research grant from the Graduate School of the University of Toledo.

ON MATRIX PROPERTIES OF CONVERGENCES

Józef Burzyk [1/]

1. In this paper we are concerned with matrix properties of convergences. We deal with relations between the following properties.

Y. If $x_{ij} \in X$ for $i,j \in N$, $x_{ij} \to x_i$ as $j \to \infty$ for $i \in N$ and $x_i \to x$, then there is a sequence $\{m_i\}$ such that $x_{in_i} \to x$ whenever $m_i \leq n_i$ for $i \in N$.

D. If $x_{ij} \in X$ for $i,j \in N$, $x_{ij} \to x_i$ as $j \to \infty$ for $i \in N$ and $x_i \to x$, then there is a sequence $\{m_i\}$ such that $x_{im_i} \to x$.

Obviously, Y implies D. The author was informed by R. Frič that P. Kratochvil gave an example of a FUSH-convergence space satisfying D but not Y. In this paper we construct an example of a FLUSH-convergence group satisfying D but not Y. By the same method one can construct an example of a FLUSH-convergence linear space with D but not with Y.

We recall that if G is a convergence in a set X then G^* is the convergence in X such that $x_n \xrightarrow{G^*} x$ iff for each subsequence $\{u_n\}$ of $\{x_n\}$ there is a subsequence $\{v_n\}$ of $\{u_n\}$ such that $v_n \xrightarrow{G} x$. If G is an F-convergence, then G^* is an extension of G.

<u>Theorem 1</u>. If G is an FLSHD-convergence in an abelian group, then G^* is an FLUSHD-convergence.

A simple prove of Theorem 1 is omitted here.

Theorem 1 says that D-property is a hereditary property under the star operation on FLSHD-convergences. In this paper we show that Y is not necessarily a hereditary property under the star operation on FLSHY-convergences.

2. In this section we are concerned with convergences in the power set P(N) of all subsets of N. We consider P(N) as an abelian group with the symmetric difference for the addition. We have $A+B=A-B=(A\setminus B)\cup(B\setminus A)$ for $A,B \in P(N)$. We note that $-A=A$ and $0=\emptyset$. As usually, I_A denotes the characteristic function of a set A. We define

$$\|A\| = \sum_{n=1}^{\infty} \frac{1}{2^n} I_A(n)$$

[1/] Institute of Mathematics, Polish Academy of Sciences
Katowice Branch, Wieczorka 8, 40-013 Katowice, Poland

The functional $\|\cdot\|$ is a norm on $P(N)$ and $(P(N),\|\cdot\|)$ is a normed group. We write $A_n \to A$ iff $\|A_n - A\| \to 0$.

In the sequel we shall deal with restrictions of the norm convergence in $P(N)$. The restrictions will be obtained by imposing some conditions on unions of convergent sequences. To this end we adopt the following definitions.

Assume that $\mathcal{R}$ is a subideal of $P(N)$, i.e., $\mathcal{R} \subset P(N)$, $A,B \in \mathcal{R}$ implies $A \cup B \in \mathcal{R}$ and $B \subset A$, $A \in \mathcal{R}$ implies $B \in \mathcal{R}$. By $G(\mathcal{R})$ we denote the convergence in $P(N)$ such that $A_n \xrightarrow{G(\mathcal{R})} A$ iff $\|A_n - A\| \to 0$ and $\bigcup_{n=1}^{\infty} (A_n - A) \in \mathcal{R}$. If $\mathcal{R} = P(N)$, then $G(\mathcal{R})$ is the norm convergence in $P(N)$. If $\mathcal{R}$ is the family of all finite subsets of N, then $A_n \xrightarrow{G(\mathcal{R})} A$ iff $A_n = A$ for almost all $n \in N$.

<u>Theorem 2.</u> For each subideal $\mathcal{R}$ of $P(N)$, $G(\mathcal{R})$ is an FLSH-convergence.

A simple proof of the theorem is omitted here.

For any positive integers m and n such that $m \leqslant n$, $[m,n]$ denotes the set of all integers p with $m \leqslant p \leqslant n$.

We note that $A_n \to A$ in $P(N)$ iff for each $k \in N$, there exists $n_0 \in N$ such that for each $n \geqslant n_0$ we have

$$A_n \cap [1,k] = A \cap [1,k].$$

By $\mathcal{F}$ we denote the family of all finite subsets of N. For any sequence of sets $A_n \in P(N)$ we assume that $V(\{A_n\})$ is the family of all sets of the form

$$\bigcup_{n=1}^{\infty} (A_n \setminus C_n)$$

where $C_n \in \mathcal{F}$ for $n \in N$.

<u>Theorem 3.</u> For each subideal $\mathcal{R}$ of $P(N)$ the following conditions are equivalent:

(i) $G(\mathcal{R})$ satisfies Y,

(ii) $G(\mathcal{R})$ satisfies D,

(iii) $\mathcal{R}$ satisfies the condition:

For each sequence $\{A_n\}$ in $\mathcal{R}$ we have $V(\{A_n\}) \cap \mathcal{R} \neq \emptyset$.

<u>Proof</u>: The implication (i)$\Rightarrow$(ii) is obvious. To prove that (ii)$\Rightarrow$(iii) we take a sequence $\{A_n\}$ with $A_n \in \mathcal{R}$. Next, we consider the matrix

$$A_{ij} = A_i \setminus [1,j]$$

for $i,j \in N$. We note that $A_{ij} \xrightarrow{G(\mathcal{R})} 0$ as $j \to \infty$ for $i \in N$. By D there exists a sequence of positive integers $\{m_i\}$ such that

$$A_{im_i} \xrightarrow{G(\mathcal{R})} 0.$$

Assume that $C_i = [1,m_i]$ for $i \in N$. Then we have

$$\bigcup_i (A_i \setminus C_i) = \bigcup_i A_{im_i} \in \mathcal{R}.$$

which proves that

$$V(\{A_n\}) \cap \mathcal{R} \neq \emptyset.$$

Now we are going to prove that (iii) $\Longrightarrow$ (i). To this end we take a matrix $\{A_{ij}\}$ of elements in 2^N such that $A_{ij} \xrightarrow{G(\mathcal{R})} 0$ as $j \to \infty$ for $i \in N$. We put

$$A_i = \bigcup_{j=1}^{\infty} A_{ij}$$

for $i \in N$. Then $A_i \in \mathcal{R}$ for $i \in N$. By (iii), there are finite subsets C_i of N such that

$$\bigcup_{i=1}^{\infty} (A_i \setminus C_i) \in \mathcal{R}.$$

Since $\|A_{ij}\| \to 0$ as $j \to \infty$ for $i \in N$ and C_i are finite sets, there exists an increasing sequence of positive integers $\{p_i\}$ such that $\|A_{ij}\| < \frac{1}{i}$ and $A_{ij} \cap C_i = \emptyset$ for $j \geqslant p_i$. Let $\{q_i\}$ be an arbitrary sequence of positive integers such that $p_i \leqslant q_i$ for $i \in N$. Then we have $\|A_{iq_i}\| < \frac{1}{i}$, $A_{iq_i} \subset A_i \setminus C_i$ and

$$\bigcup_{i=1}^{\infty} A_{iq_i} \subset \bigcup_{i=1}^{\infty} (A_i \setminus C_i) \in \mathcal{R}.$$

Since $\mathcal{R}$ is an ideal, we have

$$\bigcup_{i=1}^{\infty} A_{iq_i} \in \mathcal{R}.$$

Consequently, $A_{iq_i} \xrightarrow{G(\mathcal{R})} 0$ which proves that (iii) $\Rightarrow$ (i). Thus proof of the theorem is complete.

3. We construct two examples of subfamilies $\mathcal{R}$ of P(N) satisfying the conditions:

(a) $\mathcal{F} \subset \mathcal{R}$, $N \notin \mathcal{R}$;

(b) If $A, B \in \mathcal{R}$, then $A \cup B \in \mathcal{R}$;

(c) If $A \subset B$ and $B \in \mathcal{R}$, then $A \in \mathcal{R}$;

(d) If $A_n \in \mathcal{R}$ for $n \in N$, then $V(\{A_n\}) \cap \mathcal{R} \neq \emptyset$.

The conditions (b) and (c) say that $\mathcal{R}$ is a subideal of P(N).

For each $A \in P(N)$, the symbol $|A|$ denotes the cardinality of A. If A is a finite set, then $|A|$ is the number of elements of A. For each $A \in P(N)$ and each $n \in N$ we assume that

$$m_n(A) = \frac{|A \cap [1,n]|}{n}$$

and

$$m(A) = \limsup_{n \to \infty} m_n(A).$$

By $\mathcal{O}$ we denote the family of all subsets A of N with $m(A) = 0$.

<u>Lemma 1</u>. The family $\mathcal{O}$ satisfies the conditions (a) - (d).

Proof. It is obvious that $\mathcal{O}$ satisfies (a), (b) and (c). To show (d) we take $A_n \in \mathcal{O}$ for $n \in N$. Since $m(A_i) = 0$, there exists an integer p_1 such that for each $n \geq p_1$ we have

$$m_n(A_1) < \frac{1}{2^2}.$$

There exists a positive integer p_2 such that $p_1 < p_2$ and

$$m_n(A_i) < \frac{1}{2^{2+i}}$$

for $i=1,2$ and $n \geq p_2$. By induction we construct an increasing sequence of positive integer $\{p_m\}$ such that

$$m_n(A_i) < \frac{1}{2^{m+i}}$$

for $i=1,\dots,m$ and $n \geq p_m$. This implies that

$$m_n(A_i \setminus [1,p_i]) < \frac{1}{2^{m+i}}$$

for each $n \geq p_m$ and each $i \in N$. Consequently, by the subadditivity of m_n, we have

$$m_n\Big(\bigcup_{i=1}^{\infty} (A_i \setminus [1,p_i])\Big) < \sum_{i=1}^{\infty} \frac{1}{2^{m+i}} = \frac{1}{2^m}$$

whenever $n \geq p_m$ for $m \in N$. This implies that

$$v(\{A_n\}) \cap \mathcal{O} \neq \emptyset.$$

From Lemma 1 and Theorem 2, it follows that $G(\mathcal{O})$ is a FLSHD-convergence.

Lemma 2. If $\|B_m\| \to 0$ and $\{|B_m|\}$ is a bounded sequence, then there is a subsequence $\{B_{p_n}\}$ such that

$$\bigcup_{n=1}^{\infty} B_{p_n} \in \mathcal{O}.$$

Proof. Assume that $|B_m| < a$ for $m \in N$. Set $\{p_n\}$ be an increasing sequence of positive integers such that $[1,n^2] \cap B_{p_n} = \emptyset$. Then we have

$$\Big|\Big(\bigcup_{i=1}^{\infty} B_{p_i}\Big) \cap [1,n]\Big| = \Big|\Big(\bigcup_{i=1}^{\infty} B_{p_i}\Big) \cap [1,n^2]\Big| \leq n.a.$$

Hence, we get

$$m_n\Big(\bigcup_{i=1}^{\infty} B_{p_i}\Big) = \frac{\Big|\Big(\bigcup_{i=1}^{\infty} B_{p_i}\Big) \cap [1,n]\Big|}{n} \leq \frac{na}{n^2} = \frac{a}{n}$$

Consequently, we get

$$m\Big(\bigcup_{i=1}^{\infty} B_{p_i}\Big) = 0$$

or equivalently,

$$\bigcup_{i=1}^{\infty} B_{p_i} \in \mathcal{O}.$$

The next subideals of P(N) are induced by increasing sequences. Suppose that $\{p_k\}$ is an increasing sequence of positive integers. By $\Delta(\{p_k\})$ we denote the family of all subsets A of N such that

$$\sum_{k=1}^{\infty} \frac{|A \cap [p_k, p_k+k]|}{k} < \infty$$

For any set $\mathcal{J}$ of increasing sequences we assume that

$$\Delta(\mathcal{J}) = \bigcap_{\{p_k\} \in \mathcal{J}} \Delta(\{p_k\})$$

Lemma 3. For any countable family $\mathcal{J}$ of increasing sequences of positive integers, the family $\Delta(\mathcal{J})$ satisfies the conditions (a)-(d).

Proof. It is evident that $\Delta(\mathcal{J})$ satisfies the conditions (a), (b) and (c). To prove (d), we arrange all elements of $\mathcal{J}$ into a sequence of sequences $\{(p_1^n, p_2^n, \ldots)\}_{n=1}^{\infty}$ and we take a sequence of elements A_m in $\Delta(\mathcal{J})$. We note that

$$\sum_{k=1}^{\infty} \frac{|A_m \cap [p_k^n, p_k^n+k]|}{k} < \infty$$

for every m and n in N. By induction we construct a sequence of finite sets C_m such that

$$\sum_{k=1}^{\infty} \frac{|(A_m \setminus C_m) \cap [p_k^n, p_k^n+k]|}{k} < \frac{1}{2^m}$$

for $n=1,\ldots,m$. We put

$$A = \bigcup_{m=1}^{\infty} (A_m \setminus C_m).$$

For every $n \in N$ we have

$$\sum_{k=1}^{\infty} \frac{|A \cap [p_k^n, p_k^n+k]|}{k} \leq \sum_{k=1}^{\infty} \sum_{m=1}^{\infty} \frac{|(A_m \setminus C_m) \cap [p_k^n, p_k^n+k]|}{k} \leq$$

$$\sum_{m<n} \sum_{k=1}^{\infty} \frac{|(A_m \setminus C_m) \cap [p_k^n, p_k^n+1]|}{k} + \sum_{m=n}^{\infty} \sum_{k=1}^{\infty} \frac{|(A_m \setminus C_m) \cap [p_k^n, p_k^n+k]|}{k} \leq$$

$$C + \sum_{m=n}^{\infty} \frac{1}{2^m} < \infty$$

*) For the definition see the next page.

The last inequality shows that $A \in \Delta(\mathcal{J})$ which completes the proof.

By Lemma 2 and Theorem 2, $G(\Delta(\mathcal{J}))$ is a FLSHY-convergence.

Lemma 4. Suppose that $\mathcal{J}$ is a countable family of strongly increasing sequences*) of positive integers and $\{B_m\}$ is a sequence of subsets of N such that $\|B_m\| \to 0$ and the sequence $\{|B_m|\}$ is bounded. Then there is a subsequence $\{B_{q_m}\}$ such that $\bigcup_{m=1}^{\infty} B_{q_m} \in \Delta(\mathfrak{J})$.

Proof. We arrange all members of $\mathfrak{J}$ into a sequence of sequences $\{(p_1^n, p_2^n, \ldots)\}$ and we assume that $|B_m| < a$ for $m \in N$. By induction we select a subsequence $\{B_{q_m}\}$ such that

$$B_{q_m} \cap [1, p_{2^m}^n + 2^m] = \emptyset$$

for $n=1,\ldots,m$. We note that if $n<m$ and

$$(1) \qquad B_{q_m} \cap [p_k^n, p_k^n+k] \neq \emptyset$$

then $k > 2^m$. Since $|B_m| < a$ for $m \in N$ and intervals $[p_k^n, p_k^n+k]$ are disjoint, the number of all k-th such that (1) holds, is not greater than a. Hence we get

$$\sum_{k=1}^{\infty} \frac{B_{q_m} \cap [p_k^n, p_k^n+k]}{k} < \frac{a^2}{2^m}$$

if $n<m$. We put

$$B = \bigcup_{m=1}^{\infty} B_{q_m}.$$

Then for every $n \in N$ we have

$$\sum_{k=1}^{\infty} \frac{|B \cap [p_k^n, p_k^n+k]|}{k} < \sum_{m<n} \sum_{k=1}^{\infty} \frac{|B_{q_m} \cap [p_k^n, p_k^n+k]|}{k}$$

$$+ \sum_{m=n}^{\infty} \sum_{k=1}^{\infty} \frac{|B_{q_n} \cap [p_k^n, p_k^n+k]|}{k} < c_n + \sum_{m=n}^{\infty} \frac{a^2}{2^m} < \infty.$$

This shows that $B \in \Delta(\mathfrak{J})$ which proves the lemma.

A sequence of positive integers $\{p_k\}$ is said to be strongly increasing if $p_{k+1} > p_k + k$ for $k=1,2,\ldots$. By $\mathfrak{Y}$ we denote the set of all strongly increasing sequences.

Lemma 5. For every countable subfamily $\mathcal{T}$ of $\mathcal{O}$ and every sequence of positive integers $\{p_k\}$ there is a strongly increasing sequence of positive integers $\{q_k\}$ such that $p_k \leq q_k$ for $k \in N$ and

$$\mathcal{T} \subset \Delta(\{q_k\}).$$

Proof. We arrange all members of $\mathcal{T}$ into a sequence $T_1, T_2, \ldots$ and we assume that

$$A_k = T_1 \cup \ldots \cup T_k$$

for $k \in N$. We note that $A_k \in \mathcal{O}$ for $k \in N$. Consequently, there are positive integers q_k such that $p_k \leq q_k$, $q_k + k < q_{k+1}$ and

$$A_k \cap [q_k, q_k+k] = \emptyset .$$

For every $n \in N$ we have

$$\sum_{k=1}^{\infty} \frac{|T_n \cap [q_k, q_k+k]|}{k} = \sum_{k<n} \frac{|T_n \cap [q_k, q_k+k]|}{k} +$$

$$+ \sum_{k=n}^{\infty} \frac{|T_n \cap [q_k, q_k+k]|}{k} = \sum_{k<n}^{\infty} \frac{|T_n \cap [q_k, q_k+k]|}{k} < \infty .$$

This means that $\mathcal{T} \subset \Delta(\{q_k\})$.

Theorem 4. There is a subfamily $\mathcal{R}$ of $\mathcal{O}$ and a family $\mathfrak{J}$ of strongly increasing sequences such that the following conditions hold:

(i) $\mathcal{R}$ satisfies conditions (a) - (d).

(ii) If $\|B_m\| \to 0$ and $\{|B_m|\}$ is a bounded sequence, then there is a subsequence $\{B_{p_m}\}$ such that

$$\bigcup_{m=1}^{\infty} B_{p_m} \in \mathcal{R} .$$

(iii) For each sequence of positive integers $\{p_n\}$, there exists a sequence $\{q_n\}$ in $\mathfrak{J}$ such that $p_n \leqslant q_n$ for $n \in N$.

(iv) $\mathcal{R} \subset \Delta(\mathfrak{J})$.

Proof. We note that the cardinality of the set of all sequences $\{T_m\}$ in $\mathcal{O}$ and the cardinality of the set of all sequences $\{p_m\}$ in N are ω_1. We arrange all sequences in $\mathcal{O}$ into a transfinite sequence $\{\{T_m^{\gamma}\}_{m=0}^{\infty};\ 1 \leqslant \gamma < \omega_1\}$ and we arrange all sequences in N into a transfinite sequence $\{\{p_m^{\gamma}\}_{m=1}^{\infty};\ 1 \leqslant \gamma < \omega_1\}$. By transfinite induction we shall construct a transfinite sequence $\{\mathcal{T}_\gamma\}_{\gamma=0}^{\omega_1}$ of subfamilies $\mathcal{T}_\gamma$ of $\mathcal{O}$ and a transfinite sequence $\{\mathfrak{J}_\gamma\}_{\gamma=0}^{\omega_1}$ of subsets $\mathfrak{J}_\gamma$ of $\mathcal{Y}$ such that for each γ with $0 \leqslant \gamma < \omega_1$ the following conditions hold:

(1) $\mathcal{F} \subset \mathcal{T}_\gamma \subset \mathcal{O}$, $N \notin \mathcal{T}_\gamma$;

(2) $|\mathcal{T}_\gamma| \leqslant \aleph_0$ and $|\mathfrak{J}_\gamma| < \aleph_0$;

(3) if $\xi < \gamma$, then $\mathcal{T}_\xi \subset \mathcal{T}_\gamma$ and $\mathfrak{J}_\xi \subset \mathfrak{J}_\gamma$;

(4) $\mathcal{T}_\gamma$ satisfies (a) and (b),

(5) $\mathcal{T}_\gamma \subset \Delta(\mathfrak{J}_\gamma)$;

(6) if $\|T_m^{\gamma}\| \to 0$ and $\{|T_m^{\gamma}|\}$ is a bounded sequence, then there is a subsequence $\{T_{p_m}^{\gamma}\}$ such that

$$\bigcup_{m=1}^{\infty} T_{p_m}^{\gamma} \in T_\gamma ;$$

(7) there exists a sequence $\{q_n\} \in \mathfrak{J}_\gamma$ such that $p_n^{\gamma} \leqslant q_n$ for $n \in N$;

(8) if $\varkappa(\gamma) = \inf \{\beta :\ T_n^{\beta} \in \bigcup_{\xi<\gamma} \mathcal{T}_\xi$ for $n \in N$, $V(\{T_n^{\beta}\}) \cap \bigcup_{\xi<\eta} \mathcal{T}_\xi = \emptyset\}$, then $V(\{T_n^{\varkappa(\gamma)}\}) \cap \mathcal{T}_\gamma = \emptyset$.

We start our construction of $\mathcal{T}_\gamma$ and $\mathfrak{J}_\gamma$ for $\gamma = 0$. We assume that $\mathcal{U} = \mathcal{F}$. By Lemma 5, there exists a sequence $\{q_n\}$ in $\mathcal{Y}$ such that $p_m^{0} \leqslant q_n$

for $n \in N$ and $\mathcal{U} \subset \Delta(\{q_n\})$. We assume that $\mathfrak{I}_0$ is the set containing the only sequence $\{q_n\}$. To define $\mathcal{T}_0$ we consider the sequence $\{T^0_m\}$. If the sequence does not satisfy the assumption of (6), we define $\mathcal{U}_0 = \mathcal{U}$. Assume that $\{T^0_m\}_{m=1}^{\infty}$ satisfies the assumptions of (6). Then applying lemmas 2 and 4 we find a subsequence $\{T^0_{p_m}\}$ such that

$$\bigcup_{m=1}^{\infty} T^0_{p_m} \in \Delta(\mathfrak{I}_0) \cap \mathcal{O}.$$

In this case we put $\mathcal{U}_0 = \mathcal{U}[B]$ with

$$B = \bigcup_{m=1}^{\infty} T^0_{p_m}$$

We note that $\varkappa(0)$ does not exist. In this case we define $\mathcal{T}_0 = \mathcal{U}_0$. Now it is easy to see that the families $\mathcal{T}_0$ and $\mathfrak{I}_0$ satisfy conditions (1) - (8) with $\gamma = 0$.

Suppose that $\mathcal{T}_\xi$ and $\mathfrak{I}_\xi$ for $\xi < \gamma$ have been constructed and satisfy (1) - (8). To construct $\mathcal{T}_\gamma$ and $\mathfrak{I}_\gamma$ we put

$$\mathcal{U} = \bigcup_{\xi < \eta} \mathcal{T}_\xi \quad \text{and} \quad \mathfrak{I} = \bigcup_{\xi < \gamma} \mathfrak{I}_\xi .$$

Then, by (3) and (5), we have $\mathcal{U} \subset \Delta(\mathfrak{I})$. By (2), $\mathcal{U}$ is a countable family. By Lemma 5, there exists a sequence $\{q_n\}$ in $\mathcal{Y}$ such that

$$p^\gamma_n < q_n \text{ for } n \in N \quad \text{and} \quad \mathcal{U} \subset \Delta(\{q_n\}).$$

We now define $\mathfrak{I}_\gamma = \mathfrak{I} \cup \{\{q_n\}_{n=1}^{\infty}\}$ and we note that $\mathcal{U} \subset \Delta(\mathfrak{I}_\gamma)$. Now consider sequence $\{T^\gamma_m\}$. If the sequence does not satisfy the assumptions of (6), we put $\mathcal{U}_\gamma = \mathcal{U}$. Assume that $\{T^\gamma_m\}$ satisfies the assumptions of (6). Applying Lemma 2 and Lemma 4, we find a subsequence $\{T^\gamma_{p_m}\}$ such that

$$\bigcup_{m=1}^{\infty} T^\gamma_{p_m} \in \Delta(\mathfrak{I}_\gamma) \cap \mathcal{O}.$$

In this case we put $\mathcal{U}_\gamma = \mathcal{U}[B]$ with

$$B = \bigcup_{m=1}^{\infty} T^\gamma_{p_m} .$$

If $\varkappa(\gamma)$ does not exist, we define $\mathcal{T}_\gamma = \mathcal{U}_\gamma$.

If $\varkappa(\gamma)$ exists, we consider sequence $\{T^{\varkappa(\gamma)}_m\}$. Then $T^{\varkappa(\gamma)}_m \in \mathcal{U}$ for $m \in N$. Then from Lemma 1 and Lemma 2 it follows that

$$V(\{T^{\varkappa(\gamma)}_m\}) \cap \Delta(\mathfrak{I}_\gamma) \cap \mathcal{O} \neq \phi.$$

Assume that

$$C \in V(\{T^{\varkappa(\gamma)}_m\}) \cap \Delta(\mathfrak{I}_\gamma) \cap \mathcal{O}$$

and

$$\mathcal{T}_\gamma = \mathcal{U}_\gamma[C].$$

It is easy to see that $\mathcal{T}_\gamma$ and $\mathfrak{I}_\gamma$ satisfy conditions (1) - (8).

Finally, we put

$$\mathcal{R} = \Big(\bigcup_{\xi<\omega_1} \mathcal{T}_\xi\Big)^* \quad \text{and} \quad \mathfrak{I} = \bigcup_{\xi<\omega_1} \mathfrak{I}_\xi$$

We explain that if $\mathcal{A}$ is a family of sets, then $\mathcal{A}^*$ is the family of all subsets of sets in $\mathcal{A}$.

It can be proved that $\mathcal{R}$ and $\mathfrak{I}$ satisfy the conditions (i) - (iv).

4. An example of an FLUSHD-convergence in $P(N)$ which does not satisfy Y'. Assume that $\mathcal{R}$ and $\mathfrak{I}$ are families satisfying the condition (i) - (iv) in Theorem 4. By (i) and Theorem 2, $G(\mathcal{R})$ is an FLSHY-convergence in $P(N)$. By Theorem 1, $G(\mathcal{R})^*$ is an FLUSHD-convergence. We claim that $G(\mathcal{R})^*$ does not satisfy Y-condition. In fact, consider matrix $A_{ij}=[j,j+i]$ for $i,j \in N$. By (ii), we have $A_{ij} \xrightarrow{G(\mathcal{R})^*} 0$ as $j \to \infty$ for $i \in N$. We assert that for each sequence $\{p_i\}$ of positive integers, there exists a sequence $\{q_i\}$ such that $p_i \leq q_i$ and the sequence A_{iq_i} is not $G(\mathcal{R})^*$ -convergent. Otherwise there exists a sequence $\{p_i\}$ of positive integers such that $A_{iq_i} \xrightarrow{G(\mathcal{R})^*} 0$ whenever $p_i \leq q_i$ for $i \in N$. By (iii) there is a strongly increasing sequence $\{q_i\}$ in $\mathfrak{I}$ such that $p_i \leq q_i$ and $\mathcal{R} \subset \Delta(\{q_i\})$. In particular, for this sequence, we have $A_{iq_i} \xrightarrow{G(\mathcal{R})^*} 0$. Thus there exists an increasing sequence of positive integers $\{p_i\}$ such that $A_{p_i q_p} \xrightarrow{G(\mathcal{R})} 0$. This implies that

$$\bigcup_{i=1}^{\infty} A_{p_i q_{p_i}} \in \mathcal{R}. \tag{1}$$

On the other hand, for each $i \in N$, we have

$$\frac{\left|\left(\bigcup_{i=1}^{\infty} A_{p_i q_{p_i}}\right) \cap [q_{p_n}, q_{p_n}+p_n]\right|}{p_n} = 1$$

Consequently,

$$\sum_{n=1}^{\infty} \frac{\left|\left(\bigcup_{i=1}^{\infty} A_{p_i q_{p_i}}\right) \cap [q_n, q_n+n]\right|}{n} = \infty$$

Hence, we get

$$\bigcup_{i=1}^{\infty} A_{p_i q_{p_i}} \notin \Delta(\{q_n\})$$

and

$$\bigcup_{i=1}^{\infty} A_{p_i q_i} \notin \mathcal{R} \tag{2}$$

for $\mathcal{R} \subset \Delta(\{q_n\})$. The relations (1) and (2) are contradictary. Therefore $G(\mathcal{R})$ does not satisfy Y.

SOME RELATIONS BETWEEN SEQUENTIAL AND FILTER CONVERGENCE

H.-P. Butzmann[1] and R. Beattie[2]

With very few exceptions, the theories of sequential and filter convergence are developped independently from each other. It was not until 1977 when R. Frič gave a relation between the two ([2]). This turns into an embedding of a very large class of sequential convergence spaces into that of filter convergence spaces. Later he studied together with D.C. Kent filter convergence spaces which are derived in this way from sequential ones. But their correspondence does not carry over algebraic compatibilities and the necessary modification, e.g. for convergence groups, is not easy to handle. Therefore we introduce yet another correspondence which behaves in some respect nicer than Frič's one.

In this paper we present both approaches together with some basic results. Finally we show that our procedure links in a very natural way the completion theories developped by Novák and Frič/Kent for sequential and filter convergence groups, respectively.

In order to make this paper reasonably self-contained, we start with the two fundamental definitions:

Definition 1 The pair (X,L) is called a sequential convergence space if X is a set and L a mapping from X into the power set of $X^{\mathbb{N}}$, the set of all sequences in X, such that the following are satsified for all $x \in X$:

($L1$) If $(x_n) \in X^{\mathbb{N}}$ and $x_n = x$ for all but finetely many $n \in \mathbb{N}$, then $(x_n) \in L(x)$

($L2$) If $(x_n) \in L(x)$, then $(x_{n_k}) \in L(x)$ for every subsequence (n_k) of (n).

1) Fakultät für Mathematik und Informatik der Universität Mannheim, Schloß, D-6800 Mannheim

2) Dept. of Mathematics, Mount Allison University, Sackville, N.B., EOA 3CO , Canada

If $(x_n) \in L(x)$ for some $(x_n) \in X^{\mathbb{N}}$ and some $x \in X$, we write

$$(x_n) \longrightarrow x \in L$$

and say (x_n) converges to x (with respect to L).

Definition 2 The pair (X,λ) is called a filter convergence space if X is a set and λ is a mapping from X into the power set of $F(X)$, the set of all filters on X, satisfying for all $x \in X$ the following:

(F1) $\dot{x} = \{A \subset X : x \in A\} \in \lambda(x)$

(F2) If $F \in \lambda(x)$, $G \in F(X)$ and $F \subset G$, then $G \in \lambda(x)$.

Again, we write $F \longrightarrow x \in \lambda$

and say F converges to x (with respect to λ)

if $F \in \lambda(x)$ for some $F \in \lambda(x)$ and some $x \in X$.

Whereas Definition 1 is a slight restriction of the usual requirements for a sequential convergence space (see e.g. [5]), Definition 2 gives a generalisation of the term "convergence space" in the sense of Binz and Fischer (see e.g. [1]) which is called L-space by Poppe ([7]). Nevertheless, the usual terminology for "convergence spaces" carries easily over to filter convergence spaces and we shall use it in this sense.

Given a sequential convergence space (X,L), we follow Frič and define

$$\varkappa(L) : X \longrightarrow P(F(X))$$

by stating that a filter $F \in F(X)$ belongs to $\varkappa(L)(x)$ for $x \in X$ iff there is a sequence $(x_n) \in L(x)$ such that $F \supset F(x_n)$, where $F(x_n)$ is the Fréchet filter generated by (x_n).

Evidently $(X,\varkappa(L))$ is a locally compact, first countable filter convergence space. As Frič/Kent proved in [4], it is unter very mild conditions pseudotopological, i.e. a filter converges to a point if every finer ultrafilter does. Finally $(X,\varkappa(L))$ is under these conditions c-embedded in the sense of Binz ([1]) if and only if (X,L) admits point-separating continuous, real-valued functions.

Whereas there are some reasonable ways to associate a filter convergence structure with a sequential one, the opposite seems to be unique: For every filter convergence space (X,λ) we define

$$L(\lambda) : X \longrightarrow P(X^{\mathbb{N}}) \qquad \text{by}$$

$$(x_n) \in L(\lambda)(x) \Longleftrightarrow F(x_n) \in \lambda(x)$$

Clearly $(X,L(\lambda))$ is a sequential convergence space. If we now want to treat sequential convergence spaces in the framework of filter convergence, the sequential one should not distinguish between sequences which generate the same filter. With this in mind one defines:

<u>Definition 3</u> A sequential convergence space (X,L) satisfies the condition (FL) if for any pair of sequences (x_n) and (y_n) in X with $F(x_n) = F(y_n)$ the sequence (x_n) converges to a point $x \in X$ iff (y_n) does.

<u>Proposition 1</u> Let (X,L) be a sequential convergence space. Then we have $L(\varkappa(L)) = L$ if and only if (X,L) satisfies condition (FL).

The result of Proposition 1 sets up a one-to-one correspondence between the class of all sequential convergence spaces satisfying condition (FL) and a subclass of that of all filter convergence spaces. Now we turn our attention to the algebraic aspect. Notice that the following definition of a sequential convergence group differs from the one given by J. Novak (cf. e.g. [2]) but coincides with it in the case considered in [6] .

<u>Definition 4</u> Let $(X,+)$ be a commutative group.

(i) The triple $(X,+,L)$ is called a sequential convergence group if (X,L) is a sequential convergence space and for all $x,y \in X$ $(x_n) \in L(x)$ and $(y_n) \in L(y)$ imply $(x_n-y_n) \in L(x-y)$.

(ii) The triple $(X,+,\lambda)$ is called a filter convergence group if (X,λ) is a filter convergence space and for all $x,y \in X$ $F \in \lambda(x)$ and $G \in \lambda(y)$ imply $F - G \in \lambda(x-y)$.

Now, given a filter convergence group $(X,+,\lambda)$, then $(X,+,L(\lambda))$ is a sequential convergence group, but the converse is not true. To be more precise: If $(X,+,\lambda)$ is a first countable topological group, then, as just pointed out, $(X,+,L(\lambda))$ is a sequential convergence group, whereas $(X,+,\varkappa(L(\lambda)))$ is a filter convergence group if and only if (X,λ) is discrete. So even for the reals $\mathbb{R}$ endowed with the natural convergence N we have that $(\mathbb{R},\varkappa(N))$ is not a filter convergence group.
In order to motivate our following definition we notice that for every filter F with a countable basis one has

$$F = \bigcap \{F(x_n) : (x_n) \in X^{\mathbb{N}} \text{ and } F(x_n) \supset F\} ,$$

and so in a topological space such a filter converges to a point iff every finer sequence does. Furthermore, in a first countable topological space a filter converges to a point iff it contains a filter with a countable basis with the above property. This is precisely the announced

Definition 5 Let (X,L) be a sequential convergence space. Then we define a mapping

$$\rho(L) : X \longrightarrow P(F(X))$$

by stating that a filter $F \in F(X)$ belongs to $\rho(L)(x)$ iff there is a filter $G \subset F$ with a countable basis such that $(x_n) \in L(x)$ for all sequences $(x_n) \in X^{\mathbb{N}}$ with $F(x_n) \supset G$.

One can easily see that $(X,\rho(L))$ is a filter convergence space. Moreover we have:

Proposition 2 Let (X,L) be a sequential and (X,λ) a filter convergence space. Then

(i) $L(\rho(L)) = L$ if and only if (X,L) satisfies (FL) .

(ii) $\rho(L(\lambda)) = \lambda$ if λ is a first countable (pre-)topology.

(iii) If $(X,+,L)$ is a sequential convergence group, then $(X,+,\rho(L))$ is a filter convergence group.

So the condition in (i) is the same one that appeared for Frič's modification, whereas (ii) shows that many "good" spaces arise in this correspondence. The space $(X,\rho(L))$ has many nice properties, so it is e.g. countably pseudotopological - i.e. every filter with a countable base converges if every finer ultrafilter does - if L is a maximal convergence structure in the sense of Novák (see e.g. [5]).
Along with convergence groups one has always the concepts of Cauchy and completeness. A filter F in a filter convergence group is called a Cauchy filter if $F - F$ converges to zero (the neutral element) and a sequence (x_n) in a sequential convergence group is called a Cauchy sequence if $(x_{n_i} - x_{k_i})$ converges to zero for every pair of subsequences (n_i) and (k_i) of (n). Finally a filter (sequential) convergence group is complete if and only if every Cauchy filter (sequence) converges. In order to avoid technical difficulties, we shall assume in the following that the involved sequential structures are maximal.

Proposition 3 Let $(X,+,L)$ be a maximal sequential convergence group. Then we have:

(i) A sequence $(x_n) \in X^{\mathbb{N}}$ is a Cauchy sequence in $(X,+,L)$ if and only if $F(x_n)$ is a Cauchy filter in $(X,+,\rho(L))$

(ii) The sequential convergence group $(X,+,L)$ is complete if and only if the filter convergence group $(X,+,\rho(L))$ is complete.

Finally we want to apply this concept to completions: Given a convergence group - sequential or filter - one wants to know whether one can embed it into a complete one. In the abelian case both problems have been solved:

Theorem (Novák [6]) Every abelian, maximal Hausdorff sequential convergence group (X,L) can be densely embedded into a complete, abelian, maximal Hausdorff sequential convergence group $(\tilde{X},\tilde{L})$.

Theorem (Frič/Kent [3]) Every abelian Hausdorff filter convergence group (X,λ) can be densely embedded into a complete abelian Hausdorff filter convergence group $(\hat{X},\hat{\lambda})$.

Both constructions are very natural and very similar and so Frič and Kent asked the question whether there is a relationship between the two. We are now in a position to answer this question:

Theorem Let $(X,+,L)$ be an abelian, maximal Hausdorff sequential convergence group. Then

$$(\tilde{X},\tilde{L}) = (\hat{X},L(\widehat{\rho(L)})) .$$

References

[1] Binz, E., Continuous Convergence on C(X), Lecture Notes in Mathematics 469, Springer Verlag, Berlin-Heidelberg-New York 1975.

[2] Frič, R., On Convergence spaces and groups, Math. Slovaca 27 (1977), 323-330.

[3] Frič, R. and Kent, D.C., Completion of pseudotopological groups, Math. Nachr. 99 (1980), 99-103.

[4] - , On c-embedded sequential convergence spaces, in: Convergence Structures and Applications to Analysis, Abh. Akad. Wiss. DDR, Abt. Math.-Naturwiss.-Technik, 1979, Nr. N4, Akademie-Verlag, Berlin 1980, 33-36.

[5] Frič, R. and Koutník, V., Sequential structures, Convergence Structures and Applications to Analysis, Abh. Akad. Wiss. DDR, Abt. Math.-Naturwiss.-Technik, 1979, Nr. N4, Akademie-Verlag Berlin 1980, 37-56.

[6] Novák, J., On completions of convergence commutative groups, General Topology and its Relations to Modern Analysis and Algebra III (Proc. Third Prague Topological Sympos., 1971), Academia, Praha 1972, 335-340.

[7] Poppe, H., Compactness in general function spaces, VEB Deutscher Verlag der Wissenschaften, Berlin 1974.

AN ALTERNATIVE DEFINITION OF PSEUDORADIAL SPACES

Mario Dolcher[1)]

The notion of a pseudoradical space (A. V. ARHANGEL'SKII [1]; H. HERRLICH [2], calling them chain-net spaces) is here presented in a form different from the usual ones, by means of what I call <u>co-limits</u> of filters; this notion may appear to have a somehow constructive character.

To show concretely how the procedure works, I add the description of some tree-spaces, whence one can draw, according to the topology chosen for the tree, in a standard way, a variety of situations of convergence.

All what is here said for pseudoradiality can be repeated for sequentiality: it suffice to restrict all considerations to the case of ordinary sequences (i.e. of length ω) and their limits.

TERMINOLOGY, NOTATIONS

An <u>η-sequence</u> of elements of a set E is any mapping $(x_\alpha)_{\alpha<\eta}$ of the set of ordinals $<\eta$ into E. A <u>sequence</u> is any η-sequence (ordinary sequences are ω-sequences). If $\beta<\eta$, $(x_\alpha)_{\beta\leq\alpha<\eta}$ is said to be a <u>rest</u> of $(x_\alpha)_{\alpha<\eta}$.

If $\mathcal{F}, \mathcal{G}$ are filters on a set E, $\mathcal{F}$ is said to be coarser than $\mathcal{G}$ (and $\mathcal{G}$ finer than $\mathcal{F}$) if $\mathcal{F}\subset\mathcal{G}$. We denote by $[\mathcal{B}]$ the filter having $\mathcal{B}$ as a basis. We write $[A]$, $[x]$ (if $\emptyset\neq A\subset E$, $x\in E$) instead of $[\{A\}]$, $[\{\{x\}\}]$.

In a topological space X, $\mathcal{U}_x$ is the neighborhood filter of x. A point x is said to be <u>accessible</u> from A $(\subset E)$ if it is a limit of a sequence of points of A; <u>accessible from outside A</u> if it is accessible from $\complement A$.

COLIMIT

Let be given, on a set E, a filter $\mathcal{G}$ and a family Φ of filters $\mathcal{F}_x$, indexed in E. We then denote by $\mathcal{F}_M$ (for $M\subset E$) the intersection filter $\bigcap_{x\in M}\mathcal{F}_x$, (generated by sets $\bigcup_{x\in M} F_x$, $F_x\in\mathcal{F}_x$) Then the set family

$$\{\bigcup_0^\infty{}_n A_n : A_0\in\mathcal{G}, \quad A_{n+1}\in\mathcal{F}_{A_n} \quad \text{for } n\in N\}$$

is a filter basis; we call this filter the <u>colimit of $\mathcal{G}$ according to Φ</u>, and we denote it by

$$\mathrm{Col}_\Phi\,\mathcal{G}\ .$$

1) Via Tagliapietra 4 - 34123 Trieste, Italy

Apparently, this filter is coarser than $\mathcal{G}$.

The particular situation for our purposes is: E is a topological space; $\mathcal{G}$ is a set-filter (i.e. $[A]$ with $\emptyset \neq A \subset E$, possibly $[x]$ with $x \in E$); $\mathcal{F}_x$ is the sequence-converging filter at x, i.e. the intersection of all sequence filters converging to x; Φ is as above.

PSEUDORADIAL SPACES

We recall that a topological space X is said to be pseudoradial iff for every non-closed set $C \subset X$ there is a point of $\overline{C} \setminus C$ accessible from C; or, equivalently, iff every non-open set A has a point accessible from outside A.

(Sequential spaces can be defined by the same words, if only accessibility is understood in the sense of ω-sequences).

THEOREM

A topological space X is pseudoradial if and only if for every point x the filter of its neighborhoods is the colimit of the ultrafilter x.

PROOF. First, let X be pseudoradial. For every $x \in X$, let A be any set of the basis of $\mathrm{Col}[x]$ given in the definition. Then no point of A is accessible from outside A: in fact, if $x \in A$, then $x \in A_m$ for some m; and consequently every sequence converging to x has a rest in A_{m+1} ($\subset A$). So, X being pseudoradial, A is an open neighborhood of x: hence, $\mathrm{Col}[x]$ is coarser than $\mathcal{U}_x$. On the other hand, independently of our assumption on X, every open neighborhood U of x may be taken as $A_1 = A_2 = \ldots$ and therefore belongs to the colimit filter. We conclude that $\mathrm{Col}[x] = \mathcal{U}_x$.

Conversely, if X is not pseudoradial there is a set $C \neq \overline{C}$ such that no point y of $\overline{C} \setminus C$ is limit of any sequence of points of C. Then there are members of $\mathcal{F}_y$ having no point in C: in fact, taking for every $(x_\alpha)_{\alpha<\eta}$ converging to y a rest lying outside C, the union of these rests belongs to $\mathcal{F}_y$, and it is outside C: let us denote by A_1 such a union of rests. Now, we prove by induction that, starting with A_1, we can find an ω-sequence $(A_n)_{n\geq 1}$ such that $\bigcup_1^\omega{}_n A_n \cap C = \emptyset$. Suppose $A_n \cap C = \emptyset$; then no point of A_n being limit of a sequence in C, we can define A_{n+1} as the union of rests, one for every sequence converging to a point of A_n, lying outside C. Then let $A = \bigcup_0^\omega{}_n A_n$: we have $A \cap C = \emptyset$, and A is not a neighborhood of y, since $y \in \overline{C}$. (Moreover, $\mathrm{Col}[y]$ is strictly finer than $\mathcal{U}_y$, as remarked at the end of the first part of the proof.)

It is noteworthy that, as the Theorem shows, in a pseudoradial space the neighborhoods of any points can be obtained by a countable number of steps, whatever may be the length of the sequences needed to describe the convergence. As it is known, the closure of a set - even in the case of sequential spaces - requires a possibly uncountable number of steps.

ON TREES

We define a (γ,λ)-tree (γ and λ being ordinals > 0) as the set $T_{\gamma,\lambda}$ of all α-sequences with $\alpha \leq \lambda$[1] of ordinals $< \gamma$. This set is partially ordered by

$x = (\xi^\beta)_{\beta<\alpha_1} < y = (\eta^\beta)_{\beta<\alpha_2}$ iff $\alpha_1 < \alpha_2$ and $\xi^\beta = \eta^\beta$ for $\beta < \alpha_1$.

If x is any α-sequence in $T_{\gamma,\lambda}$, we say that α is the <u>length of x</u>; for each $\alpha \leq \lambda$ the set T_α of all x's length α is said to be the <u>α-th level</u> in the tree. There is one element of length 0, denoted by $()$. Each T_α can be thought of as totally ordered lexicographically. For every x of limit level α the set of all points preceding it forms (ordered by length) an α-sequence called the <u>thread (leading) to x</u>, or having x as <u>endpoint</u>. <u>Successors</u> of a point of length α ($<\lambda$) are said to be the points of length $\alpha + 1$ following it. The ordinal λ is called the <u>length of the tree</u>.

In the sequel, γ and λ are assumed to be both limit ordinals. We shall restrict our statements on convergence to the trees $T_{\omega,\omega}$ and T_{ω,ω_1}; for $T_{\omega,\lambda}$ the statements and proof are more involved, but there are no essential changes; for $T_{\gamma,\lambda}$ with $\gamma \neq \omega$, the topologies that we are going to describe seem not to be the most interesting.

TOPOLOGY τ_1 ON $T_{\gamma,\lambda}$

On a tree $T_{\gamma,\lambda}$ a set is defined as open iff

(1) with each of its elements it contains a rest of the γ-sequence of its successors,

(2) with each of its elements having as length a limit ordinal, it contains a rest of the thread to it.

The space **$T=(T_{\gamma,\lambda};\tau_1)$ is not regular**, not first-countable, disconnected. It is separable iff either γ and λ are both countable, or γ is singular with cofinality ω and λ is countable. Under the same conditions its tightness and its pseucharacter are $\aleph_0$. The subspace topology on each level T_α is discrete.

<u>The tree $T = (T_{\omega,\omega};\tau_1)$ is sequential</u>. Denote by T' the set of all points of finite length and by T'' the ω-th level. First, let $x' \in T'$. With regard to the definition of $Col[x']$, a set like A

[1] It would be more suitable "with $\alpha < \lambda$", because adopting the notation in the text there is no symbol to mean e.g. the set of all finite sequences of ordinals $< \gamma$. On the other hand since the interstring case is that of γ limit ordinal included, it would be troublesome to write always $\lambda + 1$ in place of λ.

can be taken in T' , and it satisfies (1) in the definition of $\mathcal{T}_1$ ((2) being then vacuous); it then belongs to $\mathcal{U}_{x'}$. For $x''\in T''$, a set A of $\mathrm{Col}[x'']$ must contain a rest of the thread to x'' , so it fulfils (2); and it is a neighborhood of every point in such a rest as stated above for points of T' . The tree $T = (T_{\omega,\omega_1}; \mathcal{T}_1)$ is pseudoradial, not sequential. Here, let T' be the union of all countable levels, and T'' be the ω_1-th level. Since T' is not closed though ω-sequentially closed (a point of T'' cannot be ω-accessible from T'), T is not sequential. To see that it is pseudoradial, distinguish according to whether a point x is of isolated length or of limit length, or it is in T'' . There is no dificulty to check that in each case we have $\mathrm{Col}[x] = \mathcal{U}_x$.

TOPOLOGY $\mathcal{T}_2$ ON $T_{\gamma,\lambda}$

On a tree $T_{\gamma,\lambda}$ a set defined as $\mathcal{T}_2$-open iff it satisfies:

(1), (2) as for $\mathcal{T}_1$ above;

(3) with every x of limit level it contains the end-points of all threads lying in a $\mathcal{T}_1$-neighborhood of x .

This is a topology coarser than $\mathcal{T}_1$: it is Hausdorff, not regular.

The space $T = (T_{\gamma,\lambda}; \mathcal{T}_2)$ is not first countable, disconnected. It is separable, with countable pseudocharacter under the same condition on γ, λ as given for $\mathcal{T}_1$. It has uncountable tightness, even for $\lambda = \omega$: using a standard diagonal argument we can prove that e.g. the point $x_0 = (0, 0, \ldots 0, \ldots)$ does not belong to the closure of any countable subset of T_λ excluding x_0 . The topology induced on the α-th level is not discrete if α is a limit ordinal.

In the sequel, we consider the case $\gamma = \lambda = \omega$.

No non-trivial[1] ω-sequence (y_n) in T'' converges.

First, it cannot converge to a point of T', since T'' is closed. Second, we prove that (y_n) cannot converge to a point of T'' . Let $y_n = (k_n', \ldots k_n^r, \ldots)$ and $y = (\bar{k}^1, \ldots \bar{k}^n, \ldots)$ be given; suppose $y_n \neq y$ for every n . Then a neighborhood of y containing no y_n is

$$\{(\bar{k}^1, \ldots \bar{k}^n) : n \in N\} \cup \bigcup_0 {}_n (k^1, \ldots k^n) : k^i \neq k_i^i \text{ for } 1 \leq i \leq n\} \cup X''$$

where X'' is the subset of T'' consisting of all end-points of the threads contained in the preceding two summands (hence $y \in X''$) .

It is easy to draw the validity of these statements also for α-sequences, α being any countable or singular ordinal with cofinality ω.

---

1) i.e. not having a constant rest

No ω-sequence in T' converges to a point x'' of T'' if not along the thread leading to x''.

Let $x'' = (\bar{k}^1, \dots \bar{k}^n, \dots)$ and let (x_n') be any ω-sequence in T'. If lengths of the x_n''s have a maximum, we have clearly $(x_n') \not\to x''$. Otherwise, there is no restriction assuming x_n' to have length n : let $(x_n') = (k_n^1, \dots k_n^n)$. We may assume that no x_n' belongs to the thread leading to x''. Then the set

$$\{(\bar{k}^1, \dots \bar{k}^n) : n \in N\} \cup \bigcup_{0}^{\omega}{}_n \{(k^1, \dots k^n) : k^i \neq k_i^i \text{ for } 1 \leq i \leq n\}$$

is the trace of a neighborhood U of x'' containing no point x_n'.

By an analogous argument one can prove that

A non-trivial ω-sequence in T' does not converge to a point of T' if not along the sequence of its successors.

Now, let us consider longer sequences.

In T'' an uncountable sequence converges (if and) only if it has a constant rest.

Let $(y_\alpha)_{\alpha<\eta}$, with $\eta \geq \omega_1$, $y_\alpha \in T''$ be given; suppose $\lim\limits_{\alpha<\eta} y_\alpha = y$. We are entitled to confine ourselves to the case when η is an initial ordinal. From known properties of ordinals, one can easily draw the conclusion in the general case. No loss of generality in assuming $y_\alpha \neq y$ for every α.

Since T'' is closed, we have $y \in T''$.

Consider in T'' the lexicographic ordering; then there is an order-preserving bijection φ of T'' onto the real interval $J = [0, 1)$ with its usual ordering [1], and the topology obtained by φ in J is finer than the real topology on it. So, having $(y_\alpha) \to y$ in T'' it follows that we have $(\varphi(y_\alpha)) \to \varphi(y)$ in J. If η is a regular (initial) ordinal, in $\mathbb{R}$ there is not place for convergent non-trivial η-sequences: indeed, given $y \in \mathbb{R}$ and $(y_\alpha)_{\alpha<\eta}$, if (B_n) is a countable decreasing open basis of $\mathcal{U}_y$, let α_n be the minimum ordinal such that $\alpha > \alpha_n$ implies $y_\alpha \in B_n$; if $\bar{\alpha} = \sup\limits_{n \in N} \alpha_n$, we see that for $\alpha \geq \bar{\alpha}$ we must have $y_\alpha = y$ constantly.

For a singular (initial) ordinal η, the argument above fails to hold if and only if in the set of ordinals $< \eta$ there is a countable cofinal subsequence [2]. In this case, let (α_n) be an increasing ω-sequence such that $\sup\limits_{n \in N} \alpha_n = \eta$; assuming $(y_\alpha)_{\alpha<\eta} \to y$, we have

---

1) To define such a correspondence, represent the real numbers of J in binary form (as 0.11011110010 ...) with infinitely many zeros, then, inspecting the digits of the binary sequence, from the left to the right, write down the sequence of natural numbers answering "how many 1's there are between two 0's ?". (For the number above it is (2, 4, 0, 1, ...).)

2) Consider, e.g., $\eta = \omega_\omega$ and define $y_\alpha = 1/(n+1)$ for $\omega_{n-1} \leq \alpha < \omega_n$ (read ω_{-1} as 0). Then $\lim\limits_{\alpha<\eta} y_\alpha = 0$ and $(y_\alpha)_{\alpha \leq \eta}$ has no constant rest.

$(y_{\alpha_n})_{n<\omega} \longrightarrow y$, and this, as already seen, cannot be true for ω-sequences whose terms are all distinct from y .

After these statements on convergence in $T = (T_{\omega,\omega} ; \tau_2)$, it is easy to prove that T is not pseudoradial. We are going to prove that for every $x'' \in T$ we have $\mathrm{Col}[x''] \neq \mathcal{U}_{x''}$.

Let $x'' \in T''$ be given; with reference to the definition of $\mathrm{Col}[x'']$, take x'' as A_0 . Since x'' is not the limit of any non-trivial sequence in T'', and only along its thread it is limit of sequences in T' , add to A_0 the thread leading to x'' , to obtain A_1 ; for every $x \in A_1 \cap T'$ take the ω-sequence of its successors (the only one, in some sense, converging to x); so continuing we obtain a set $A \in \mathrm{Col}[x'']$ without points of T'' other than x'' . Since in every neighborhood of x'' there are infinitely many points of T'' , we have $A \notin \mathcal{U}_x$. So, the condition of colimit being not satisfied, T is not pseudoradial.

REFERENCES

[1] A. V. ARHANGEL'SKIĬ, Structure and classification of topological spaces and cardinal invariants. Russian Math. Surveys 33 (1978), 33-95.

[2] H. HERRLICH, Quotienten geordneter Räume and Folgenkonvergenz. Fund. Math. 61 (1967), 79-81.

DIAGONAL CONDITIONS IN SEQUENTIAL CONVERGENCE

Roman Frič and Peter Vojtáš[1)]

1. INTRODUCTION

Throughout the paper, each ordinal number is identified with the set of its predecessors, cardinal numbers are initial ordinals, ω denotes the first infinite cardinal, ${}^{\omega}\omega$ denotes the set of all mappings of ω into ω partially ordered "modulo finite" (i.e. $f <^* g$ whenever $f(n) < g(n)$ for all but finitely many $n \in \omega$), each $f \in {}^{\omega}\omega$ is identified with its graph $\{(n, f(n)); n \in \omega\}$, Mon denotes the set of all strictly increasing mappings of ω into ω, a sequence S of points of a set X is considered as a mapping of ω into X and for $s \in$ Mon by $S \circ s$ we denote the composition of s and S; it is the subsequence of S the n-th term of which is $S(s(n))$. If $\langle S_n \rangle$ is a sequence of sequences of points, then to each $f \in {}^{\omega}\omega$ there corresponds a diagonal sequence S_f the n-th term of which is $S_n(f(n))$. Subsequences of diagonals are subdiagonals.

Let X be an infinite set equipped with a sequential convergence structure (e.g., X is a topological space and a sequence $\langle x_n \rangle$ of points converges in X to a point x whenever each neighborhood of x contains x_n for all but finitely many indexes, or a suitable subset $\mathcal{L}$ of $X^{\omega} \times X$ is given and we say than a sequence $\langle x_n \rangle$ $\mathcal{L}$-converges to x whenever $(\langle x_n \rangle, x) \in \mathcal{L}$). We usually assume all four axioms of sequential convergence (FUSH in the Katowice notation or, equivalently, L_i, $i = 0, 1, 2, 3$, in the Prague notation) but in most cases only two of them (viz., (S) For each $x \in X$ the constant sequence $\langle x \rangle$ converges to x; (F) Each subsequence of a convergent sequence converges to the same limit) are relevant. A reader interested in the topological considerations only can assume that X is a sequential space (sequentially open sets are open).

Let V be a sequence of points converging in X to a point v and for each $n \in \omega$ let S_n be a sequence of points converging to the point $V(n)$. Then $(\langle S_n \rangle, V, v)$ is said to be an s-system. Moreover, if V is the constant sequence generated by v (i.e. $V(n) = v$ for all $n \in \omega$), then $(\langle S_n \rangle, v)$ is said to be a p-system. The point v is said to be the vertex of

1) Matematický ústav SAV, Jesenná 5, 041 54 Košice, Czechoslovakia

$(\langle s_n \rangle, V, v)$, resp. $(\langle s_n \rangle, v)$. By a <u>diagonal condition</u> (abbreviated to DC) for $(\langle s_n \rangle, V, v)$, resp. for $(\langle s_n \rangle, v)$, we understand a condition asserting that for a specified set of diagonals of $\langle s_n \rangle$ specified subdiagonals converge in X to v. If a given DC holds for each s-system, resp. p-system, in X, then we say that <u>X satisfies the DC</u> in question.

As an example consider the following well-known condition: For each p-system $(\langle s_n \rangle, v)$ there exist $f \in {}^{\omega}\omega$ and $s \in$ Mon such that the subdiagonal $S_f \circ s = \langle s_{s(n)}(f(s(n))) \rangle$ converges to v. The condition will be denoted by (PSD); P stands for p-system, SD stands for subdiagonal.

DC's were studied by various authors in connection with: the relationship between a sequential convergence and the induced closure operators ([19], [18], [9], [20]), convergence in products ([19], [18], [13], [2], [7]), topological games ([8], [22], [7]), and analysis ([12]). Some of them were introduced and investigated independently and some of the related results were rediscovered, albeit sometimes in a different setting. In [18], [4], [2] and [21] systems of DC's were considered and, in a sense, all papers cited at the end of the present paper deal with mutual relations between various DC's.

Our aim is to provide a survey on DC's and their applications. First, we present sample results to illustrate the nature and role of DC's. Secondly, we discuss classification schemes for DC's introduced in [18], [4] and [2]. Thirdly, we show that some results obtained earlier under CH hold under weaker (or without any) additional set-theoretical assumptions. Finally, we mention several problems and areas in which the future research on DC's seems to be promising.

It is to be stressed that our survey is far from being complete. Among other results, many elaborated constructions and interesting problems from [2], [12] and [21] are not mentioned here.

We have tried to narrow the gap between the sequential convergence theory (e.g. [18]) and its topological counterpart (e.g. [2], [21]). The idea was to point out results and problems the two sides have in common.

Throughout the literature, the terminology and notation concerning DC's is not uniform (fortunately, it is rather suggestive). To make the cross-reference easier, we usually recall the notation used in the cited papers. For further information on the space ${}^{\omega}\omega$ and its applications to the sequential spaces theory the reader is referred to [6].

2. SAMPLE RESULTS

2.1. "$cl_s = cl_s^2$"

Let X be a set equipped with a sequential convergence satisfying axioms FS. The sequential closure operator cl_s is defined by $cl_s A = \{x \in X ; x = \lim x_n , x_n \in A , n \in \omega\}$. It is known that the following (subdiagonal) condition

(SD) For each s-system $(\langle S_n \rangle, V, v)$ there exists $f \in {}^\omega\omega$ and $s \in$ Mon such that the subdiagonal $S_f \circ s$ converges to v,

guarantees that $cl_s^2 = cl_s$, i.e. cl_s is idempotent and hence a topological closure. Note that the so-called sequential fan (the quotient of a disjoint sum of countably many convergent sequences with their limit points identified) satisfies $cl_s^2 = cl_s$ but it fails to satisfy (SD). Clearly (SD) implies (PSD). As proved by J. NOVÁK and L. MIŠÍK in [19], the situation for sequential convergence groups is much more convenient.

THEOREM 2.1. ([19], see also [18]) Let X be a commutative group equipped with a compatible sequential convergence (FLUSH). Then $cl_s = cl_s^2$ iff X satisfies (PSD).

REMARK 2.2. It can be readily seen that virtually the same proof works also for noncommutative FLS-convergence groups. P. NYIKOS in [20] proved that if X is a topological group the topology of which is sequential and (PSD) holds, then X is a Fréchet space (i.e. $cl = cl_s$). He also proved that every topological group the topology of which is Fréchet satisfies (PSD). Note that for commutative groups having unique sequential limits both results follow from Theorem 2.1.

2.2. Topological convergences

Let X be a set equipped with a sequential convergence satisfying axioms FS. In [9] among conditions guaranteeing that the convergence of sequences in the topology $cl_s^{\omega_1}$ coincides with the original sequential convergence in X (such convergences are said to be topological) the following two have been considered:

(D) For each s-system $(\langle S_n \rangle, V, v)$ there exists $f \in {}^\omega\omega$ such that the diagonal S_f converges to v;

(Y) For each s-system $(\langle S_n \rangle, V, v)$ there exists $f \in {}^\omega\omega$ such that for each $g \in {}^\omega\omega$, $g > f$, the diagonal S_g converges to v.

A. KAMIŃSKI asked whether in FUSH-convergence spaces (D) implies (Y) (the converse implication holds trivially). Independently, P. KRATOCHVÍL ([11]) and J. BURZYK ([3]) have constructed under CH a FUSH-convergence space, resp. a FLUSH-convergence group, satisfying (D) but not (Y). It will be shown in Section 4 that there exists a FUSH-convergence space satisfying (D) but not (Y) without any additional set-theoretical assumptions.

For a characterization of topological convergences, as well as for further information concerning the relationship between closure operators and sequential convergences, the reader is referred to [10].

2.3. Properties of products

Let X and Y be first-countable topological spaces. Then their product $X \times Y$ is again a first-countable space. Since first-countable spaces are Fréchet, in this case we have two Fréchet spaces the topological product of which is again a Fréchet space. This is not true in general (see Example 2.3). Hence a natural question arises: "Under what conditions the product of two Fréchet spaces is again a Fréchet space?" The problem attracted considerable attention (cf. [2], [13], [23], [7]) and, in fact, it is a special case of the following more general problem: "Under what conditions the closure product of two sequential convergence (FUSH) spaces coincides with their convergence product?" This more general approach to convergence in product is followed in [16], [17], [18]. Most of the DC's considered in our paper have been considered in connection with convergence properties of products (cf. [2]). We start with an illustrative example belonging to folklore (see e. g. [17], [13], [7]). A similar example can be found in [16] on page 22. Note that solving a problem from [13], P. SIMON has constructed in ZFC a compact Fréchet space whose square fails to be Fréchet ([23]).

EXAMPLE 2.3. Let X be the sequential fan (X can be visualized as $(\omega \times \omega) \cup \{v\}$, where points of $\omega \times \omega$ are isolated and a neighborhood base at v is formed by sets $U_f = \{v\} \cup \{(n, m) \in \omega \times \omega \;;\; m > f(n) \;,\; f \in {}^{\omega}\omega\}$ and let Y be the space $\omega + 1$. Both X and Y are Fréchet spaces with unique sequential limits and hence also FUSH-convergence spaces. Their topological product (the same as closure product; a neighborhood base at (x, y) in the closure product is formed by product of closure neighborhoods of x in X and of y in Y) fails to be a Fréchet space. In their convergence product a sequence $\langle(x_n, y_n)\rangle$ converges to (x, y) iff $\langle x_n\rangle$ converges in X to x and $\langle y_n\rangle$ converges in Y to y. Note that X does not satisfy (PSD) and Y is a first-countable compact space.

THEOREM 2.4. ([18]) Let X and Y be FUSH-convergence spaces. Assume that X satisfies (PSD) and Y satisfies the following condition:

(α) For each $x \in X$ there exists a decreasing sequence $\langle U_n(x)\rangle$ of closure neighborhoods of x such that each one-to-one sequence $\langle x_n\rangle$, $x_n \in U_n(x)$, contains a subsequence converging to x.

Then the closure product and the convergence product of X and Y coincide.

Clearly, each first-countable space satisfies condition (α). Combining Example 2.3 and Theorem 2.4 we get the following characterization of FUSH-convergence spaces satisfying condition (PSD).

COROLLARY 2.5. Let X be a FUSH-convergence space. Then the closure product of X with the closed unit interval I and the convergence product of X and I coincide iff X satisfies (PSD).

LEMMA 2.6. (cf. Lemma 3 in [18]) Let X be a FUSH-convergence space. Then (PSD) is equivalent to the following condition

(δ) If $\langle A_n\rangle$ is a decreasing sequence of subsets of X and $x \in X$ is a point of X such that $x \in \mathrm{cl}_s A_n$ for all $n \in \omega$, then there exists a sequence $\langle x_n\rangle$ such that for each $n \in \omega$ we have $x_n \in A_n$ and $\langle x_n\rangle$ converges to x.

REMARK 2.7. Note that if X is a topological space and in (δ) the sequential closure cl_s is replaced by the topological closure cl, then we get the definition of a countably bi-sequential space (sometimes called also a strongly Fréchet space). A. V. ARCHANGELSKIJ proved (Theorem 5.23 in [2]) that if X is a Fréchet space, then X is countably bi-sequential iff X satisfies a certain condition which is equivalent to our (PSD) (ARCHANGELSKIJ considers only one-to-one sequences S_n in p-systems). The same result is in [20] attributed to G. GRUENHAGE. Since, if X is a FUSH-convergence space and $\mathrm{cl}_s = \mathrm{cl}$, then X is a Fréchet space, this characterization of countably bi-sequential spaces via (PSD) follows from Lemma 2.6. E. MICHAEL proved (Proposition 4.D.5 in [13]) that a space X is countably bi-sequential iff the (topological) product space $X \times I$ is Fréchet. Again, if we restrict ourselves to Fréchet spaces, this characterization of countably bi-sequential spaces follows from Corollary 2.5. This, together with Remark 2.2, shows that in some cases the idempotency of the closure operator in sequential convergence theory is superfluous.

2.4. Analysis

A linear space equipped with a partial order in which each two elements possess a supremum is said to be a Riesz space. As an example, consider a measure space $(X, \mathcal{S}, \mu)$, take the set of all real μ-almost everywhere finite μ-measurable functions on X, equip it with the μ-almost everywhere pointwise partial ordering and identify two functions whenever they are equal μ-almost everywhere. The resulting space $M(X, \mu)$ is a Riesz space. Roughly, the classical Egoroff's theorem states that if in $M(X, \mu)$ a sequence $\langle f_n \rangle$ converges to f pointwise, then this convergence is uniform outside a set of small μ-measure. The theorem has a generalization to abstract Riesz spaces. In this generalization, called in [12] the abstract Egoroff-type theorem, the so-called Egoroff property plays a fundamental role.

Let L be a Riesz space. We say that an element $f \in L$ has the Egoroff property if, given a double sequence $\langle u_{nk} \rangle_{n,k \in \omega}$ in L such that for each $n \in \omega$ we have $|f| \geqq u_{nk} \downarrow 0$, there exists a sequence $\langle v_m \rangle$ such that $v_m \downarrow 0$ and for every pair (m, n) of natural numbers there exists a natural number $k(m, n)$ such that $v_m \geqq u_{n, k(m, n)}$ holds. The space L is said to have theEgoroff property if each f in L has the Egoroff property. We say that L has the strong Egoroff property if, given a double sequence $\langle u_{nk} \rangle_{n,k \in \omega}$ in L such that for each $n \in \omega$ we have $u_{nk} \downarrow 0$, there exists a sequence $\langle v_m \rangle$ such that $v_m \downarrow 0$ and for every pair (m, n) of natural numbers there exists a natural number $k(m, n)$ such that $v_n \geqq u_{n,k(m,n)}$ holds. Evidently, the strong Egoroff property implies the Egoroff property.

The sequence space l_∞ is an example of a space possessing the Egoroff property but not the strong Egoroff property. If μ is σ-finite, then $M(X, \mu)$ has the strong Egoroff property and hence also the Egoroff property.

Let L be a Riesz space. By convergence in L we understand the order convergence, i.e. $\langle f_n \rangle \to f$ whenever there exists in L a sequence $\langle p_n \rangle$ such that $p_n \downarrow 0$ and $|f - f_n| \leqq p_n$ holds for all $n \in \omega$. Then the following are equivalent (cf. [12]):

(i) L has the strong Egoroff property;

(ii) L satisfies (D);

(iii) L satisfies (SD) (-called the "diagonal gap property" in [12]).

Further, the strong Egoroff property is equivalent to the fact that the order closure coincides with the sequential closure in L.

An interesting fact about the Riesz spaces theory as presented in [12] is that many fundamental properties of particular Riesz spaces are proved under CH.

2.5. Topological games

Consider the following infinite game introduced by G. GRUENHAGE ([8]). Let X be a topological space and p a point of X. In the n-th step, $n \in \omega$, the first player chooses a neighborhood $\mathcal{N}_n$ of p and the second player chooses a point $p_n \in \mathcal{N}_n$. The first player wins if the sequence $\langle p_n \rangle$ converges to p, otherwise the second player wins. The space X is said to be a <u>w-space</u> if for each $p \in X$ the second player does not have a winning strategy. As shown by P. L. SHARMA ([22]), X is a w-space iff the following condition holds:

(C) If $\langle A_n \rangle$ is a sequence of subsets of X and $p \in \mathrm{cl} A_n$ for each $n \in \omega$, then there exists a sequence $\langle p_n \rangle$ such that $p_n \in A_n$ for each $n \in \omega$ and $\langle p_n \rangle$ converges to p.

As pointed out by G. GRUENHAGE (cf. [20]), w-spaces are precisely Fréchet spaces satisfying the following DC:

(PD) For each p-system $(\langle S_n \rangle, v)$ there exists $f \in {}^\omega\omega$ such that the diagonal $S_f = \langle S_n(f(n)) \rangle$ converges to p.

3. CLASSIFICATION OF DC's

In [18], J. NOVÁK introduced the following five conditions for FUSH-convergence spaces (double sequences $\{x_{mn}\}_{m,n=1}^{\infty}$ are considered instead of sequences $\langle S_n \rangle$ of sequences S_n and the term cross-sequence instead of diagonal is used in [18]):

(α) - see Theorem 2.4;

(β) For each p-system $(\langle S_n \rangle, v)$ there exists $f \in {}^\omega\omega$ such that for each $g > f$, $g \in {}^\omega\omega$, there exists $s \in \mathrm{Mon}$ such that the subdiagonal $S_g \circ s = \langle S_{s(n)}(g(s(n))) \rangle$ converges to v;

(γ) - the same as condition (PSD);

(δ) - see Lemma 2.6;

(ρ) $\equiv \neg(\gamma)$ - included for historical reasons only (cf. [19]).

J. NOVÁK proved that $(\alpha) \Rightarrow (\beta) \Rightarrow (\gamma) \Leftrightarrow (\delta)$ and observed that each FUSH-convergence space having a countable (closure) neighborhood base at each point satisfies condition (α). Further, he pointed out that the one-point compactification of a discrete space having uncountable cardinality fulfils (α) without being first-countable, and gave an example of a FUSH-convergence space satisfying (β) but not (α) (Example 3 in [18]). In Section 4 we shall construct (under the assumption that there is a scale in $^{\omega}\omega$) a Fréchet space satisfying (γ) but not (β).

In [4], M. CONTESSA and F. ZANOLIN considered (in our notation) conditions (D), (PD), (SD) and (PSD) for FUSH-convergence spaces. Clearly, (D) $\Rightarrow$ (SD) $\Rightarrow$ (PSD) and (D) $\Rightarrow$ (PD) $\Rightarrow$ (PSD). They asked whether (SD) implies (PD). There are simple examples from which it follows that if the answer is "no", then no two of the four conditions are equivalent. P. MIKUSIŃSKI ([14]) constructed under CH a FUSH-convergence space satisfying (SD) but not (PD). In Section 4 it will be shown that the same result can be obtained without any additional set-theoretical assumption; further, it will be pointed out that for Fréchet spaces conditions (PSD) and (SD) are equivalent.

Yet another system of diagonal conditions was considered by A. V. ARCHANGELSKIJ in [2]. These conditions involve Fréchet (i.e. Fréchet-Urysohn) spaces and p-systems with one-to-one sequences - such systems will be called p^*-systems. The conditions are studied mainly in connection with product spaces. Note that beyond some self-evident cases, he did not investigate the mutual relationship between classes of spaces satisfying his conditions $\langle i - FU \rangle$, $i = 1, \ldots, 5$. In a slightly different notation, four of these conditions appear in the unpublished manuscript [21] by P. NYIKOS, who lectured on this topic at the Colloquium on Topology, held in Eger in 1983. Here we present some of his results and also some of his interesting observations.

Let X be a set equipped with a sequential convergence satisfying the Urysohn axiom

(U) If a sequence S does not converge to $x \in X$, then there exists $s \in$ Mon such that for each $t \in$ Mon the subsequence $S \circ s \circ t$ of S does not converge to x.

It is known that if a one-to-one sequence $S = \langle x_n \rangle$ converges in X to a point $x \in X$, then for each one-to-one mapping $s \in {}^{\omega}\omega$ the one-to-one sequence $S \circ s$ also converges to x. Hence in a sequential convergence space satisfying (U), instead of saying that a one-to-one sequence $\langle x_n \rangle$ converges to a point $x \in X$, we can say that the set $S = \{x_n ; n \in \omega\}$ converges to x. From this point of view we shall make no distinction between S

and the corresponding sequence. Now we are ready to recall the conditions studied by A. V. ARCHANGELSKIJ and P. NYIKOS (generalized to sequential convergence spaces; note that only the Urysohn axiom is relevant), and to reformulate them into DC's.

Let X be a FUS-convergence space (e.g. a Fréchet space). Consider the following conditions:

(α_1) For each p^*-system $(\langle S_n \rangle, v)$ there exists a countable set S converging in X to v such that for each $n \in \omega$ we have $|S_n \setminus S| < \omega$ (i.e. all but finitely many elements of the one-to-one sequence S_n belong to S);

(α_2) For each p^*-system $(\langle S_n \rangle, v)$ there exists a countable set S converging in X to v such that for each $n \in \omega$ we have $|S_n \cap S| = \omega$;

(α_3) For each p^*-system $(\langle S_n \rangle, v)$ there exists a countable set S converging in X to v such that for infinitely many $n \in \omega$ we have $|S_n \cap S| = \omega$;

(α_4) For each p^*- system $(\langle S_n \rangle, v)$ there exists a countable set S converging in X to v such that for infinitely $n \in \omega$ we have $S_n \cap S \neq \emptyset$.

Conditions $\langle i - FU \rangle$ are related to (α_j) as follows: $(\alpha_1) \equiv \langle 1 - FU \rangle$, $(\alpha_2) \equiv \langle 5 - FU \rangle$, $(\alpha_3) \equiv \langle 3 - FU \rangle$, $(\alpha_4) \equiv \langle 4 - FU \rangle$. Clearly, for $i = 1, 2, 3$ we have $(\alpha_i) \Rightarrow (\alpha_{i+1})$. On the other hand, under the assumption that there is no scale in ${}^{\omega}\omega$, P. NYIKOS has constructed a Fréchet space satisfying (α_2) but not (α_1). For $i = 3, 4$ he has examples of Fréchet spaces satisfying (α_i) but not (α_{i-1}) without any additional set-theoretical assumptions.

REMARK 3.1. (see e.g. [14], [21]) Notice that in (α_2) the condition $|S_n \cap S| = \omega$ can be equivalently replaced by $S_n \cap S \neq \emptyset$. Hint. Each p^*-system $(\langle S_n \rangle, v)$ can be replaced by a p^*-system $(\langle S_n' \rangle, v)$ in which $\langle S_n' \rangle$ is obtained by repeating each S_n infinitely many times and each time it appears we leave out from the sequence $\langle S_n(k) \rangle_{k=1}^{\infty}$ larger and larger initial segments. From $S_n' \cap S \neq \emptyset$ we get $|S_n \cap S| = \omega$.

REMARK 3.2. (cf. [21]) Assume that X is a space satisfying condition (α_i) but not (α_j). Then there exists a countable Fréchet space Y all points but one of which are isolated and Y satisfies (α_i) but not (α_j). Indeed, let $(\langle S_n \rangle, v)$ be a p^*-system in which (α_j) fails. Let Y be the underlying set of $(\langle S_n \rangle, v)$. Then Y can be equipped with a sequential convergence in which a sequence S converges to x provided $x = v$ and

S converges in X to v. Clearly, Y has the desired properties. Hence Y is a topological space of the form $\omega \cup \{\mathcal{F}\}$, where $\mathcal{F}$ is a suitable filter on ω and a neighborhood base at $\{\mathcal{F}\}$ is formed by sets $F \cup \{\mathcal{F}\}$, $F \in \mathcal{F}$. Several deep results concerning conditions (α_i) were obtained in [21] along these lines.

REMARK 3.3. For FUSH-convergence spaces, condition (α_1) is equivalent to the following one:

(PY) For each p-system $(\langle S_n \rangle, v)$ there exists $f \in {}^{\omega}\omega$ such that for each $g \in {}^{\omega}\omega$, $g > f$, the diagonal $S_g = \langle S_n(g(n)) \rangle$ converges to v.

Hint. First, observe that "it makes no difference whether we consider p-systems or p^*-systems". Secondly, if $(\langle S_n \rangle, v)$ is a p^*-system and (α_1) holds, then there exists $f \in {}^{\omega}\omega$ such that for each $g \in {}^{\omega}\omega$, $g > f$, the diagonal S_g is a subsequence of S. Thirdly, if $(\langle S_n \rangle, v)$ is a p^*-system and (PY) holds, then for a suitable $f \in {}^{\omega}\omega$ the set $S = \{S_n(k) ;\ u \in \omega$ and $k > f(n)\}$ converges (according to (U)) to v. Clearly $|S_n \setminus S| < \omega$ for each $n \in \omega$.

REMARK 3.4. Similarly, for FUSH-convergence spaces, condition (α_2) is equivalent to (PD) - see Remark 3.1, condition (α_4) is equivalent to (PSD), and condition (α_3) can be transformed into a suitable PDC.

4. IMPROVEMENTS

4.1. (D) versus (Y)

As mentioned in Section 2, condition (D) follows from condition (Y) and, under CH, there are examples of sequential convergence spaces satisfying (D) but not (Y) ([11], [3]); we are going to show that CH can be avoided completely.

P. NYIKOS in [21], under the assumption that there is no scale in ${}^{\omega}\omega$ (recall that a scale in ${}^{\omega}\omega$ is a well-ordered and dominating subset of ${}^{\omega}\omega$), gave an example of a countable Fréchet space, with unique sequential limits, satisfying condition (α_2) but not condition (α_1). Since the space has exactly one nonisolated point, each s-system can be replaced in this space by an "equivalent" p^*-system. Consequently, it follows from Remark 3.1, Remark 3.2 and Remark 3.3 that the space in question satisfies (D) but not (Y).

We are going to construct, under the assumption that there is a scale in ${}^{\omega}\omega$, a Fréchet space X, with unique sequential limits, satisfying (D) but not (Y).

The two examples together imply that, in ZFC, there always exists a Fréchet space with unique sequential limits (i.e. a FUSH-convergence space such that $cl_s = cl_s^2$) which satisfies condition (D) but not (Y).

Our construction is a modification of KRATOCHVÍL's example from [11]. When constructing a sequential convergence space, we have to make a list of convergent sequences in a given set. To kill (Y), we have to construct an s-system $(\langle S_n \rangle, V, v)$ so that for a large (e.g. unbounded) set $\mathcal{F}$ of functions in ${}^{\omega}\omega$, each diagonal $S_f = \langle S_n(f(n)) \rangle$, $f \in \mathcal{F}$, does not converge to the vertex v. A natural candidate for the underlying set of our space X is $((\omega + 1) \times (\omega + 1)) \setminus \{(\omega, n) ; n \in \omega\}$. The convergence in X will be defined in such a way that: the k-th column $C_k = \{(k, n) ; n \in \omega\}$ will converge to (k, ω), $k \in \omega$; the top line $\{(n, \omega) ; n \in \omega\}$ will converge to (ω, ω); for a suitable scale $\mathcal{F}$ in ${}^{\omega}\omega$, each $f = \{(n, f(n)) ; n \in \omega\} \in \mathcal{F}$ will be totally divergent; every one-to-one sequence which is almost disjoint with each C_k, $k \in \omega$, and each $f \in \mathcal{F}$ will converge to (ω, ω).

REMARK 4.1. (cf. [1], [11]) Let X be a set and for each $x \in X$ let $\mathcal{N}_x$ be a system of subsets of X such that $x \in N$ for each $N \in \mathcal{N}_x$. Then a FUS-convergence for X is defined as follows: $\langle x_n \rangle$ converges to x whenever for each $N \in \mathcal{N}_x$ we have $x_n \in N$ for all but finitely many $n \in \omega$. Further, assume that whenever x and y are distinct points of X, then there are almost disjoint sets $M \in \mathcal{N}_x$ and $N \in \mathcal{N}_y$. Then the convergence satisfies also axiom (H) of the uniqueness of sequential limits. Note that if $S = \{x_n \in X ; n \in \omega\}$ is a countable subset of X such that for each $x \in X$ there is a set $N \in \mathcal{N}_x$ such that S and N are almost disjoint, then no subsequence of S converges in X.

Using the just described method, we shall equip X with a FUSH-convergence having the properties mentioned before. It remains to take care of (D). Note that from (D) it will follow that in X we have $cl_s = cl_s^2$ (cf. Section 2.1), and hence X will be a Fréchet space. To verify (D), it suffices to consider only s-systems with vertex (ω, ω). Indeed, since each point $x \in X \setminus \{(\omega, \omega)\}$ has a countable neighborhood base, (D) holds for each s-system with vertex x different from (ω,ω) trivially. Moreover, as we shall see, it suffices to verify that (D) holds only for two basic types of s-systems with vertex (ω,ω) (in fact, one of these two is a p-system).

Our construction is based on two technical lemmas, in which we use the assumption that there is a scale in ${}^{\omega}\omega$. It is known (cf. [5], [6]) that this assumption is independent of axioms of ZFC and it is equivalent to the assumption that the following two cardinal invariants of the space ${}^{\omega}\omega$ are equal: $\underline{b}$ - the least cardinality of an unbounded family of functions in ${}^{\omega}\omega$ and $\underline{d}$ - - the least cardinality of a dominating family of functions in ${}^{\omega}\omega$. Further, under $\underline{b} = \underline{d}$ there is in ${}^{\omega}\omega$ a scale of the cardinality $\underline{b} = \underline{d}$.

Recall that we identify each $f \in {}^{\omega}\omega$ with its graph $\{(n, f(n)): n \in \omega\} \subset \omega \times \omega$, dom (f) denotes the domain of f , and $f <^* g$ means that $f(n) < g(n)$ for all but finitely many $n \in \omega$.

The first lemma guarantees that if $\langle R_n \rangle$ is a sequence such that each R_n is an infinite subset of the n-th column $C_n = \{(n, k) ; k \in \omega\}$, then there exists a selector $g \in {}^{\omega}\omega$ of the sequence $\langle R_n \rangle$ such that g (i.e. its graph) is almost disjoint with each $f \in \mathcal{F}$ (i.e. with its graph). This will in fact guarantee that in our space X condition (D) holds for each s-system with vertex (ω,ω) each "base" sequence of which is a one-to-one sequence in some column C_n . The selector g will be the required diagonal converging to (ω,ω) .

LEMMA 4. 2. Let $\mathcal{F} = \{f_\alpha ; \alpha \in \underline{b} = \underline{d}\}$ be a scale in ${}^{\omega}\omega$. For each $n \in \omega$, let R_n be an infinite subset of ω . Then there exists $g \in {}^{\omega}\omega$ such that:

(i) For each $n \in \omega$ we have $g(n) \in R_n$; and

(ii) For each $f_\alpha \in \mathcal{F}$ the set $g \cap f_\alpha$ is finite.

PROOF. Since each R_n is infinite, by induction it is easy to construct a sequence $\langle g_n \rangle$ in ${}^{\omega}\omega$ such that for each $k \in \omega$ we have $g_n(k) \in R_k$ and an increasing sequence $\langle \alpha_n \rangle$ in $\underline{b} = \underline{d}$ such that for each $n \in \omega$ we have

$$g_n <^* f_{\alpha_n} < g_{n+1} .$$

Put $\alpha = \sup \alpha_n$. Then there exists a strictly increasing sequence $\langle k_n \rangle$ in ω such that for each $n \in \omega$ and each $m > k_n$ we have $f_\alpha(m) > g_n(m)$. Define

$$g(m) = g_0(m) \text{ if } m \leq k_1 \text{ , and}$$

$$g(m) = g_n(m) \text{ if } k_n < m \leq k_{n+1} .$$

Clearly, $g <^* f_\alpha$ and $f_\beta <^* g$ whenever $\beta < \alpha$. This completes the proof.

Denote by $\mathcal{P}$ the set $\bigcup\{{}^{M}\omega ;\ M \in [\omega]^{\omega}\}$ of all partial functions in ${}^{\omega}\omega$, i.e. all subsets of $\omega \times \omega$ having the form $g \upharpoonright M = \{(k, g(k)) ;\ k \in M\}$, where g belongs to ${}^{\omega}\omega$ and M is an infinite subset of ω .

The second lemma asserts that if $\langle g_n \rangle$ is a sequence in $\mathcal{P}$ and each g_n (i.e. its graph) is almost disjoint with each $f \in \mathcal{F}$ (i.e. with its graph), then there is a selector $g \in \mathcal{P}$ of $\langle g_n \rangle$ which is also almost disjoint with each $f \in \mathcal{F}$. This will guarantee that in our space X condition (D) holds for each p-system with vertex (ω, ω) each "base" sequence of which is a one-to-one subdiagonal of the sequence $\langle C_n \rangle$ of columns C_n . Ideas similar to that used in the proof of the lemma were already employed by other authors, see e.g. [5], [21], [15].

LEMMA 4.3. Let $\mathcal{F} = \{f_\alpha ;\ \alpha \in \underline{b} = \underline{d}\}$ be a scale in ${}^{\omega}\omega$. Let $\langle g_n \rangle$ be a sequence in $\mathcal{P}$ such that for each $\alpha \in \underline{b} = \underline{d}$ and each $n \in \omega$ the set $g_n \cap f_\alpha$ is finite. Then there exists $g \in \mathcal{P}$ such that:

(i) For each $n \in \omega$ we have $g \cap g_n \neq \emptyset$; and

(ii) For each $f_\alpha \in \mathcal{F}$ the set $g \cap f_\alpha$ is finite.

PROOF. For each $n \in \omega$ there exists $\beta_n \in \underline{b} = \underline{d}$ such that $g_n <^* f_{\beta_n}$. Put $\beta = \sup \beta_n$. Then for each $\alpha < \beta$ there exists $d_\alpha \in {}^{\omega}\omega$ such that for each $n \in \omega$ we have $\mathrm{dom}\,(g_n \cap f_\alpha) \subset d_\alpha(n)$. Since $\beta < \underline{b} = \underline{d}$, there exists $d \in {}^{\omega}\omega$ such that for each $\alpha < \beta$ we have $d_\alpha <^* d$. Put

$$k_0 = \min\{k \in \mathrm{dom}(g_0) ;\ g_0(k) < f_\beta(k) \text{ and } k > d(0)\},$$

and inductively

$$k_{n+1} = \min\{k \in \mathrm{dom}(g_n) ;\ g_n(k) < f_\beta(k) \text{ and } k > d(n+1) + k_n\} .$$

Define

$$g(k_n) = g_n(k_n) ,\quad n \in \omega .$$

Now, let $\alpha \in \underline{b} = \underline{d}$ be a fixed ordinal. If $\alpha \geq \beta$, then from $g < f_\beta$ it follows that the set $g \cap f_\alpha$ is finite. If $\alpha < \beta$, then there exists $n_0 \in \omega$ such that for each $n \geq n_0$ we have $d(n) > d_\alpha(n)$. Consequently, since for each $n \geq n_0$ we have $\mathrm{dom}(g_n \cap f_\alpha) \subset d_\alpha(n) \subset d(n)$ and $k_n > d(n)$, we get $\mathrm{dom}(g \cap f_\alpha) \subset \subset \{k_0, k_1, \ldots, k_{n_0}\}$. This completes the proof.

Now we are ready to describe our space X formally.

EXAMPLE 4.4. Let $\mathcal{F} = \{f_\alpha ;\ \alpha \in \underline{b} = \underline{d}\}$ be a scale in ${}^{\omega}\omega$. Put $X = ((\omega + 1) \times (\omega + 1)) \setminus \{(\omega, n) ;\ n \in \omega\}$. For each $x \in X$ define a system $\mathcal{N}_x$ of subsets of X as follows:

(i) For each $m, n \in \omega$, $\mathcal{N}_{(m,n)}$, consists of the singleton $\{(m, n)\}$;

(ii) For each $n \in \omega$, $\mathcal{N}_{(n,\omega)}$ consists of all sets of the form $\{(n,\omega)\} \cup (C_n \setminus F)$, where F is a finite subset of C_n ;

(iii) $\mathcal{N}_{(\omega,\omega)}$ consists of all sets of the form $\{(\omega,\omega)\} \cup (((\omega \times (\omega + 1)) \setminus (C_n \cup f_\alpha)) \setminus F)$, where $n \in \omega$, $f_\alpha \in \mathcal{F}$ and F is a finite subset of $(\omega \times (\omega + 1))$.

Define a sequential convergence for X as follows:

(iv) $\langle x_n \rangle$ converges to x whenever for each $N \in \mathcal{N}_x$ we have $x_n \in N$ for all but finitely many $n \in \omega$.

It is easy to verify that, according to Remark 4.1, X is a FUSH--convergence space.

We are going to prove that X satisfies condition (D). So, let $(\langle S_n \rangle, V, v)$ be an s-system in X . If $v \neq (\omega,\omega)$, then it follows directly from the construction of X that $(\langle S_n \rangle, V, v)$ satisfies (D) . If $v = (\omega,\omega)$, then a straightforward but tedious calculation shows that $(\langle S_n \rangle, V, (\omega,\omega))$ can be split into a finite number of s-systems, each of which is either trivial or it is of the type described in Lemma 4.2 or Lemma 4.3. Since each of these systems possesses a diagonal converging to (ω,ω) , by gluing them together we get a diagonal of the original s-system $(\langle S_n \rangle, V, (\omega,\omega))$ and, by the Urysohn axiom (U), the diagonal converges in X to (ω,ω) (recall that, assuming (U), if $\langle x_n \rangle$ and $\langle y_n \rangle$ converge to x , then their conjunction $\langle z_n \rangle = \langle x_n \rangle \wedge \langle y_n \rangle$, defined by $z_{2n} = x_n$ and $z_{2n+1} = y_n$, also converges to x and, if $s \in {}^{\omega}\omega$ is a one-to--one mapping, then also $\langle z_{f(n)} \rangle$ converges to x). Consequently, X satisfies condition (D) .

Observe that from (D) it follows that X is a Fréchet space (cf. Section 2.1).

Finally, since $\mathcal{F}$ is a scale in ${}^{\omega}\omega$ and each $f_\alpha \in \mathcal{F}$ is a totally divergent sequence, the s-system $(\langle C_n \rangle, \langle (n,\omega) \rangle, (\omega,\omega))$ does not satisfy condition (Y) .

4.2. (β) versus (γ)

As we pointed out in Section 3, the mutual relationship between conditions (β) and (γ) remained in [18] unsolved. We are going to construct, under the assumption that there is a scale in ${}^{\omega}\omega$, a Fréchet space satisfying (γ) but not (β). The space is, in fact, a modification of Example 4.4 and its properties follow from Lemma 4.2 and Lemma 4.3.

EXAMPLE 4.5. Let $\mathcal{F} = \{f_\alpha ; \alpha \in \underline{b} = \underline{d}\}$ be a scale in ${}^{\omega}\omega$.

Let Y be a quotient of the space X from Example 4.4, with all points (n, ω) identified to the point (ω, ω). Then Y is the set $(\omega \times \omega) \cup \{v\}$ equipped with a FUSH-convergence determined by systems $\mathcal{N}_y$ of subsets of Y (cf. Remark 4.1) defined as follows:

(i) For each $(m, n) \in \omega \times \omega$, $\mathcal{N}_{(m,n)}$ consists of the singleton $\{(m, n)\}$;

(ii) The system $\mathcal{N}_v$ consists of sets $\{v\} \cup (((\omega \times \omega) \setminus f_\alpha) \setminus F)$, where $f_\alpha \in \mathcal{F}$ and F is a finite subset of $\omega \times \omega$.

Because v is the only nonisolated point in Y, Y is a Fréchet space.

Since $\mathcal{F}$ is a scale in $^\omega\omega$ and, by (ii), each $f_\alpha \in \mathcal{F}$ is a totally divergent one-to-one sequence, the space Y does not satisfy condition (β) at the p-system $(\langle C_n \rangle, v)$.

To show that Y satisfies $(\gamma) \equiv$ (PSD), it suffices to split each p-system $(\langle S_n \rangle, v)$ into simple p-systems and use Lemma 4.2 and Lemma 4.3 to construct a suitable subdiagonal converging in Y to v. This involves no difficulty and is omitted.

4.3. (SD) versus (PD)

Answering a problem posed in [4], P. MIKUSIŃSKI has constructed in [14] a FUSH-convergence space satisfying (SD) but not (PD). In his construction he has employed the Continuum hypothesis. We are going to show that such space exists in ZFC.

THEOREM 4.6. Let X be a Fréchet space with unique sequential limits (i.e. a FUSH-convergence space in which $cl_s = cl_s^2$). Then conditions (SD) and (PSD) are equivalent in X.

PROOF. Clearly, (SD) implies (PSD). To prove the converse implication, assume that $(\langle S_n \rangle, V, v)$ is an s-system. Then there are three possibilities.

Case I. For some $s \in \mathrm{Mon}$, $V \circ s$ is a constant sequence. Then $V \circ s = \langle v \rangle$ and the existence of a subdiagonal of $(\langle S_n \rangle, V, v)$ converging to v follows from (PSD).

Case II. For some $s \in \mathrm{Mon}$, $V \circ s$ is a one-to-one sequence and for each $n \in \omega$ there exists $k_{s(n)}$ such that $S_{s(n)}(k_{s(n)}) = V(s(n))$. Then $\langle S_{s(n)}(k_{s(n)}) \rangle$ is the required subdiagonal converging to v.

Case III. There exists an s-system $(\langle T_n \rangle, U, v)$ such that U is a one-to-one sequence and $U = V \circ s$ for some $s \in \mathrm{Mon}$, further, each T_n is a one-to-one subsequence of $S_{s(n)}$ and

$v \neq T_n(k)$ for all $k \in \omega$. For $A = \bigcup_{n\in\omega} \bigcup_{k\in\omega} \{T_n(k)\}$ we have $v \in \mathrm{cl}A \setminus A$. Since X is a Fréchet space, there is in A a sequence T converging to v, and, by the uniqueness of sequential limits, T catches at most finitely many points from each T_n, $n \in \omega$. Hence, for some $t \in \mathrm{Mon}$, $T \circ t$ is a subdiagonal of $(\langle S_n \rangle, V, v)$ and it converges to v. This completes the proof.

P. NYIKOS in [21] gave, in ZFC, an example of a Fréchet space with unique sequential limits, satisfying condition (α_4) but not condition (α_3) and hence also not condition (α_2). As shown in Section 3, (α_4) is equivalent to (PSD) and (α_2) to (PD). Hence by Theorem 4.6, the space satisfies (SD) but not (PD).

REMARK 4.7. It might be in place to observe that instead of Theorem 4.6 we could use a similar argument as in Section 4.1. Namely, assuming a set-theoretical assumption A - possibly void, let X be a FUSH-convergence space satisfying (α_i) but not (α_j). Then, by Remark 3.2, there exists a (countable) Fréchet space Y such that: it has unique sequential limits, it has exactly one non-isolated point, it satisfies (α_i) but not (α_j). Since in such space each s-system can be replaced by an "equivalent" p^*-system, we have, under the assumption A, a space satisfying the DC equivalent to (α_i) but not the DC equivalent to (α_j).

5. CONCLUDING REMARKS

At the end of our paper we would like to mention several areas in which there are interesting unsolved problems concerning DC's.

1. Gaps in the classification of DC's. We have mentioned several examples (showing that one DC does not imply another one) constructed under additional set-theoretical assumptions. It would be nice to have the corresponding absolute results. Here we recall the problem from [21] whether there is in ZFC a Fréchet space satisfying (α_2) but not (α_1). Also the exact position of condition (β) in the hierarchy of other DC's seems to be an interesting problem.

2. DC's in sequential convergence groups. Even though there are some isolated results concerning the mutual relationship between various DC's in FLUSH-convergence groups (cf. [18], [12], [21], [3]) the situation is far from being satisfactory. Problems in this area seem to be particularly hard.

3. Analysis. In [12] there are several interesting theorems and examples using CH. Is it possible to avoid this assumption?

4. DC's in special classes of spaces. As pointed out in [2] and [21], very little is known about special classes of spaces, such as compact, regular, sequentially compact, homogeneous, etc., in which a particular DC holds.

It is our feeling that there are also other areas in mathematics where DC's play an important role. But that is another story.

REFERENCES

[1] Antosik, P.: On a topology of convergence. Colloq. Math. 21 (1970), 205-209.

[2] Archangelskij, A. V.: The frequency spectrum of a topological space and the product operation. (Russian.) Trudy Moskovskogo matematičeskogo obščestva 40 (1979), 171-206.

[3] Burzyk, J.: The example of FLUSHD convergence without Y. These proceedings.

[4] Contessa, M. and F. Zanolin: On a question of Josef Novák about convergence spaces. Rend. Sem. Mat. Univ. Padova, 58 (1977), 155-161.

[5] van Douwen, E. K.: The integers and topology. In: J. Vaughan and K. Kunen, ed., Handbook of Set-theoretical Topology. North-Holland. (To appear.)

[6] Frič, R. and P. Vojtáš: The space ω_ω in sequential convergence. These proceedings.

[7] Gerlits, J. and Zs. Nagy: Products of convergence properties. Comment. Math. Univ. Carolin. 23 (1982), 747-756.

[8] Gruenhage, G.: Infinite games and generalizations of first countable spaces. Gen. Top. Appl. 6 (1976), 339-352.

[9] Kamiński, A.: On characterization of topological convergence. Proc. Conf. on Convergence, Szczyrk, 1979. Katowice, 1980, 50-70.

[10] Koutník, V.: Closure and topological sequential convergence. These proceedings.

[11] Kratochvíl, P.: Sequential convergences generated by coneighborhoods and an example of D not Y space. Czechoslovak Math. J. (To appear.)

[12] Luxemburg, W. A. J. and A. C. Zaanen: Riesz spaces. Vol. 1. North-Holland, Amsterdam - London 1971.

[13] Michael, E.: A quintuple quotient quest. Gen. Top. Appl. 2 (1972), 91-138.

[14] Mikusiński, P.: Bases of convergence and diagonal conditions. Rend. Ist. Mat. Univ. Trieste 15 (I-II)(1983), 32-38.

[15] Mišík, L., Jr.: Sequential completeness and {0, 1}-sequential completeness are different. Czechoslovak Math. J. (To appear.)

[16] Novák, J.: Sur les espaces ($\mathcal{L}$) et sur les produits cartesiens ($\mathcal{L}$). Publ. Fac. Sciences Univ. Masaryk, Brno, fasc. 273 (1939).

[17] Novák, J.: On convergence spaces and their sequential envelopes. Czechoslovak Math. J. 15 (1965), 74-100.

[18] Novák, J.: On convergence groups. Czechoslovak Math. J. 20 (1970), 357-374.

[19] Novák, J. and L. Mišík: On L-spaces of continuous functions. (Slovak.) Matematicko-fyzikálny sborník 1 (1951), 1-17.

[20] Nyikos, P.: Metrizability and the Fréchet-Urysohn property

in topological groups. Proc. Amer. Math. Soc. 83 (1981), 793-801.
[21] Nyikos, P.: More about (ω_1, ω_1)-gaps; More about $(\varkappa,\lambda)$-gaps and α_i-spaces. Handwritten notes, 1983.
[22] Sharma, P. L.: Some characterizations of W-spaces nad w-spaces. Gen. Top. Appl. 9 (1978), 289-293.
[23] Simon, P.: A compact Fréchet space whose square is not Fréchet. Comment. Math. Univ. Carolin. 21 (1980), 749-753.

THE SPACE $^{\omega}\omega$ IN SEQUENTIAL CONVERGENCE

Roman Frič and Peter Vojtáš[1)]

ABSTRACT. We survey cardinal invariants and combinatorial phenomena connected with the space $^{\omega}\omega$ of all mappings of ω into ω equipped with the partial order "modulo finite" and the algebra $\mathcal{P}(\omega)/_{fin}$ of all subsets of ω "modurlo finite". We discuss recent constructions of peculiar sequential spaces in which infinite combinatorics has been exploited.

1. INTRODUCTION

The space $^{\omega}\omega$ of all mappings of natural numbers into natural numbers equipped with the partial order "modulo finite" (i.e., for $f, g \in {^{\omega}\omega}$ we put $f <^* g$ whenever $f(n) < g(n)$ for all but finitely many $n \in \omega$) and the algebra $\mathcal{P}(\omega)/_{fin}$ are extremely useful tools for specialists in sequential convergence. As it is known for long, some of the properties of $^{\omega}\omega$ and $\mathcal{P}(\omega)/_{fin}$ depend on the particular model of set theory we are working with. An excellent account of the properties of $^{\omega}\omega$ and $\mathcal{P}(\omega)/_{fin}$ with applications to topology was written by E. K. van DOUWEN ([11]). In our paper we would like to stress the importance of some combinatorial ideas in sequential convergence. We make some remarks on constructions of peculiar sequential spaces based on $^{\omega}\omega$, survey cardinal invariants and combinatorial phenomena connected with $^{\omega}\omega$ and $\mathcal{P}(\omega)/_{fin}$ and present some old and new results around $^{\omega}\omega$ and $\mathcal{P}(\omega)/_{fin}$.

1.1. Definitions, notation, basic facts.

For the reader s convenience, in this section we recall the basic terminology and notation. Ordinals are von Neumann ordinal numbers (i.e., the sets of all predecessors), cardinals are initial ordinals, ω is the set of all natural numbers - the least infinite ordinal (cardinal) number.

The set theory we are working with is ZFC (the Zermelo-Fraenkel set theory with the axiom of choice); CH denotes the Continuum hypothesis $2^{\omega} = \omega_1$. Our standard reference book on set theory is [23].

Functions in $^{\omega}\omega$ are identified with their graphs in $\omega \times \omega$. A set $\mathcal{B} \subset {^{\omega}\omega}$ is said to be unbounded if $(\forall g \in {^{\omega}\omega})\ (\exists f \in \mathcal{B})$

1) Matematický ústav SAV, Jesenná 5, 041 54 Košice, Czechoslovakia

$(f \not<^* g)$. Denote by $\underline{b}$ the minimal size of an unbounded set in ${}^\omega\omega$. It is a folklore that there is a family $\mathcal{F} = \{f_\alpha : \alpha \in \underline{b}\} \subset {}^\omega\omega$ such that: $\mathcal{F}$ is unbounded, $\alpha < \beta$ implies $f_\alpha <^* f_\beta$ (i.e., $\mathcal{F}$ is well-ordered), each f_α is increasing (note that if M is an infinite subset of ω , then $\mathcal{F}$ restricted to M has the same properties in ${}^M\omega$, the space of all mappings of M into ω , with respect to the order inherited from ${}^\omega\omega$) . A set $\mathcal{D} \subset {}^\omega\omega$ is said to be <u>dominating</u> if $(\forall g \in {}^\omega\omega)(\exists f \in \mathcal{D})(g <^* f)$. Denote by $\underline{d}$ the minimal size of a dominating family of functions in ${}^\omega\omega$. A set $\mathcal{S} \subset {}^\omega\omega$ is said to be a <u>scale</u> if it is dominating and well-ordered under $<^*$. The existence of a scale in ${}^\omega\omega$ is independent of the axioms of ZFC, it follows from CH , and it is equivalent to the condition $\underline{b} = \underline{d}$.

Let X be an infinite set. Denote by $[X]^\omega$, resp. by $[X]^{<\omega}$, the set of all countable, resp. finite, subsets of X . For $A, B \in [X]^\omega$, we say that they are <u>almost disjoint</u> if $A \cap B$ is finite, and we write $A \subset^* B$ if $A \setminus B$ is finite. A set $\mathcal{A} \subset [X]^\omega$ is said to be an <u>almost disjoint family</u>, abbr. ADF, if $(\forall A, B \in \mathcal{A}, A \neq B)$ $(A \cap B$ is finite), resp. a <u>maximal almost disjoint family</u>, abbr. MADF, if $\mathcal{A}$ is an ADF and it is not contained properly in any other ADF in $[X]^\omega$.

Consider the Boolean algebra $\mathcal{P}(\omega)$ of all subsets of ω . Then $\mathcal{P}(\omega)/_{fin}$ is the factor algebra of $\mathcal{P}(\omega)$ modulo the ideal of all finite subsets of ω ; instead of a factor class $[M]$, $M \subset \omega$, we sometimes work with its representative M . We recall the Stone duality between $\omega^* = \beta\omega \setminus \omega$ (the remainder of the Čech-Stone compactification $\beta\omega$ of the discrete space ω) and $\mathcal{P}(\omega)/_{fin}$. Points of ω^* correspond to free ultrafilters on ω , clopen sets to subsets of ω , open sets to ideals, closed sets to filters, regular open dense sets to maximal almost disjoint families, nowhere dense sets to free filters (cf. [40], [12]).

We often work with spaces of the type $N \cup \mathcal{N}$ (denoted also by $\Psi(\omega, \mathcal{N})$), where N is a countable infinite set and $\mathcal{N} \subset [N]^\omega$ is an AD family; points in N are isolated and a neighborhood base at $A \in \mathcal{N}$ is formed by sets $\{A\} \cup (A \setminus F)$, F is a finite subset of A .

2. MOTIVATION

Mathematicians usually work in ZFC but sometimes use certain additional set-theoretical assumptions (CH , $\underline{b} = \underline{d}$, etc.) and their consequences. Of course, the weaker assumption - the better. To illustrate the nature of ${}^\omega\omega$ and its role in sequential convergence, we start with three closely related examples of Fréchet spaces each of which is sequentially regular but fails to be regular. Recall that a space X is said to be sequentially regular if the convergence of sequences in X is projectively generated by the set $C(X)$ of all continuous functions on X, i.e., $x_n \longrightarrow x$ in X iff for each $f \in C(X)$ we have $f(x_n) \longrightarrow f(x)$. Note that for sequential convergence spaces (cf. [33]) the sequential regularity plays an analogous role as the complete regularity does for topological spaces.

EXAMPLE 1. Assuming CH, let $\mathcal{F} = \{f_\alpha : \alpha \in \omega_1\}$ be a scale in ${}^\omega\omega$. Consider the set $Y_a = (\omega \times (\omega + 1)) \cup \omega_1$ equipped with the following topology: all points in $\omega \times \omega$ are isolated; for each $\alpha \in \omega_1$, a neighborhood base at α is formed by sets $\{\alpha\} \cup (f_\alpha \setminus F)$, where F is a finite subset of $f_\alpha = \{(n, f_\alpha(n)) : n \in \omega\}$; for each $n \in \omega$, a neighborhood base at (n, ω) is formed by sets $\{(n, \omega)\} \cup (C_n \setminus F)$, where $C_n = \{(n, m) : m \in \omega\}$ and F is a finite subset of C_n. Clearly, Y_a is a space of the type $\Psi(\omega, \mathcal{N})$. Finally, let X_a be the quotient space obtained from Y_a by identifying points (n, ω) to one point p. Then X_a is a Fréchet space, homeomorphic to the space L_9 in [33]. Using the fact that $\mathcal{F}$ is well-ordered, it can be shown that X_a is sequentially regular. (Hint. Assume $x_n \not\longrightarrow x$. We have to find a continuous function φ such that $\varphi(x_n) \not\longrightarrow \varphi(x)$. The case $x \neq p$ is trivial. Let $x = p$. For each $\alpha \in \omega_1$, define a function φ_α on X_a as follows: $\varphi_\alpha((n, m)) = 0$ whenever $m < f_\alpha(n)$, $\varphi_\alpha(\beta) = 0$ whenever $\beta < \alpha$, and $\varphi_\alpha(y) = 1$ otherwise. For each $\alpha \in \omega_1$, $\varphi_\alpha(p) = 1$ and, since $\mathcal{F}$ is well-ordered, φ_α is a continuous function. Clearly, there exists $\alpha \in \omega_1$ such that $\varphi_\alpha(x_n) = 0$ for infinitely many $n \in \omega$.) On the other hand, using the fact that $\mathcal{F}$ is dominating, it can be shown that X_a fails to be regular. (Hint. Sets $O_f = \{p\} \cup \{(m, n) \in \omega \times \omega : m > f(n)\}$, $f \in {}^\omega\omega$, form a neighborhood base at p. Since $\mathcal{F}$ is a dominating family, $(\forall f \in {}^\omega\omega)(\exists \alpha \in \omega_1)(f <^* f_\alpha)$. Further, $f <^* f_\alpha$ implies that $f_\alpha \setminus O_f$ is a finite set, and hence $\alpha \in \operatorname{cl} O_f$. Consequently, the point p and the closed set ω_1 cannot be separated by open sets.)

Note that CH in Example 1 can be trivially weakened to the assumption that there is a scale in ${}^\omega\omega$. Further, as pointed out in [17], Example 1 can be modified so that even this assumption can be completely avoided, viz. the scale $\mathcal{F}$ can be replaced by a well-ordered unbounded family $\mathcal{F}$ in ${}^\omega\omega$ (such family always exists in ZFC).

EXAMPLE 2. Let $\{f_\alpha \in {}^\omega\omega : \alpha \in \underline{b}\}$ be a well-ordered ($f_\alpha <^* f_\beta$ iff $\alpha < \beta$) unbounded family in ${}^\omega\omega$. For $Y_b = (\omega \times (\omega + 1)) \cup \underline{b}$ define a topology in the same way as for Y_a in Example 1 and, identifying all points (n, ω) to one point p, consider $X_b = (\omega \times \omega) \cup \{p\} \cup \underline{b}$ equipped with the corresponding quotient topology. Again, using virtually the same argument as in Example 1, it can be shown that X_b is a sequentially regular Fréchet space which fails to be regular (for each O_f there exists $\alpha \in \underline{b}$ such that $O_f \cap f_\alpha$ is an infinite set, and hence $\alpha \in \mathrm{cl}\, O_f$).

Spaces X_a and X_b are uncountable. As shown in [16], even a countable Fréchet space can be sequentially regular but not regular.

EXAMPLE 3. Let $\mathcal{A}$ be an ADF on ω such that the cardinality of $\mathcal{A}$ is 2^ω (such $\mathcal{A}$ always exists in ZFC). Since the cardinality of ${}^\omega\omega$ is 2^ω, there exist one-to-one correspondences $\mathcal{A} = \{A_\alpha : \alpha \in 2^\omega\}$ between $\mathcal{A}$ and 2^ω and ${}^\omega\omega = \{f_\alpha : \alpha \in 2^\omega\}$ between ${}^\omega\omega$ and 2^ω. Consider the set $Y_c = (\omega \times (\omega + 1)) \cup 2^\omega$ equipped with the following topology: all points in $\omega \times \omega$ are isolated; for each $n \in \omega$, a neighborhood base at (n, ω) is formed by sets $\{(n, \omega)\} \cup (C_n \setminus F)$, where $C_n = \{(n, m) : m \in \omega\}$ and F is a finite subset of C_n; for each $\alpha \in 2^\omega$, a neighborhood base at α is formed by sets $\{\alpha\} \cup (f_\alpha \restriction A_\alpha \setminus F)$, where F is a finite subset of $f_\alpha \restriction A_\alpha$, the restriction of f_α to A_α. Again, Y_c is a space of the type $\Psi(\omega, \mathcal{N})$. Let X_c be the quotient space obtained from Y_c by identifying all points $\alpha \in 2^\omega$ to one point p. It is not so hard to verify that X_c has the desired properties.

All three examples are quotients of spaces of the type $N \cup \mathcal{N}$ ($= \Psi(\omega, \mathcal{N})$) with $N = \omega \times \omega$. The properties of $\mathcal{N}$, and hence of the whole space $N \cup \mathcal{N}$, depend on whether or not there is in ${}^\omega\omega$ a certain family of functions. This, however, may depend on additional set-theoretical assumptions.

3. REMARKS ON CARDINAL INVARIANTS

In this section we recall the definitions of certain cardinal invariants and make some remarks concerning their applications. A mnemonic notation of these invariants (they are usually denoted by lower case German letters) is largerly due to E. van DOUWEN; $\underline{h}$ is due to P. NYIKOS and $\underline{n}$ is due to B. BALCAR, J. PELANT and P. SIMON.

Denote by $\underline{b}$ the minimal size of an unbounded family of functions in ${}^{\omega}\omega$, i.e.,

$$\underline{b} = \min\{|\mathcal{B}| : \mathcal{B} \subset {}^{\omega}\omega \ \& \ (\forall g \in {}^{\omega}\omega)(\exists f \in \mathcal{B})(f \not<^{*} g)\};$$

denote by $\underline{d}$ the minimal size of a dominating family of functions in ${}^{\omega}\omega$, i.e.,

$$\underline{d} = \min\{|\mathcal{D}| : \mathcal{D} \subset {}^{\omega}\omega \ \& \ (\forall g \in {}^{\omega}\omega)(\exists f \in \mathcal{D})(g <^{*} f)\};$$

denote by $\underline{p}$ the minimal size of a free filter with the pseudo-intersection property or, equivalently, the minimal size of a neighborhood base of a nowhere dense subset of ω^{*}, i.e.,

$$\underline{p} = \min\{|F| : F \subset \mathcal{P}(\omega)/_{fin} \ \& \ F \text{ is centred } \& \ \wedge F = 0\} =$$
$$= \min\{|\mathcal{F}| : \mathcal{F} \subset [\omega]^{\omega} \ \& \ (\forall \mathcal{G} \in [\mathcal{F}]^{<\omega})(\cap \mathcal{G} \text{ is infinite}) \ \& \ (\forall X \in [\omega]^{\omega})(\exists Y \in \mathcal{F})(X \setminus Y \text{ is infinite})\};$$

denote by $\underline{t}$ the minimal size of a nowhere dense tower of ω^{*}, i.e.,

$$\underline{t} = \min\{|T| : T \subset \mathcal{P}(\omega)/_{fin} \ \& \ (T, <) \text{ is well-ordered } \& \ \& \ \wedge T = 0\} =$$
$$= \min\{|\mathcal{T}| : \mathcal{T} \subset [\omega]^{\omega} \ \& \ (\mathcal{T}, \subset^{*}) \text{ is well ordered } \& \ \& \ (\forall X \in [\omega]^{\omega})(\exists Y \in \mathcal{T})(X \setminus Y \text{ is infinite})\};$$

denote by $\underline{h}$ the minimal size of the height of a base matrix (a tree-like π-base), i.e.,

$$\underline{h} = \min\{|\mathcal{F}| : (\forall F \in \mathcal{F})(F \text{ is a nowhere dense subset of } \omega^{*}) \ \& \ \& \ (\cup \mathcal{F} \text{ is dense in } \omega^{*})\} =$$
$$\doteq \sup\{\varkappa : (\forall \mu \in \varkappa)(\forall \text{ system } \mathbb{P} = \{\mathcal{P}_{\alpha} : \alpha \in \mu\} \text{ of MAD families on } \omega)(\exists \mathcal{P} \subset [\omega]^{\omega})(\mathcal{P} \text{ is a MAD family } \& \ \mathcal{P} \text{ refines } \mathbb{P})\},$$

where $\mathcal{P}$ refines $\mathbb{P}$ means that $(\forall \alpha \in \mu)(\forall X \in \mathcal{P})(\exists ! Y \in \mathcal{P}_{\alpha})(X \subset^{*} Y)$;

denote by $\underline{n}$ the so-called Novák number (also called the Baire number) of ω^{*}, i.e.,

$$n = \min\{|\mathcal{F}| : (\forall F \in \mathcal{F})(F \text{ is a nowhere dense subset of } \omega^{*}) \ \& \ \& \ \cup \mathcal{F} = \omega^{*}\} =$$
$$= \sup\{\varkappa : (\forall \mu \in \varkappa)(\forall \text{ system } \mathbb{P} = \{\mathcal{P}_{\alpha} : \alpha \in \mu\} \text{ of MAD families on } \omega)(\exists \mathcal{U} \in \omega^{*})(\forall \alpha \in \mu)(\exists P \in \mathcal{P}_{\alpha})(P \in \mathcal{U})\}.$$

Further, we just mention the following two cardinal invariants: $\underline{a}$ denotes the minimal size of a MAD family on ω and $\underline{s}$ denotes the minimal size of a splitting family in $\mathcal{P}(\omega)/_{fin}$.

In ZFC, the following inequalities can be proved (in the diagram $\alpha \longrightarrow \beta$ means that $\alpha \leqq \beta$):

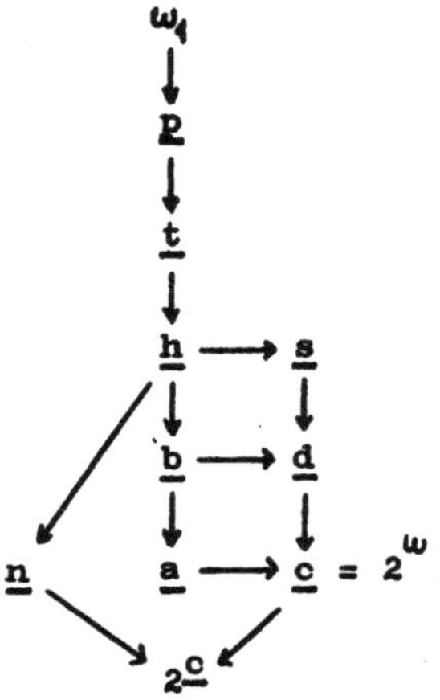

Note that B. BALCAR, J. PELANT and P. SIMON proved ([2]) that $\underline{h} \leqq \underline{b}$; R. C. SOLOMON proved ([42]) that $\underline{b} \leqq \underline{a}$; as communicated to us, P. SIMON and independently P. NYIKOS proved $\underline{s} \leqq \underline{d}$. The remaining inequalities belong to folklore. The exact relationship between $\underline{a}$ and $\underline{d}$ is not known (there are models in which $\underline{a} = \underline{d}$ or $\underline{a} < \underline{d}$, but we do not know whether there is a model in which $\underline{d} < \underline{a}$) . Also we do not know whether there is a model in which $p < t$ (in all known models we have $p = t$) . It follows from the literature ([27], [23], [38], [39]) that, except the two just mentioned unsolved problems, nothing can be added to the above diagram and it is consistent with ZFC that all other inequalities in the above diagram can be sharp.

Next, we present some results from the set-theoretical topology. In doing so, we follow two principles: - using some additional set-theoretical assumptions concerning cardinal invariants, it is possible to prove nonabsolute results; - using some combinatorial objects of nonspecified cardinalities (existing in ZFC), it is possible to get absolute results. Also, we would like to stress the usefulness of combinatorial methods in sequential topology, sequential convergence spaces, general topology, analysis, functional analysis and probability.

S. MRÓWKA ([31]) constructed a MAD family $\mathcal{N}$ such that $\beta(N \cup \mathcal{N})$ is the one-point compactification of $N \cup \mathcal{N}$. P. SIMON ([41]) constructed (in ZFC) a compact Fréchet space the square of which is not a Fréchet space. A. J. BERNER ([43]), under $\underline{a} = 2^{\omega}$, constructed a noncompact space X such that βX is Fréchet. J. NOVÁK ([32]) constructed a regular sequential space on which each (real-valued) continuous function is constant. M. HUŠEK ([15]) constructed a space $N \cup \mathcal{N}$ having different sequential and $\{0, 1\}$-sequential envelopes. Solving a problem from [15], L. MI-

ŠÍK, Jr. ([30]) constructed under $\underline{h} = 2^{\omega}$ a $\{0, 1\}$-sequentially regular space $N \cup \mathcal{N}$ which is sequentially complete but not $\{0, 1\}$-sequentially complete. G. GRUENHAGE ([20]) proved that the product $S_{\omega} \times S_{\varkappa}$ is a sequential space iff $\underline{b} \geqq \varkappa^{+}$, where S_{α} is the quotient space of a disjoint union of α-many convergent sequences with their limits identified; from this it follows, e.g., that the question "Is $S_{\omega} \times S_{\omega_1}$ a sequential space?" is undecidable in ZFC. A. BLASZCZYK and A. SZYMAŃSKI ([5]) proved that, under CH, for each $p \in \omega^{*}$ the space $\omega^{*} \setminus \{p\}$ is not normal. V. V. FEDORČUK ([13]) proved that, under $\underline{s} = \omega_1$ & $2^{\omega_1} = 2^{\omega}$, there is a compact space of the cardinality 2^{ω} in which no nontrivial sequence converges. B. BALCAR and P. VOJTÁŠ ([3]) proved that in ZFC each point in ω^{*} is a 2^{ω}-point. Using MAD families, M. CONTESSA and F. ZANOLIN ([9]) solved a problem of J. Novák concerning the classification of sequential convergence spaces. A. KAMIŃSKI in a series of papers studied conditional probabilities (cf. [25]) and, under CH, he recently obtained some interesting results in this area. Under CH, P. KRATOCHVÍL ([26]) and J. BURZYK ([7]) constructed a convergence space, resp. a convergence group satisfying condition D but not condition Y; more information about related problems and results can be find in a paper by the present authors in these Proceedings. V. I. MALYHIN and N. N. HOLŠČEVNIKOVA ([22], [29]) proved that, under $\underline{t} = 2^{\omega}$, each dominating family of convergent series (in a natural partial ordering) has the cardinality 2^{ω}. Nice applications of combinatorial methods in sequential spaces and their generalizations can be found in the classical paper [1] by A. V. ARHANGEĽSKII and in a recent paper by Z. FROLÍK, R. ISLER and G. TIRONI ([19]).

In the last decade several interesting books and survey articles on cardinal invariants and combinatorial methods have been published, e.g., by M. E. RUDIN ([35]), W. W. COMFORT ([8]), I. JUHÁSZ ([24]), G. M. REED ([34]), or are to be published, e.g., the Handbook of set-theoretic topology, edited by J. E. VAUGHAN and K. KUNEN. The classical textbook [37] by W. SIERPIŃSKI on infinite combinatorics is still motivating.

4. NEW RESULTS

In this section we discuss the role and nature of $\underline{h}$ and $\underline{n}$ - two important cardinal invariants of $\mathcal{P}(\omega)/_{fin}$. Answering a question asked by E. K. van DOUWEN in [11], we illustrate how these invariants can be used in proofs by transfinite induction and in constructions of certain spaces.

Let X be a topological space. Recall that X said to be <u>sequentially compact</u> if $(\forall A \in [X]^{\omega})(\exists B \in [A]^{\omega})(\exists x \in X)$ (B converges to x), i.e., each sequence has a convergent subsequence, and X is said to be <u>countably compact</u> if $(\forall A \in [X]^{\omega})$ $(\exists x \in X)$ (A clusters at x), i.e., each infinite set has a cluster point. Following [11] we define

$$\mu = \min\{\varkappa : \text{some product of } \varkappa\text{-many sequentially compact spaces is not sequentially compact}\}.$$

C. T. SCARBOROUGH and A. H. STONE ([36]) proved that $\mu \geq \omega_1$ and S. H. HECHLER ([21]) observed that, in fact, their proof yields $\mu \geq \underline{t}$. D. BOOTH ([63]) showed that $\mu \leq \underline{s}$. J. PELANT and P. SIMON independently showed that $\mu \leq \underline{h}$. Question 6.10 in [11] reads as follows: "Can μ be expressed as a set-theoretically defined cardinal number?" The answer is provided by the next theorem. P. SIMON informed us that the same result has been obtained by P. NYIKOS, J. PELANT and P. SIMON.

THEOREM. $\mu = \underline{h}$.

PROOF. Since we already know that $\mu \leq h$, it suffices to prove $\mu \geq h$. In order to make it more instructive, we break the proof into several steps and add some remarks about the ideas used and the strategy we follow:

1. Remark. Assume that X_α, $\alpha \in \varkappa$, are sequentially compact spaces. Following C. T. SCARBOROUGH and A. H. STONE ([36]) and S. H. HECHLER ([21]), if $\{y_n : n \in \omega\}$ is a sequence in the product space $\prod X_\alpha$, then we construct inductively a tower $\{A_\alpha : \alpha \in \varkappa\}$ in ω in such a way that $\{\mathrm{proj}_\alpha(y_n) : n \in A_\alpha\}$ is a convergent sequence in the space X_α. Now, provided $\varkappa < \underline{t}$, the properties of $\underline{t}$ guarantee the existence of a subsequence $\{y_n : n \in A\}$ of $\{y_n : n \in \omega\}$ which converges in $\prod X_\alpha$, i.e., for each $\alpha \in \varkappa$, its projection into X_α converges to some point in X_α. The idea of our proof is to replace $\underline{t}$ by $\underline{h}$. For, the properties of $\underline{h}$ will permit us to work simultaneously with a large numbers of "towers of convergent subsequences" (instead of with just one as before) and the induction "will be alive with probability 1" for $\varkappa < \underline{h}$.

2. Lemma. Let X be a sequentially compact space and let $\{y_n : n \in \omega\}$ be a sequence in X. Let $\mathcal{B}$ be a MAD family on ω. Then there is a MAD family $\mathcal{A}$ on ω such that:

(i) $(\forall A \in \mathcal{A})(\{y_n : n \in A\}$ converges in $X)$;

(ii) $\mathcal{A}$ refines $\mathcal{B}$.

(Hint. An AD family which is maximal with respect to convergence is also maximal with respect to the inclusion; property (ii) is trivial.)

3. Assume $\mathcal{H} < \underline{h}$. Let X_α, $\alpha \in \mathcal{H}$, be sequentially compact spaces and let $\prod X_\alpha$ be their product. Let $\{y_n : n \in \omega\}$ be a sequence in $\prod X_\alpha$. Then there are two families $\{\mathcal{A}_\alpha : \alpha \in \mathcal{H}\}$ and $\{x_A^\alpha : \alpha \in \mathcal{H}$ & $A \in \mathcal{A}_\alpha\}$ such that:

(iii) $(\forall \alpha \in \mathcal{H})(\mathcal{A}_\alpha$ is a MAD family on $\omega)$;

(iv) $\mathcal{A}_\beta$ refines $\mathcal{A}_\alpha$ whenever $\alpha, \beta \in \mathcal{H}$ & $\alpha < \beta$;

(v) $(\forall \alpha \in \mathcal{H})(\forall A \in \mathcal{A}_\alpha)(x_A^\alpha \in X_\alpha$ & $\{\mathrm{proj}_\alpha(y_n) : n \in A\}$ converges in X_α to $x_A^\alpha)$.

The construction of $\{\mathcal{A}_\alpha : \alpha \in \mathcal{H}\}$ and $\{x_A^\alpha : \alpha \in \mathcal{H}$ & $A \in \mathcal{A}_\alpha\}$ can be carried out by induction on α. If α is isolated, we simply use our Lemma. If α is a limit ordinal, we moreover use some refinement of the previous MAD families. The calculations are straightforward and are omitted. Now, since $\mathcal{H} < \underline{h}$, there is a MAD family $\mathcal{A}$ on ω such that $\mathcal{A}$ refines the matrix $\{\mathcal{A}_\alpha : \alpha \in \mathcal{H}\}$. Choose an arbitrary $A \in \mathcal{A}$. Then there is a unique family $\{A_\alpha : \alpha \in \mathcal{H}\}$ such that $(\forall \alpha \in \mathcal{H})$ $(A_\alpha \in \mathcal{A}_\alpha$ & $A \subset^* A_\alpha)$. In the product space $\prod X_\alpha$, denote by y the point the α-th projection $\mathrm{proj}_\alpha(y)$ of which is x_A^α, $\alpha \in \mathcal{H}$. From the construction of the family $\{A_\alpha : \alpha \in \mathcal{H}\}$ it follows readily that the subsequence $\{y_n : n \in A\}$, of the original sequence $\{y_n : n \in \omega\}$, converges to y. This completes the proof.

REMARK. Observe that if (in the same notation as in the above proof) $\mathcal{H} < \underline{n}$, then there is an ultrafilter $\mathcal{U}$ on ω such that $(\forall \alpha \in \mathcal{H})(\exists A_\alpha \in \mathcal{A}_\alpha)(A_\alpha \in \mathcal{U})$. Then the point y (defined by $\mathrm{proj}_\alpha(y) = x_A^\alpha$, $\alpha \in \mathcal{H}$) is an accumulation point of the original sequence $\{y_n : n \in \omega\}$.

COROLLARY. Every product of less than $\underline{n}$ sequentially compact spaces is countably compact.

Observe that it is not known whether in ZFC there is a family $\{X_\alpha, \alpha \in A\}$ of sequentially compact spaces such that the product space $\prod X_\alpha$ is not countably compact (cf. [11]).

It is known (cf. [10]) that there are maximal towers having cardinalities different from $\underline{t}$. Towers can be used, e.g., in constructions similar to that given by S. P. FRANKLIN and M. RAJAGOPALAN ([14]), cf. Example 7.1 in [11].

Concerning numbers $\underline{h}$ and $\underline{n}$, observe that they were used in [18] to construct a completely regular space X in which no nontrivial sequence converges but in the Čech-Stone compactification βX there is a nontrivial convergent sequence lying together with its limit in $\beta X \setminus X$.

5. CONCLUDING REMARKS

We have tried to point out some aspects of set theory which are connected with sequential convergence. On one side, they provide tools for constructions and help us to prove theorems but, on the other side, they sometimes fix limits beyond which our attempts to prove some assertions necessarily fail, e.g., if we are trying to prove in ZFC something which is, in fact, independent of ZFC.

We conclude with few remarks concerning the usage of cardinal invariants:

$\underline{b}$ and $\underline{d}$ are useful when constructing spaces of the type $N \cup \mathcal{N}$ for $N = \omega \times \omega$ and $\mathcal{F} \subset {}^{\omega}\omega$, $\mathcal{F} \subset \mathcal{N}$, is indexed by $\underline{b}$ or $\underline{d}$, see e.g. Example 2 ;

t - it helps us to keep the transfinite induction alive, see e.g. [36], [21];

$\underline{h}$ and $\underline{n}$ help us to keep the transfinite induction alive simultaneously "with probability 1", see e.g. Theorem, or they are useful when constructing spaces, see e.g. [17].

REFERENCES

[1] Arhangel'skii, A. V.: On the cardinality of bicompacta satisfying the first axiom of countability. Soviet. Math. Dokl. 10 (1969), 951-955.

[2] Balcar, B., J. Pelant, and P. Simon: The space of ultrafilters on N covered by nowhere dense sets. Fund. Math. 110 (1980), 11-24.

[3] Balcar, B. and P. Vojtáš: Almost disjoint refinement of families of subsets of N . Proc. Amer. Math. Soc. 79 (1980), 465-470.

[4] Berner, A. J.: $\beta(X)$ can be Fréchet. Proc. Amer. Math. Soc. 80 (1980), 367-373.

[5] Błaszczyk, A. and A. Szymański: Hausdorff's gaps versus normality. Bull. Acad. Polon. Sci. 30 (1982), 371-378.

[6] Booth, D.: A Boolean view of sequential compactness. Fund. Math. 85 (1974), 99-102.

[7] Burzyk, J.: The example of FLUSHD convergence without Y . These proceedings.

[8] Comfort, W. W.: Deciding some undecidable topological

statements. Ann. New York. Acad. Sc. 321 (1979), 9-26. (Slovak translation in: Pokroky Mat. Fyz. Astronom. 27 (1982), 252-272.)

[9] Contessa, M. and F. Zanolin: On a question of J. Novák about convergence spaces. Rend. Sem. Mat. Univ. Padova, 58 (1977), 155-161.

[10] Dordal, P. L.: Independence results concerning some combinatorial properties of continuum. Ph. D. Thesis, Harvard University, Mass., 1982.

[11] van Douwen, E. K.: The integers and topology. Handbook of set-theoretical topology. (Ed. J. Vaughan and K. Kunen). (North-Holland, to appear.)

[12] Engelking, R.: General topology. PWN, Warszawa, 1977.

[13] Fedorčuk, V.: A compactum having the cardinality of the continuum with no convergent sequences. Math. Proc. Camb. Phil. Soc. 81 (1977), 177-181.

[14] Franklin, S. P. and M. Rajagopalan: Some examples in topology. Trans. Amer. Math. Soc. 155 (1971), 305-314.

[15] Frič, R. and M. Hušek: Projectively generated convergence of sequences. Czechoslovak Math. J. 33 (108) (1983), 525-536.

[16] Frič, R. and J. Novák: A tiny peculiar Fréchet space. Czechoslovak Math. J. 34 (109) (1984), 22-27.

[17] Frič, R. and P. Vojtáš: Variants of complete regularity. Abstracts Amer. Math. Soc. 2 (1981), # 81T - 54 - 572.

[18] Frič, R. and P. Vojtáš: Convergent sequences in βX. Proc. 11-th Winter School on Abstract Analysis 1983. (To appear.)

[19] Frolík, Z., Isler, R. and G. Tironi: Some results on chain-net and sequential spaces. Quaderni Matematici, n. 66, 1983, Universitá Trieste, Trieste, Italy.

[20] Gruenhage, G.: k-spaces and products of closed images of metric spaces. Proc. Amer. Math. Soc. 80 (1980), 478-482.

[21] Hechler, S. H.: On some weakly compact spaces and their products. Gen. Top. Appl. 5 (1975), 83-93.

[22] Holščevnikova, N. N.: Undecidability of some questions concerning convergence of series and sequences. Mat. Zametki 34 (1983), 711-718.

[23] Jech, T.: Set theory. Academic Press, New York, 1978.

[24] Juhász, I.: Cardinal functions in topology - ten years later. Math. Centre Tracts, 123, Math. Centrum, Amsterdam, 1980.

[25] Kamiński, A.: On the Rényi theory of conditional probabilities. (To appear.)

[26] Kratochvíl, P.: Sequential convergences generated by coneighborhoods and an example of D not Y space. Czechoslovak Math. J. (To appear.)

[27] Kunen, K.: Set theory. An introduction to independence proofs. Studies in Logic and the Foundations of Mathematics, 102, North-Holland, Amsterdam-New York, 1980.

[28] Luzin, N.: On subsets of the series of natural numbers (Russian). Izv. Akad. Nauk SSSR. Ser. Mat. 11 (1947), 403-411.

[29] Malyhin, V. I. and N. N. Holščevnikova: Independence of two set-theoretic statements in summation theory (Russian). Mat. Zametki 28 (1980), 869-882.

[30] Mišík, L., Jr.: Sequential completeness and $\{0, 1\}$-sequential completeness are different. Czechoslovak Math. J. 34 (109) (1984), 424-431.

[31] Mrówka, S.: Some set-theoretic constructions in topology. Fund. Math. 94 (1977), 83-92.

[32] Novák, J.: A regular space on which every continuous function is constant (Czech). Časopis Pěst. Mat. Fys. 73 (1948), 58-68.

[33] Novák, J.: On convergence spaces and their sequential envelopes. Czechoslovak Math. J. 15 (90)(1965), 74-100.

[34] Reed, G. M., editor: Set-theoretic topology. Academic Press., New York-London, 1977.
[35] Rudin, M. E.: Lectures on set-theoretic topology. CBMS Regional Conference Series in Mathematics, No. 23, Amer. Math. Soc., Providence, R. I., 1975.
[36] Scarborough, C. T. and A. H. Stone: Products of nearly compact spaces. Trans. Amer. Math. Soc. 124 (1964), 131-147.
[37] Sierpiński, W.: Hypothése du continu. Second edit., Chelsea Publ. Comp., New York, 1956.
[38] Shelah, S.: Proper forcing. Lecture Notes in Math., 940, Springer, Berlin-New York, 1982.
[39] Shelah, S.: Cardinal invariant of the continuum: consistency results. Abstracts Amer. Math. Soc. 4 (1983), # 83T-03-510.
[40] Sikorski, R.: Boolean algebras. Springer-Verlag, Berlin, 1960.
[41] Simon, P.: A compact Fréchet space whose square is not Fréchet. Comment. Math. Univ. Carolin. 21 (1980), 749-753.
[42] Solomon, R. C.: Families of functions and sets. Czechoslovak Math. J. 27 (102)(1977), 556-559.
[43] Steen, L. A. and J. A. Seebach, Jr.: Counterexamples in topology. Second edition. Springer-Verlag, New York-Heildelberg, 1978.
[44] Vaughan, J. E. and K. Kunen, ed.: Handbook of set-theoretic topology. (North-Holland, to appear.)

COARSE CONVERGENCE GROUPS

Roman Frič[1)] and Fabio Zanolin[2)]

1. INTRODUCTION

Let P be a property of topologies. A topological space $(X, \mathcal{T})$ is said to be P-minimal if $\mathcal{T}$ has P, but no topology on X which is strictly weaker than $\mathcal{T}$ has P (cf. [22]).

Consider the following situation: X is a group and a topology $\mathcal{T}$ on X has P if it is a Hausdorff topology compatible with the group structure of X. In this case X equipped with a P-minimal topology is said to be a minimal topological group.

While there is an extensive literature on P-minimal topological spaces and, in particular, on minimal topological groups (cf. [22], [5], [3], [4], [19], [2] and the references therein), little attention has been paid so far to their sequential convergence counterparts ([11], [8]). The aim of the present paper is to initiate systematic study of minimal sequential convergence groups. Since convergence structures are defined via convergent sequences (resp. filters) and minimal means that the structure in question has as many convergent sequences (resp. filters) as possible, we prefer to call such structures coarse. Observe that the property of being P-minimal has been characterized by B. BANASCHEWSKI in purelly categorical terms. An object A in a concrete category K for which all monomorphisms $A \to B$ are embeddings is said to be a minimal object of K; this categorical definition yields the usual P-minimal topologies when applied to particular cases appearing in the literature. Future research based on this categorical approach seems to be promising.

DEFINITION. Let P be a property of convergence structures. A convergence structure $\mathcal{L}$ on X is said to be P-coarse if $\mathcal{L}$ has P, but no convergence structure on X which is strictly coarser than $\mathcal{L}$ has P. A set equipped with a P-coarse convergence structure is said to be a P-coarse convergence space.

Throughout the rest of this paper we shall deal with groups equipped with compatible sequential convergence structures. More precisely, the underlying set is always a group, say G, and it

1) Matematický ústav SAV, Karpatská 5, 040 01 Košice, Czechoslovakia
2) Istituto di Matematica, Universita, P. le Europa 1, 34100 Trieste, Italy

is equipped with a sequential convergence structure $\mathcal{G} \subset G^N \times G$ which satisfies the so-called FLUSH-axioms (Katowice notation, cf. [12]) or equivalently, $\mathcal{L}$-axioms (Prague notation, cf. [13]). Recall that H ($\equiv \mathcal{L}_0$) denotes the uniqueness of sequential limits, S ($\equiv \mathcal{L}_1$) means that for all $g \in G$ the constant sequence $\langle g \rangle$ converges to g, F ($\equiv \mathcal{L}_2$) means that each subsequence of a convergent sequence converges to the same limit, U ($\equiv \mathcal{L}_3$) denotes the Urysohn axiom, and L ($\equiv$ SG) denotes the compatibility of $\mathcal{G}$ with the group structure of G. So P stands for FLUSH and P-coarse will be condensed to coarse. We shall use the following notational convention: $\mathcal{S}$ denotes the set of all monotone mappings of the set N of all natural numbers into itself and if $S = \langle x_n \rangle$ is a sequence of points, then for $s \in \mathcal{S}$ the subsequence of S the n-th term of which is $x_{s(n)}$ is denoted by $S \circ s$.

REMARK 1. Let G be a group. A subgroup $\mathcal{H}$ of G^N is said to be normal with respect to G if $\langle gS(n)g^{-1} \rangle \in \mathcal{H}$ whenever $g \in G$ and $S \in \mathcal{H}$. Let $\mathcal{A}$ be a nonempty subset of G^N. Let $\delta\mathcal{A}$ be the set of all sequences $S \circ s$ such that $S \in \mathcal{A}$ and $s \in \mathcal{S}$ and let $\varsigma\mathcal{A}$ be the set of all sequences $S \in G^N$ such that for each $s \in \mathcal{S}$ there exists $t \in \mathcal{S}$ such that $S \circ s \circ t \in \mathcal{A}$. Finally, let $[\mathcal{A}]_G$ be the smallest subgroup of G^N containing $\mathcal{A}$ and normal with respect to G. Then the following assertions hold (cf. [9], [23]):

(i) $[\mathcal{A}]_G$ consists precisely of finite products of sequences of the form $\langle gS(n)^{\varepsilon} g^{-1} \rangle$, where $\varepsilon \in \{-1, 1\}$, $g \in G$ and $S \in \mathcal{A}$;

(ii) $\delta[\delta\mathcal{A}]_G = [\delta\mathcal{A}]_G$;

(iii) $\varsigma[\delta\mathcal{A}]_G$ is the smallest subgroup of G^N containing $\mathcal{A}$, closed with respect to δ and ς and normal with respect to G;

(iv) Let $\mathcal{G}$ be a FLUS-convergence structure for G. Then $\mathcal{G}^{\leftarrow}(e) = \{S \in G^N ;\ S \to e\}$ is a subgroup of G^N closed with respect to δ and ς and normal with respect to G;

(v) If $\mathcal{A} = \varsigma[\delta\mathcal{A}]_G$, then there exists a FLUS-convergence structure $\mathcal{G}$ for G such that $\mathcal{A} = \mathcal{G}^{\leftarrow}(e)$. Further, $\mathcal{G}$ satisfies H iff $\mathcal{A}$ contains no constant sequence except $\langle e \rangle$.

The above properties enable us to equip a group with a suitable FLUSH-convergence structure in which a prescribed set of sequences converge to its neutral element (cf. [8]).

A subtitle of our paper might be "Minimal versus coarse". For, we present our results on coarse groups paired with the corresponding results on minimal topological groups. As we shall see, the two notions have quite different properties.

2. EXISTENCE

I. PRODANOV in [17] showed that the group Q of all rational numbers does not admit minimal group topologies (but each proper subgroup of Q does). On the other hand, as shown in [8], every commutative group admits a coarse convergence structure. Our next theorem generalizes this result to the noncommutative case. We start with a lemma on which the theorem is based.

LEMMA. Let G be a group. Let $\mathcal{B}$ and $\mathcal{C}$ be nonempty subsets of G^N such that $\delta\mathcal{C} \cap [\delta\mathcal{B}]_G = \emptyset$. Consider the set Γ of all subsets $\mathcal{H}$ of G^N such that:

(i) $\mathcal{H} = [\mathcal{H}]_G$;

(ii) $\mathcal{B} \subset \mathcal{H}$ and $\delta\mathcal{C} \cap \mathcal{H} = \emptyset$;

(iii) $\mathcal{H} = \delta\mathcal{H} = \xi\mathcal{H}$.

Then Γ is nonempty and it contains a maximal element with respect to inclusion.

PROOF. Observe that $\delta\mathcal{C} \cap \xi[\delta\mathcal{B}]_G = \emptyset$ and hence Γ is nonempty. Let $\{\mathcal{H}_\iota ;\ \iota \in I\}$ be a chain in Γ . Put $\mathcal{H}_0 = \xi(\bigcup_{\iota \in I} \mathcal{H}_\iota)$. A straightforward calculation shows that $\mathcal{H}_0 \in \Gamma$ and $\mathcal{H}_\iota \subset \mathcal{H}_0$ for all $\iota \in I$. The assertion now follows by Zorn's lemma.

THEOREM 1. Let $\mathcal{G}$ be a FLUSH-convergence structure for a group G . Then there exists a coarse convergence structure $\mathcal{G}_c$ for G such that $\mathcal{G} \subset \mathcal{G}_c$.

PROOF. Put $\mathcal{B} = \mathcal{G}^{\leftarrow}(e)$ and $\mathcal{C} = \{\langle p \rangle \in G^N ;\ p \neq e\}$. Since $\delta\mathcal{C} = \mathcal{C}$ and $\mathcal{G}$ is a FLUSH-convergence structure, according to Remark 1 we have $\mathcal{B} = [\delta\mathcal{B}]_G$ and $\delta\mathcal{C} \cap [\delta\mathcal{B}]_G = \emptyset$. By Lemma, there exists a maximal subgroup $\mathcal{H}$ of G^N such that $\mathcal{B} \subset \mathcal{H}$, $\mathcal{C} \cap \mathcal{H} = \emptyset$ and $\mathcal{H} = \xi[\delta\mathcal{H}]_G$. It follows from Remark 1 that there exists a FLUS-convergence structure $\mathcal{G}_c$ for G such that $\mathcal{H} = \mathcal{G}_c^{\leftarrow}(e)$. Clearly, $\mathcal{G}_c$ satisfies H and it is a coarse convergence structure for G .

3. CHARACTERIZATION

J. JAKUBÍK in [11] characterized coarse compatible sequential convergence structures in linear spaces. Theorem 2 provides a characterization of coarse convergence structures in groups. Provided the group is commutative, the correspondingly modified criterion has analogous form as that obtained in [11].

THEOREM 2. Let $\mathcal{G}$ be a FLUSH-convergence structure for a group G. Then $\mathcal{G}$ is coarse iff the following condition (C) is satisfied:

(C) For each sequence $S \in G^N$ either:

(C_1) some subsequence $S \circ s$, $s \in \mathcal{S}$, of S converges to the neutral element e of G,

or

(C_2) for some $p \in G$, $p \neq e$, the constant sequence $\langle p \rangle$ is a product of finitely many sequences of the type $\langle gU(n)^{\varepsilon} g^{-1} \rangle$, where g is an element of G, $\varepsilon \in \{-1, 1\}$, and $U \in G^N$ either converges to e or it is a sebsequence of S.

PROOF. Clearly, $\mathcal{G}$ is coarse iff $\mathcal{G}^{\leftarrow}(e)$ is a maximal subgroup of G^N which is normal with respect to G, closed with respect to operations σ and ξ and such that for no $p \in G$, $p \neq e$, the constant sequence $\langle p \rangle$ belongs to $\mathcal{G}^{\leftarrow}(e)$, i.e., iff for each sequence $S \in G^N$, $S \notin \mathcal{G}^{\leftarrow}(e)$, there exists $p \in G$, $p \neq e$, such that $\langle p \rangle \in \xi[\sigma(\{S\} \cup \mathcal{G}^{\leftarrow}(e))]_G$. Since $\sigma(\{S\} \cup \mathcal{G}^{\leftarrow}(e)) = \sigma\{S\} \cup \mathcal{G}^{\leftarrow}(e)$ and each subsequence of $\langle p \rangle$ is again $\langle p \rangle$, the assertion of Theorem 2 follows from (i) in Remark 1.

REMARK 2. If G is a commutative group ($e \equiv 0$), then condition (C_2) in Theorem 2 has the form

(C_2') for some $p \in G$, $p \neq o$, the constant sequence $\langle p \rangle$ is a sum of a sequence converging to 0 and a finite linear combination with integral coefficients of subsequences of S,

or, equivalently,

(C_2'') for some $p \in G$, $p \neq 0$, some finite linear combination with integral coefficients of subsequences of S converges to p.

EXAMPLE 1. Consider the "rational" torus Q/Z. It is a first--countable topological (T_2) group and hence a FLUSH-convergence group. As shown by R. M. STEPHENSON, Jr. in [22], Q/Z is a minimal topological group. Using (C), it can be readily seen that Q/Z fails to be coarse.

4. OPERATIONS

D. DOITCHINOV in [5] showed that a product of two minimal topological groups need not be minimal. We do not know anything about products of coarse groups.

QUESTION 1. Is coarseness of groups preserved by products?

As proved by I. PRODANOV in [18], each closed subgroup of a minimal topological commutative group is minimal. From Remark 2 we get immediately the corresponding statement for coarse groups.

THEOREM 3. Each closed subgroup of a coarse commutative group is coarse.

U. SCHWANENGEL constructed in [21] an example of a precompact minimal topological group a closed normal subgroup of which fails to be minimal.

REMARK 3. The second author recently constructed a coarse non-commutative group a closed subgroup of which fails to be coarse. The result will be published elsewhere.

5. COARSE VERSUS SEQUENTIALLY PRECOMPACT

As shown by A. OLSHANSKIJ in [15], there exists a countable group admitting only one compatible (T_2) group topology, viz. the discrete one. In this case the discrete topology is minimal but fails to be precompact.

REMARK 4. First example (uncountable, under CH) of a group admitting only the discrete group topology was given by S. SHELAH. OLSHANSKIJ's result is based on a complicated construction by S. J. ADIAN (cf. [1]).

For commutative groups we have the following result by I. PRODANOV and L. STOJANOV ([20]): Every minimal commutative group is precompact.

Let G be a FLUSH-convergence group. Consider the following two versions of sequential precompactness:

(i) G is a subgroup of a sequentially compact FLUSH-convergence group H;

(ii) Each sequence in G contains a (two-sided) Cauchy subsequence.

Clearly, (i) implies (ii). We do not know whether (ii) implies (i). Note that not every FLUSH-convergence group has a completion (cf. [7]). As shown by J. NOVÁK in [14], every commutative FLUSH-convergence group has a completion but it can have several nonhomeomorphic completions (e. g. the group Q of rational numbers).

QUESTION 2. Is every commutative coarse convergence group sequentially precompact with respect to (i) or (ii)?

In the noncommutative case the situation is the same as for topological groups.

EXAMPLE 2. As shown by A. OLSHANSKIJ, there exists a natural number m, a countable noncommutative group G and elements $a_1, \ldots, a_m \in G$, $a_i \neq e$, such that each $x \in G$, $x \neq e$, solves at least one of the following $2(m-1)$ equations: $x^m = a_i$ or $x = a_i$, $i = 1, \ldots, m-1$. From this fact we easily infer that no one-to-one sequence in G can converge to e in any FLUSH-convergence structure for G.

6. FREE GROUPS

Let X be a completely regular space and let $F(X)$ be the free topological group over X. It was proved by V. PESTOV ([16]) that $F(X)$ fails to be a minimal topological group.

The free sequential convergence groups were introduced in ([9]). Further information is provided by [8] and [10].

THEOREM 4. Let $F(X)$ be the free FLUSH-convergence group over a FUSH-convergence space X. Then $F(X)$ fails to be coarse.

PROOF. It follows from the construction of $F(X)$ that for each $x \in X$ the sequence $\langle x^{s(n)} \rangle$, with $s \in \mathcal{S}$ defined by $s(n) = 2^{n!}$, kills condition (C) in Theorem 2.

COROLLARY 1. Let $F(X)$ be the free (algebraic) group over a finite set X of generators. Then $F(X)$ can be equipped with a nondiscrete FLUSH-convergence structure.

PROOF. The finite set X can be considered as a discrete FUSH-convergence space and hence the free FLUSH-convergence group $F(X)$ is discrete. By Theorem 4, $F(X)$ is not coarse and hence, by Theorem 1, there exists a coarse (nondiscrete) FLUSH-convergence structure for $F(X)$.

This solves a problem posed in [6].

7. CONCLUDING REMARKS

We have already mentioned some of the open questions concerning coarse convergence groups. Further, it would be interesting to find out more about coarse convergence structures for subgroups of powers R^m of the real line R, resp. T^m of the torus $T = R/Z$. We conclude our paper by mentioning some of the related results.

EXAMPLE 3. None of the groups Z, Q, R equipped with the usual metric convergence is coarse. (Hint. Apply Theorem 2.)

The proof of the following theorem will appear elsewhere.

THEOREM 5. Let G be a divisible torsion-free commutative group equipped with a coarse convergence structure $\mathcal{G}$. Then G is complete.

COROLLARY 2. Let Q (resp. R) be equipped with a coarse convergence structure $\mathcal{G}$ which is coarser than the usual metric convergence structure. Then it is complete.

EXAMPLE 4. The group R is commutative divisible and torsion-free. It is complete in the usual metric convergence but it fails to be coarse.

EXAMPLE 5. Let $\mathcal{G}$ be a coarse convergence structure for R which is coarser than the usual metric convergence structure. Then $\mathcal{G} \upharpoonright Q$ fails to be a coarse convergence structure for Q. Indeed, consider an irrational number x and a sequence S of rational numbers converging in the usual metric convergence to x. Assume, to the contrary, that $\mathcal{G} \upharpoonright Q$ is coarse. By Remark 2, some finite linear combination $\sum_{i=1}^{m} z_i (S \circ s_i)$, where z_i are integers and $S \circ s_i$, $i = 1, \ldots, m$, are subsequences of S, would $\mathcal{G} \upharpoonright Q$-converge to a nonzero rational y. Hence $x \cdot \sum_{i=1}^{m} z_i = y$, which is impossible.

REFERENCES

[1] Adian, S. I.: Classification of periodic words and their application in group theory. Burnside groups (Ed. J. L. Mennicke), Lecture Notes in Mathematics 806, Springer-Verlag, Berlin Heidelberg 1980, 1-46.

[2] Banaschewski, B.: Minimal topological algebras. Math. Ann. 211 (1974), 107-114.

[3] Comfort, W. W. and D. Grant: Cardinal invariants, pseudocompactness and minimality: some recent advances in the topological theory of topological groups. Topology Proc. 6 (1981), 227-265.

[4] Dikranjan, D.: Minimal topologies on Abelian groups. Seminar notes, Universita Padova, Padova 1983.
[5] Doitchinov, D.: Produits de groupes topologiques minimaux. Bull. Sci. Math. 97 (1972), 59-64.
[6] Frič, R.: Four questions concerning constructions of convergence groups. To appear.
[7] Frič, R. and F. Zanolin: A sequential convergence group having no completion. To appear.
[8] Frič, R. and F. Zanolin: Remarks on sequential convergence in free groups. To appear.
[9] Frič, R. and F. Zanolin: Sequential convergence in free groups. To appear.
[10] Frič, R. and F. Zanolin: Fine convergence in free groups. To appear.
[11] Jakubík, J.: On convergence in linear spaces. (Slovak. Russian summary.) Mat. - Fyz. Časopis Slovensk. Akad. Vied 6 (1956), 57-67.
[12] Mikusiński, P.: Problems posed at the conference. Proc. Conf. on Convergence, Szczyrk, 1974. Katowice, 1980, 110-112.
[13] Novák, J.: On convergence groups. Czechoslovak Math. J. 20 (1970), 357-374.
[14] Novák, J.: On completions of convergence commutative groups. General Topology and its Relations to Modern Analysis and Algebra III (Proc. Third Prague Topological Sympos., 1971). Academia, Praha 1972, 335-340.
[15] Olshanskij, A.: A note on a countable group admitting no non-isolated group topology (Russian). Vestnik. Moskov. Univ. Ser. I Mat. Meh. (1980), No. 3, 103.
[16] Pestov, V.: On the structure and embeddings of topological groups. To appear.
[17] Prodanov, I.: Precompact minimal group topologies and p-adic numbers. Ann. de l'Univ. de Sofia 66 (1971/1972), 249-266.
[18] Prodanov, I.: Some minimal group topologies are precompact. Math. Ann. 227 (1979), 117-125.
[19] Prodanov, I. and L. Stoyanov: Minimal group topologies. To appear.
[20] Prodanov, I. and L. Stoyanov: Every minimal Abelian group is precompact. C. R. Acad. Bulgare Sci. 37 (1984), 23-26.
[21] Schwanengel, U.: An example of a q-minimal precompact topological group containing a non-minimal closed normal subgroup. Manuscripta Math. 27 (1979), 323-327.
[22] Stephenson, R.: Minimal topological groups. Math. Ann. 192 (1971), 193-195.
[23] Zanolin, F.: Solution of a problem of Josef Novák about convergence groups. Boll. Un. Mat. Ital. 5, 14-A, (1977), 375-381.

CONVERGENCE VECTOR SPACES FOR ANALYSIS

Alfred Frölicher[1) and Andreas Kriegl[2)

Independently the authors introduced classes of vector spaces which they considered as the right ones for calculus in infinite dimension. These classes seemed quite different : certain locally convex spaces in [8;9] and certain vector spaces with a smooth structure in [3;4]. Later it was proved that the two classes are actually identical : the locally convex topology and the smooth structure are two among many possible structures which can be used to characterize these spaces which in short were called convenient vector spaces; cf. [5].

Here we give another characterization by means of convergence structures. Mackey convergence can be considered for arbitrary bornological vector spaces and gives, as we shall see, an identification of the category of bornological vector spaces with a full subcategory of the category of convergence vector spaces. By restriction we obtain, considering the convenient vector spaces with their bornologies, an imbedding of the category of convenient vector spaces into that of convergence vector spaces, i.e. the mentioned characterization.

The Mackey convergence structure plays an important role : derivatives are obtained as Mackey limits ; and Mackey-completeness which is fundamental for calculus is just completeness in the usual sense of convergence vector spaces. We shall also consider the topology associated to Mackey convergence, called the Mackey closure topology, and we shall show that it is the same as the C^∞-topology of [8] where one finds various other characterizations showing its role in differentiation theory.

Mackey convergent sequences are of course convergent for the Mackey closure topology, but the converse fails in general. In fact we will show that in general there exists no topology having as convergent sequences the Mackey convergent sequences. This is proved by giving an

1) Section de Mathématiques, Université de Genève, 2-4, rue du Lièvre CH-1211 GENEVE 24

2) Institut für Mathematik, Universität Wien, Strudlhofgasse 4 A-1090 WIEN

example which lacks the Urysohn property for Mackey convergent sequences.

§1. Mackey convergence in bornological vector spaces

Let BVS be the category of bornological vector spaces (cf. [6]) and CVS the category of convergence vector spaces (cf.[2]), the morphisms being the linear bornological (i.e. bounded on bounded sets) respectively the linear continuous maps. All vector spaces considered shall be over $\mathbb{R}$, and $\mathbb{V}$ shall denote the zero-neighborhood filter on $\mathbb{R}$. We shall consider functors μ : BVS → CVS and ν : CVS → BVS which do not modify the underlying vector spaces and act as identity on the morphisms. The convergence vector space μE, where $E \in \mathrm{Obj}(\underline{BVS})$, has by definition as zero-converging filters those which are finer ($\leq$) than the filters of the form $\mathbb{V}\cdot B$ where B is bounded in E ($\mathbb{V}\cdot B$ has as filter basis the sets $V\cdot B$ with $V \in \mathbb{V}$). One easily verifies that this determines a unique compatible convergence structure and that for a morphism $l : E_1 \to E_2$ of BVS, $l : E_1 \to E_2$ is a morphism of CVS. The so defined convergence is called the Mackey convergence. The functor ν is defined as follows : For $F \in \mathrm{Obj}(\underline{CVS})$, a subset B is bounded in νF iff $\mathbb{V}\cdot B$ converges to zero on F. Again, verification that this is a vector bornology and that functoriality holds is easy.

Lemma 1. ν is right adjoint to μ and $\nu\mu = \mathrm{Id}$.

Proof. As immediate consequences of the definitions one has :

a) For any $F \in \mathrm{Obj}(\underline{CVS})$, $\mathrm{Id}_F : \mu\nu F \to F$ is continuous ;

b) For any $E \in \mathrm{Obj}(\underline{BVS})$, $\mathrm{Id}_E : E \to \nu\mu E$ is bornological.

c) Let B be bounded in $\nu\mu E$. So there exists B' bounded in E with $\mathbb{V}\cdot B \leq \mathbb{V}\cdot B'$. Hence there exists $\delta > 0$ such that $I_1\cdot B' \supset I_\delta \cdot B$ where $I_\delta = [-\delta,\delta]$. Therefore $I_1\cdot B' \supset \delta\cdot B$ or $B \subset I_{1/\delta}\cdot B'$ and so B is bounded in E, since with B' also $I_\lambda\cdot B'$ is bounded in E for any $\lambda \in \mathbb{R}^+$. This together with b) proves $\nu\mu = \mathrm{Id}$. The adjunction is implied by a) and b).

Definition 2 . By BCVS we denote the full subcategory of CVS formed by the bornological convergence vector spaces, i.e. by those which satisfy $\nu\mu F = F$.

Remark. If in particular F is a locally convex space, then $\nu\mu F = F$ implies that F is a bornological locally convex space in the usual sense. Since the converse fails, one has, for locally convex spaces, to distinguish between bornological as locally convex space and bornological as convergence vector space. Nevertheless our terminology seems justified since as an immediate consequence of Lemma 1 one has :

Proposition 3. The category BVS of bornological vector spaces is isomorphic to the category BCVS of bornological convergence vector spaces, and BCVS is a coreflective subcategory of CVS.

With respect to limits and colimits formed in CVS one has the following permanence properties : being bornological (in the sense of Definition 2) is inherited by finite products, subspaces and (as consequence of the adjunction) by any colimit, so in particular by coproducts and quotients.

Definition 4. A convergence vector space is called locally convex iff for any zero-converging filter F also the filter $\hat{F}$ generated by the convex hulls of the sets of F converges to zero.

This notion coincides with the usual one in case F is a topological vector space. The locally convex convergence vector spaces form a coreflective subcategory of CVS; the retraction functor γ is the identity on morphisms and on objects F it is characterized as follows : F converges to zero on γF iff $F \leq \hat{G}$ for some G which is converging to zero on F.

Since for a locally convex convergence vector space F the space νF is a convex bornological space (cf [6]), and for a convex bornological space E the space μE is a locally convex convergence vector space, the isomorphism of Proposition 3 yields by restriction the

Proposition 5. The category CBS of convex bornological spaces is isomorphic to the category of bornological locally convex convergence vector spaces.

2. Topological convex bornological spaces

The standard description of topological convex bornological spaces uses the functor b : LCS → CBS from locally convex spaces to convex bornological spaces and its left adjoint t : CBS →LCS. The functor b associates to a locally convex space L the underlying vector space with its usual bornology (also called its von Neumann bornology) : a set is bounded in bL iff for every zero-neighborhood U in L there exists $\lambda \in \mathbb{R}$ with $B \subset \lambda \cdot U$, or equivalently iff $\mathbb{V} \cdot B$ converges to zero in L. Hence b can also be described as restriction of the previous functor ν to LCS . For a convex bornological space E, a basis for the zero-neighborhood filter of tE is formed by the absolutely convex bornivorous subsets U of E. (U bornivorous means : for each bounded subset B of E there exists $\lambda \in \mathbb{R}$ with $B \subset \lambda \cdot U$). The topological convex bornological spaces E are those which satisfy E = btE. They form a reflective subcategory of CBS which is isomorphic to the coreflective subcategory of LCS formed by the so-called bornological locally convex spaces L, i.e. by those satisfying L = tbL.

Proposition 6. A convex bornological space E is topological if and only if the bounded subsets B are exactly those for which l(B) is bounded for all $l \in E^x$. (E^x denotes the bornological dual of E; i.e. $E^x = \{l : E \to \mathbb{R};$ l linear and bornological$\}$).

Proof. We use the theorem which says that a subset B of a locally convex space L is bounded (in the von Neumann bornology, i.e. in bL) iff l(B) is bounded for all $l \in L'$ (by L' we denote the topological dual of L). For a separated space L, a proof of this can be found in almost any book on locally convex spaces; one then reduces the general case to the separated one by going to the quotient L/N where N is the closure of $\{0\}$ in L for which one shows that $N = \{x \in L; l(x) = 0$ for all $l \in L'\}$.

Suppose now first that E is topological. Then, with L = tE, one has E = bL and $E^x = L'$; hence the property follows from the quoted theorem.

Suppose, conversely, that E has the property of Proposition 6. Let L be the locally convex space obtained by putting on E the weak topology with respect to E^x. Then $E^x = L'$ and by the theorem quoted above, a subset B is bounded in bL iff l(B) is bounded for all $l \in L'$, so according to the assumption, iff it is bounded in E. This proves that E = bL and the assertion E = btE follows from the property btb = b.

In order to describe those convergence vector spaces which correspond, via the isomorphism of Proposition 3, to the topological convex bornological spaces we use a functor λ : CVS → CBS characterized as follows : a subset B of F (where $F \in$ Obj(CVS)) is bounded in λF iff l(B) is bounded for all $l \in F' = \{l : F \to \mathbb{R};$ l linear and continuous$\}$. We remark that λF is always a topological convex bornological space.

Proposition 7. The category TCBS of topological convex bornological spaces is isomorphic to the full subcategory of CVS formed by the objects F satisfying $F = \mu\lambda F$.

Proof

a) Let F be a convergence vector space with $F = \mu\lambda F$. Applying ν and using $\nu\mu$ = Id we get $\nu F = \lambda F$. Hence by the above remark : $\nu F \in$ Obj(TCBS).

b) Let E be a topological convex bornological space. One first verifies that $E^{\times} = (\mu E)'$. By Proposition 6 it follows that $E = \lambda\mu E$, so $F := \mu E$ satisfies $F = \mu\lambda F$.

3. The Mackey closure topology

It is well known that the inclusion functor from the category TOP of topological spaces to the category CS of convergence spaces has a left adjoint τ : CS $\to$ TOP which preserves the underlying sets. The Mackey closure topology of a bornological vector space E is obtained by applying τ to the Mackey convergence structure. Since the Mackey convergence structure has obviously the property that for any filter which converges to a point there exists a coarser filter with countable base converging to the same point, the closed sets for the Mackey closure topology are exactly those which are closed under Mackey-convergent sequences. Hence the Mackey closure topology is final with respect to the Mackey-convergent sequences considered as maps $s : \overline{\mathbb{N}} \to E$ where $\overline{\mathbb{N}} = \mathbb{N} \cup \{\infty\}$ is the one point compactification of $\mathbb{N}$ and $s(\infty)$ is the Mackey limit of $s(n)$ for $n \to \infty$. As proved in [8] for separated spaces, this topology is also final with respect to the smooth curves of E, a curve $c : \mathbb{R} \to E$ being smooth iff $l \cdot c : \mathbb{R} \to \mathbb{R}$ is smooth for all $l \in E^{\times}$.

According to the functoriality, every $l \in E^{\times}$ is continuous with respect to the Mackey closure topology. Conversely, let $l : E \to \mathbb{R}$ be linear but not an element of $E^{\times}$. Then there exists a bounded subset B of E such that $l(B)$ is unbounded, and one can choose, for $n \in \mathbb{N}$, points $b_n \in B$ with $|l(b_n)| \geq n$. Then the points $a_n := 1/n \cdot b_n$ are Mackey convergent to zero and hence also convergent to zero with respect to the Mackey closure topology. Since $|l(a_n)| \geq 1$ one concludes that l is not continuous with respect to the Mackey closure topology. This shows that $E^{\times} = \{l : E \to \mathbb{R};\ l$ linear and continuous with respect to the Mackey closure topology$\}$. Combining with Proposition 6 we obtain therefore :

Proposition 8. The bornology of a topological convex bornological space is completely determined by the associated Mackey closure topology.

Remark. There are examples which show that the Mackey closure topology is often not a vector space topology. Of course it is invariant with respect to translations and homothetic transformations.

§4. Convenient vector spaces.

Proposition 9. For a topological convex bornological space E the following properties are equivalent :

1) E is regular, i.e. E^x separates points of E;
2) The Mackey closure topology of E is separated;
3) The Mackey convergence structure of E is separated;
4) The bornology of E is separated, i.e. $\{0\}$ is the only bounded linear subspace of E.

Proof.

1) ⇒ 2) is trivial, since the elements of E^x are continuous with respect to the Mackey closure topology.

2) ⇒ 3) is trivial, since the Mackey convergence structure is finer than the Mackey closure topology.

3) ⇒ 4). Suppose that $a \in E$ belongs to a bounded linear subspace of E. Then the constant sequence a,a,... is Mackey-convergent to a and also to 0; hence $a = 0$.

4) ⇒ 1). Let $0 \neq a \in E$. Then the linear subspace <a> spanned by a is unbounded in E, hence (since E is topological) also unbounded in the locally convex space $L = tE$. So there exists a zero-neighborhood U in L which does not absorb <a> and hence there exists $t \neq 0$ with $t \cdot a \notin U$. Therefore $1/t \cdot U$ is a zero-neighborhood in L which does not contain a. So L is a separated locally convex space and by the theorem of Hahn-Banach there exists $l \in L'$ with $l(a) \neq 0$. Since $L' = E^x$ this proves 1).

Proposition 10 . For a separated topological convex bornological space E the following properties are equivalent :

1) Every Mackey-Cauchy sequence of E is Mackey-convergent;
2) With the Mackey convergence structure, E is a complete convergence vector space;
3) The bornology of E is complete.

Proof.

1) ⇒ 2) is a consequence of the countability property of the Mackey convergence structure which was mentioned at the beginning of §3.

2) ⇒ 1) is trivial. For the definition of complete bornologies and for the equivalence of 3) with 1) in the case of topological convex bornological spaces we refer to [6].

Since a convenient vector space (in the sense of [5]) is a topological convex bornological space which is separated and Mackey-complete, we obtain from Propositions 7, 9 and 10 the

Proposition 11. The category of convenient vector spaces is isomorphic to the category of those separated complete convergence vector spaces which satisfy $F = \mu\lambda F$.

Remark. In order to develop a calculus for a class of vector spaces yielding with the smooth maps as morphisms a cartesian closed category, convergence vector spaces were used in [2]. Though a much better and more natural way to achieve this goal is given in [5], using convenient vector spaces, one might be interested in a comparison of the used spaces. The answer is the following : a convenient vector space with its Mackey convergence structure is equable and admissible in the sense of [2] . The proof uses the fact that the locally convex topology of tE (cf. §2) coincides with the locally convex topology associated in the sense of §2 to the Mackey convergence structure of E.

For convenient vector spaces E_1 and E_2 the space of (linear) morphisms $E_1 \to E_2$ becomes a convenient vector space in a natural way by means of the following bornology : a subset B is bounded in $L(E_1,E_2)$ iff $B(A)$ is bounded in E_2 for each A which is bounded in E_1. The Mackey convergence structure of $L(E_1,E_2)$ can be expressed by those of E_1 and E_2 as follows : one applies the functor $\mu\nu$ to the following auxiliary convergence structure of $L(E_1,E_2)$: a filter H on $L(E_1,E_2)$ is converging to zero iff for any filter F on E_1 with $\mathbb{V}\cdot F$ zero-converging on E_1 the filter $H(F)$ is zero-converging on E_2. So the function space structure is similar but not identical with the one used in [2] .

§5. Mackey convergent sequences and the Urysohn property

The Urysohn property says : if every subsequence of a sequence s has a subsequence converging towards x, then the sequence s itself converges towards x. It is well known and trivial, that the Urysohn property holds for any topological convergence (i.e. convergence with respect to some topology). Before showing that it fails for Mackey convergent sequences we compare different convergences of a sequence $x_1, x_2, \ldots$ in a convenient vector space E. As before we consider E as bornological vector space and we write :

$x = \text{M-}\lim_{n\to\infty} x_n$ if convergence holds in μE, i.e. with respect to the Mackey convergence structure;

$x = \text{M-cl-}\lim\limits_{n\to\infty} x_n$ if convergence holds in $\tau\mu E$, i.e. with respect to the Mackey closure topology;

$x = \lim\limits_{n\to\infty} x_n$ if convergence holds in tE, i.e. with respect to the locally convex topology.

Because the structures of μE, $\tau\mu E$ and tE are translation-invariant, we can restrict our considerations to zero-converging sequences. The Mackey convergence of a sequence is a special case of that for filters which was defined in §1, and can equivalently be discribed as follows : $\text{M-}\lim\limits_{n\to\infty} x_n = 0$ iff there exist reals μ_n with $\lim\limits_{n\to\infty} \mu_n = \infty$ and $\{\mu_n \cdot x_n; n \in \mathbb{N}\}$ bounded in E.

<u>Lemma 12</u>. a) $Id_E : \mu E \to \tau\mu E$ is continuous;

b) $Id_E : \tau\mu E \to tE$ is continuous.

<u>Proof</u>.

a) Since τ is left adjoint to the inclusion functor of <u>TOP</u> into <u>CS</u> (convergence spaces), $Id_X : X \to \tau X$ is continuous for any $X \in \text{Obj}(\underline{CS})$.

b) Let F be a filter which is Mackey convergent to zero. Then $F \le \mathbb{V}.B$ for some B which is bounded in E. But $E = btE$; so B is bounded in btE which means that $\mathbb{V}\cdot B$ is zero-converging in tE. This proves that $Id_E : E \to tE$ is continuous. One now applies τ and uses $\tau tE = tE$.

<u>Corollary 13</u>. a) $\text{M-}\lim\limits_{n\to\infty} x_n = 0 \Rightarrow \text{M-cl-}\lim\limits_{n\to\infty} x_n = 0$.

b) $\text{M-cl-}\lim\limits_{n\to\infty} x_n = 0 \Rightarrow \lim\limits_{n\to\infty} x_n = 0$.

<u>Proposition 14</u>. For a sequence $x_1, x_2, \ldots$ in a convenient vector space E the following properties are equivalent :

1) $\text{M-cl-}\lim\limits_{n\to\infty} x_n = 0$;

2) Every subsequence has a subsequence which is Mackey-convergent to zero.

<u>Proof</u>.

1) $\Rightarrow$ 2). It is enough to show that the sequence itself has a subsequence which is Mackey-convergent to zero. The set $A = \{x_n;\ n \in \mathbb{N} \text{ and } x_n \neq 0\}$ is not closed in the Mackey closure topology (unless in the case $A = \emptyset$ which is trivial), hence (cf. §3) not closed under Mackey-convergent sequences. So there exist $x_{n_1}, x_{n_2}, \ldots$ with $x = \text{M-}\lim\limits_{k\to\infty} x_{n_k}$ existing but $x \notin A$. Since infinitely many of the values n_k must be different (otherwise the limit would belong to A) we can suppose $n_1 < n_2 < \ldots$

From $\text{M-cl-}\lim_{k\to\infty} x_{n_k} = 0$ and $\text{M-}\lim_{k\to\infty} x_{n_k} = x$ we deduce, using Corollary 13 a), that $x = 0$.

2) $\Rightarrow$ 1) follows from Corollary 13 a) and the Urysohn property of topological convergence.

In order to show that the implications in Corollary 13 cannot be reversed we use the space L which consists of the product of $2^{\aleph_0}$ copies of the real line $\mathbb{R}$ together with the product topology. It is known (cf. [7], p. 282) that this locally convex space L is bornological. Since it is also separated and complete, the space $E := bL$ is a convenient vector space. Furthermore, a set B is bounded in E iff all its projections on the factors $\mathbb{R}$ are bounded (cf. [7], p.35).

For such a space one finds in [6], page 122, the description of a sequence with $\lim_{n\to\infty} x_n = 0$ though the sequence is not Mackey-converging to zero. By the same arguments one shows that no subsequence is Mackey-convergent to zero. So, according to Proposition 14, we obtain the part b) of the following

<u>Proposition 15</u>. a) $\text{M-cl-}\lim_{n\to\infty} x_n = 0 \not\Rightarrow \text{M-}\lim_{n\to\infty} x_n = 0$

b) $\lim_{n\to\infty} x_n = 0 \not\Rightarrow \text{M-cl-}\lim_{n\to\infty} x_n = 0$

<u>Proof of a)</u>. An example given by Mikusinski and Pochciał in [10] yields this result. That example, adapted to our notation, is as follows. We use as index set for the space L mentioned before a set $\mathcal{A}$ of subsets of $\mathbb{N}$ with the following properties :

i) $\operatorname{card}\mathcal{A} = 2^{\aleph_0}$;

ii) for $A, B \in \mathcal{A}$: $A \neq B \Rightarrow A \cap B$ finite ;

iii) $X \subset \mathbb{N}$ and X infinite $\Rightarrow A \cap X$ infinite for some $A \in \mathcal{A}$. The existence of such a set $\mathcal{A}$ was proved in [1], using Zorn's lemma. By c_0 we denote the set of zero-converging sequences in $\mathbb{R}$. Let furthermore $\psi : \mathcal{A} \to c_0$ be a bijection (which exists, since the sets have the same cardinality). We consider the sequence $x_1, x_2, \ldots$ in L for which x_n is defined by its components $(x_n)_A$, $A \in \mathcal{A}$, as follows :

$$(x_n)_A = \begin{cases} (\psi(A))(n) & \text{if } n \in A; \\ 0 & \text{if } n \notin A. \end{cases}$$

We first show, using Proposition 14, that $\text{M-cl-}\lim_{n\to\infty} x_n = 0$.

So let $x_{n_1}, x_{n_2}, \ldots$ be a subsequence. Then $X := \{n_k; k \in \mathbb{N}\}$ is an infinite subset of $\mathbb{N}$ and by iii) there exists $A_0 \in \mathcal{A}$ with $A_0 \cap X$ infinite. So we obtain $k_1 < k_2 < \ldots$ such that $A_0 \cap X = \{n_{k_i}; i \in \mathbb{N}\}$. $\psi(A_0)$ is a zero-converging sequence of $\mathbb{R}$ and we can choose reals μ_i with $\lim_{i\to\infty} \mu_i = \infty$ and $\{\mu_i \cdot \psi(A_0)(n_{k_i}); i \in \mathbb{N}\}$ bounded. Since $\psi(A)(n) = (x_n)_A$ we have : $B_A := \{(\mu_i \cdot x_{n_{k_i}})_A; i \in \mathbb{N}\}$ is bounded for $A = A_0$. For $A \neq A_0$, $(\mu_i \cdot x_{n_{k_i}})_A = 0$ except for finitely many i according to property ii).

So B_A is also bounded for $A \neq A_0$ and we conclude that $\{\mu_i \cdot x_{n_{k_i}}; i \in \mathbb{N}\}$ is bounded in E, i.e. that the subsequence $x_{n_{k_i}}$ is Mackey-convergent to zero. But the whole sequence $x_1, x_2, \ldots$ is not Mackey-convergent to zero. In fact, otherwise there exist reals μ_n with $\lim_{n\to\infty} \mu_n = \infty$ and $\{\mu_n \cdot x_n; n \in \mathbb{N}\}$ bounded in E. One chooses $\lambda \in c_0$ with $\lim_{n\to\infty} \lambda_n \cdot \mu_n = \infty$ and considers $A := \psi^{-1}(\lambda)$. Then $(\mu_n \cdot x_n)_A = \mu_n \cdot \psi(A)(n) = \mu_n \cdot \lambda_n$ is not bounded with respect to n, in contradiction to $\{\mu_n \cdot x_n; n \in \mathbb{N}\}$ bounded in E.

Corollary 16. The Urysohn property fails for Mackey convergence of sequences in a convenient vector space.

Corollary 17. There exists in general no topology on a convenient vector space E having as convergent sequences exactly the Mackey-convergent sequences of E.

REFERENCES

[1] Contessa M. and Zanolin F. : On a question of Josef Novák about convergence spaces. Rend.Sem. Mat. Padova 58 (1977), 151-161.

[2] Frölicher A. and Bucher W. : Calculus in vector spaces without Norm. Lecture Notes in Mathematics Vol. 30, Springer 1966.

[3] Frölicher A. : Smooth Structures. Lecture Notes in Mathematics Vol 962, 69-81, Springer 1982.

[4] Frölicher A. : Cartesian closed Categories and Analysis of smooth Maps. As preprint in Seminarberichte Fernuniversität Hagen Nr. 16 (1982), 2-16. To appear also in the Proceedings of the Buffalo Conference of May 1982.

[5] Frölicher A., Gisin B. and Kriegl A. : General differentiation theory. Aarhus Universitet, Various Publications Series No 35, Aarhus 1983, 125-153.

[6] Hogbe-Nlend H. : Bornologies and functional analysis. North Holland 1977.

[7] Jarchow H. : Locally convex spaces. Teubner 1981.

[8] Kriegl A. : Die richtigen Räume für Analysis im Unendlich-Dimensionalen. Monatshefte f. Math. 91 (1982),109-124.

[9] Kriegl A. : Eine kartesisch abgeschlossene Kategorie glatter Abbildungen zwischen beliebigen lokalkonvexen Vektorräumen. Monatshefte f. Math. 95 (1983), 287-309.

[10] Mikusiński P. and Pochciał J. : On Mackey Convergence. Bull. Polish Acad. of Sci. 31 (1983), 151-155.

DISCRETE CONVERGENCE

Siegfried Gähler [1)]

I. Introduction. The theory of discrete convergence is a branch of numerical analysis which from the viewpoint of topology is very interesting. Roughly spoken, discrete convergence is a convergence $x_n \to x$ where, in general, all points x_n and x belong to different sets. Discrete convergence spaces appear in papers of several authors under different names and in different degrees of generality. For instance, the approximation schemes used by Browder [1], Petryshyn [8], [9] and others yield discrete convergence spaces. Special types of discrete convergence spaces are used also in papers of Grigorieff [7], Stummel [11] and Vainikko [13]. Under restrictions to the family of index sets the notion of a discrete limit space introduced and studied by Stummel in [12] is identical with the notion of a separated discrete convergence space which is Urysohn.

The underlying paper presents a survey on certain essentials of the theory of discrete convergence. One of the basic problems is the following: Let f be a mapping of a set E into a set E', x' be a point of E', and let us ask for a solution x of $f(x) = x'$. If x cannot be obtained directly, one can try to use a sequence of mappings $f_n: E_n \to E'_n$ and a sequence of points $x'_n \in E'_n$ which in a certain sense converge to f or x', respectively, where for every n the equation $f_n(x_n) = x'_n$ has a solution x_n. If all turns well, the sequence of points x_n in a certain sense converges to a point $x \in E$ which is a solution of $f(x) = x'$. Obviously in these considerations "convergence in a certain sense" means a certain discrete convergence. Later (in section V) we shall take up the problem and give conditions under which a solution really is possible by the method described.

II. Basic notions. Let $\mathbb{N}$ be the set of all positive integers, $\mathfrak{N}$ be the set of all infinite subsets of $\mathbb{N}$ and let $\leq$ denote the partial ordering in $\mathfrak{N}$ given by the set inclusion. Let $E, E_1, E_2, \ldots$ be arbitrary non-empty sets and τ be a mapping of E into $P(\bigcup_{\mathbb{N}' \in \mathfrak{N}} \prod_{n \in \mathbb{N}'} E_n)$, where $P(U)$ denotes the power set of U. Let $(x_n)_{n \in \mathbb{N}'} \in \tau(x)$ (with $\mathbb{N}' \in \mathfrak{N}$ and with $x_n \in E_n$ for every $n \in \mathbb{N}'$) and $\mathbb{N}^+ \leq \mathbb{N}'$ imply $(x_n)_{n \in \mathbb{N}^+} \in \tau(x)$. Then $X = ((E, E_1, E_2, \ldots), \tau)$ is called a discrete convergence space and τ its discrete convergence. Often instead of $(x_n)_{n \in \mathbb{N}'} \in \tau(x)$ we write $(x_n)_{n \in \mathbb{N}'} \xrightarrow[\tau]{} x$ and say that $(x_n)_{n \in \mathbb{N}'}$ converges discretely (to x).

---

1) Akademie der Wissenschaften der DDR, Institut für Mathematik, DDR - 1086 Berlin, Mohrenstraße 39

A discrete convergence space X is called a discrete approximation if for every $x \in E$ there exists a sequence $(x_n)_{n \in N} \underset{\tau}{\rightarrow} x$. X and τ are said to be weakly separated or to be separated if $(x_n)_{n \in N} \underset{\tau}{\rightarrow} x, y$ respectively $(x_n)_{n \in N'} \underset{\tau}{\rightarrow} x, y$ ($N' \in \mathfrak{N}$) implies $x = y$. X and τ are said to be Urysohn if for a given $(x_n)_{n \in N'} \in \prod_{n \in N'} E_n$, $N' \in \mathfrak{N}$, and $x \in E$ the following holds: If for every $N^+ \leq N'$ there exists an $N^* \leq N^+$ such that $(x_n)_{n \in N^*} \underset{\tau}{\rightarrow} x$, then we have $(x_n)_{n \in N'} \underset{\tau}{\rightarrow} x$.

Examples, special cases: 1. Let E be an L-space in the sense of Fréchet and let $E_1 = E_2 = \dots = E$. On $(E, E_1, E_2, \dots)$ there is given a discrete convergence τ by

$$(x_n)_{n \in N'} \in \tau(x),\ N' \in \mathfrak{N}, \iff (x_n)_{n \in N'} \text{ converges to } x \text{ in the sense of the L-space } E.$$

$X = ((E, E_1, E_2, \dots), \tau)$ is a discrete approximation and is separated. X is Urysohn if and only if E is an L^*-space.

2. Let E be a set and for every $n \in N$, let E_n be a metric space with metric δ_n and p_n be a mapping of E into E_n. On $(E, E_1, E_2, \dots)$ there is given a discrete convergence τ by

$$(x_n)_{n \in N'} \in \tau(x),\ N' \in \mathfrak{N}, \iff (\delta_n(x_n, p_n(x)))_{n \in N'} \rightarrow 0.$$

$X = ((E, E_1, E_2, \dots), \tau)$ is called a metric discrete convergence space. A metric discrete convergence space X is a discrete approximation with $(p_n(x))_{n \in N} \underset{\tau}{\rightarrow} x$ for every $x \in E$ and, moreover, is Urysohn. X is weakly separated or is separated if and only if for every $x, y \in E$ with $x \neq y$, $\limsup_{n \to \infty} \delta_n(p_n(x), p_n(y)) > 0$ ([3], Lemma 2) respectively $\liminf_{n \to \infty} \delta_n(p_n(x), p_n(y)) > 0$ (proof analogous to the proof of [3], Lemma 2).

The mapping $\delta : E \times E \rightarrow [0, \infty]$ given by $\delta(x, y) = \limsup_{n \to \infty} \delta_n(p_n(x), p_n(y))$ is a pseudo-metric on E (i.e. one has $\delta(x, x) = 0$, $\delta(x, y) = \delta(y, x)$ and $\delta(x, z) \leq \delta(x, y) + \delta(y, z)$).

3. A discrete convergence space $X = ((E, E_1, E_2, \dots), \tau)$ is said to be linear if E and every E_n, $n \in N$, are linear and if τ is compatible with the linear structures of E and E_n, $n \in N$, in the sense that for every $N' \in \mathfrak{N}$, every $(x_n)_{n \in N'} \underset{\tau}{\rightarrow} x$, $(y_n)_{n \in N'} \underset{\tau}{\rightarrow} y$ and every $(\alpha_n)_{n \in N'} \rightarrow \alpha$, $(\beta_n)_{n \in N'} \rightarrow \beta$ (where α_n, β_n ($n \in N'$), α, β are reals) we have $(\alpha_n x_n + \beta_n y_n)_{n \in N'} \underset{\tau}{\rightarrow} \alpha x + \beta y$. A discrete convergence space X which is both a metric and a linear discrete convergence space where every δ_n is generated by a norm $\|.\|_n$ on E_n, is called a linear normed discrete convergence space.

It can be easily seen that a linear normed discrete convergence space X is weakly separated or separated if and only if for every $x \in E$ with

$x \neq 0$, $\limsup_{n\to\infty} \| p_n(x)\|_n > 0$ respectively $\liminf_{n\to\infty} \| p_n(x)\|_n > 0$. If $X = ((E,E_1,E_2, \dots),\tau)$ is a linear normed discrete convergence space with

$$(\| p_n(\alpha x + \beta y) - \alpha p_n(x) - \beta p_n(y)\|_n)_{n\in \mathbb{N}} \to 0$$

for every $x, y \in E$ and every real α, β, the mapping $\|.\|:E \to [0,\infty]$ given by $\|x\| = \limsup_{n\to\infty} \| p_n(x)\|_n$ is a pseudo-norm on E (i.e. one has $\|0\| = 0$, $\|\alpha x\| = |\alpha|\|x\|$ for every real α and $\|x + y\| \leq \|x\| + \|y\|$) with the above δ as the corresponding pseudo-metric. $\{x \in E | \|x\| < \infty\}$ is a linear subspace of E.

III. "Topological" properties. In the theory of discrete convergence spaces several notions analogous to notions of topology can be introduced and are the base of fruitful and interesting investigations. The underlying section is devoted to some of them.

For this, for every $n \in \mathbb{N}$ let M_n be a (possibly empty) subset of E_n. The set

$$\overline{(M_1,M_2, \dots)} = \{x \in E | \exists (x_n)_{n\in \mathbb{N}'} \xrightarrow[\tau]{} x \text{ with } x_n \in M_n \text{ for all } n \in \mathbb{N}'\}$$

is called the adherence of $(M_1,M_2, \dots)$ and the set

$$\underline{(M_1,M_2, \dots)} = E \setminus \overline{(E_1\setminus M_1, E_2\setminus M_2, \dots)}$$

is called the interior of $(M_1,M_2, \dots)$. The interior of $(M_1,M_2, \dots)$ can also be expressed by

$$\underline{(M_1,M_2, \dots)} = \{x \in E | \forall (x_n)_{n\in \mathbb{N}'} \xrightarrow[\tau]{} x : x_n \in M_n \text{ for almost all } n \in \mathbb{N}'\}.$$

$\overline{(M_1,M_2, \dots)}$ and $\underline{(M_1,M_2, \dots)}$ remain unchanged if a finite number of sets M_n is replaced by other sets of E_n. Obviously $E = \overline{(E_1,E_2, \dots)}$ if X is a discrete approximation.

$(M_1,M_2, \dots)$ is said to be a discrete neighbourhood of a point $x \in E$ if $x \in \underline{(M_1,M_2, \dots)}$. The set $\mathfrak{M}(x)$ of all discrete neighbourhoods of x has the property:

(*) $(M_1,M_2, \dots)$ belongs to $\mathfrak{M}(x)$ if for every sequence $(x_n)_{n\in \mathbb{N}'}$ the following holds: If for every $(M_1',M_2', \dots) \in \mathfrak{M}(x)$ we have $x_n \in M_n'$ for almost all $n \in \mathbb{N}'$, then $x_n \in M_n$ for almost all $n \in \mathbb{N}'$.

To every mapping $\mathfrak{M}':E \to P(\prod_{n\in\mathbb{N}} P(E_n))$ there is associated a mapping $\tau':E \to P(\bigcup_{\mathbb{N}'\subset\mathbb{N}} \prod_{n\in\mathbb{N}'} E_n)$ by means of

$$\tau'(x) = \{(x_n)_{n\in \mathbb{N}'} \in \prod_{n\in\mathbb{N}'} E_n | \forall (M_1,M_2, \dots) \in \mathfrak{M}'(x): x_n \in M_n \text{ for almost all } n \in \mathbb{N}'\}$$

which is a discrete convergence on $(E,E_1,E_2, \dots)$ and is Urysohn.

Theorem 1 (6). Let $\mathfrak{M}' = \mathfrak{M}$. Then $\tau(x) \subseteq \tau'(x)$ for every $x \in E$ and, moreover, $\tau = \tau'$ if and only if τ is Urysohn.

P r o o f. $(M_1, M_2, \dots)$ belongs to $\mathfrak{M}(x)$ if and only if for every $(x_n)_{n \in \mathbb{N}'} \overrightarrow{\tau'} x$, we have $x_n \in M_n$ for almost all $n \in \mathbb{N}'$. Hence, we have $\tau(x) \subseteq \tau'(x)$ for every $x \in E$. If $\tau = \tau'$, τ is Urysohn since τ' is Urysohn. Now, let us assume that τ is Urysohn and that $\tau(x) \neq \tau'(x)$ for a certain $x \in E$. Then there exists a sequence $(x_n)_{n \in \mathbb{N}'} \in \prod_{n \in \mathbb{N}'} E_n$, $\mathbb{N}' \in \mathfrak{N}$, which converges discretely to x with respect to τ', but not with respect to τ. There exists an $\mathbb{N}^+ \subseteq \mathbb{N}'$ such that for every $\mathbb{N}^* \subseteq \mathbb{N}^+$, $(x_n)_{n \in \mathbb{N}^*}$ does not converge discretely to x with respect to τ. Let $M_n = \{x_n\}$ if $n \in \mathbb{N}^+$, $= \emptyset$ else. Then $x \notin \overline{(M_1, M_2, \dots)}$ where the adherence is taken with respect to τ. This implies $(E_1 \setminus M_1, E_2 \setminus M_2, \dots) \in \mathfrak{M}(x)$ which is a contradiction to $(x_n)_{n \in \mathbb{N}'} \overrightarrow{\tau'} x$.

<u>Theorem 2</u> ([6]). There exists a one-to-one relation between the set of all discrete convergences τ on $(E, E_1, E_2, \dots)$ which are Urysohn and the set of all mappings $\mathfrak{M} : E \to P(\prod_{n \in \mathbb{N}} P(E_n))$ with the property $(*)$ which is given by

$$(1) \quad \mathfrak{M}(x) = \Big\{(M_1, M_2, \dots) \in \prod_{n \in \mathbb{N}} P(E_n) \mid \forall (x_n)_{n \in \mathbb{N}'} \overrightarrow{\tau} x : x_n \in M_n \text{ for almost all } n \in \mathbb{N}'\Big\}$$

and

$$(2) \quad \tau(x) = \Big\{(x_n)_{n \in \mathbb{N}'} \in \prod_{n \in \mathbb{N}'} E_n \mid \forall (M_1, M_2, \dots) \in \mathfrak{M}(x) : x_n \in M_n \text{ for almost all } n \in \mathbb{N}'\Big\}.$$

P r o o f. Because of above considerations we have to verify only that for a given $\mathfrak{M}$, the discrete convergence τ associated to $\mathfrak{M}$ by (2), by means of (1) leads to an $\mathfrak{M}'$ which is equal to $\mathfrak{M}$. Obviously, $\mathfrak{M}(x) \subseteq \mathfrak{M}'(x)$ for every $x \in E$. Because of the property $(*)$ of $\mathfrak{M}$, we also get $\mathfrak{M}'(x) \subseteq \mathfrak{M}(x)$. Thus the theorem is proven.

A sequence $(M, M_1, M_2, \dots)$ with $M \subseteq E$ is said to be open or closed if $M \subseteq \underline{(M_1, M_2, \dots)}$ or $M \supseteq \overline{(M_1, M_2, \dots)}$, respectively. A set $\mathcal{L}$ of open sequences is called a base of X if

$$(3) \quad x \in \underline{(M_1, M_2, \dots)} \iff \exists (M', M_1', M_2', \dots) \in \mathcal{L} : x \in M' \text{ and } M_n' \subseteq M_n \text{ for almost all } n \in \mathbb{N}.$$

The set of all open sequences is a base of X.

<u>Theorem 3</u> ([6]). Let X be a metric discrete convergence space. For every $n \in \mathbb{N}$, $x \in E$ and $\varepsilon > 0$ let $K_n(x, \varepsilon) = \{x_n \in E_n \mid \delta_n(x_n, p_n(x)) < \varepsilon\}$.

$$\mathcal{L} = \Big\{(\underline{(K_1(x,\varepsilon), K_2(x,\varepsilon), \dots)}, K_1(x,\varepsilon), K_2(x,\varepsilon), \dots) \mid x \in E,\ \varepsilon > 0\Big\}$$

is a base of X.

P r o o f. Obviously, every sequence of $\mathcal{L}$ is open. Let $x \in \underline{(M_1, M_2, \dots)}$. Then there exists an $\varepsilon > 0$ such that $K_n(x, \varepsilon) \subseteq M_n$ for almost all $n \in \mathbb{N}$. Otherwise we could find a sequence $(x_n)_{n \in \mathbb{N}'} \underset{\tau}{\rightarrow} x$ with $x_n \in M_n$ for every $n \in \mathbb{N}'$ which is impossible. Since, moreover, we have $x \in \underline{(K_1(x,\varepsilon),K_2(x,\varepsilon), \ldots)}$, $\Rightarrow$ in (3) is verified. To prove $\Leftarrow$, let $(M_1,M_2, \ldots)$ be arbitrarily given. If there exists a $y \in E$ and an $\varepsilon > 0$ such that $x \in \underline{(K_1(y,\varepsilon),K_2(y,\varepsilon), \ldots)}$ and $K_n(y,\varepsilon) \subseteq M_n$ for almost all $n \in \mathbb{N}$, then $x \in \underline{(M_1,M_2, \ldots)}$. This proves the theorem.

A sequence $(\hat{M},M_1,M_2, \ldots)$ is said to be compact if for every $\mathbb{N}' \in \mathfrak{N}$ and every sequence $(x_n)_{n \in \mathbb{N}'}$ with $x_n \in M_n$, $n \in \mathbb{N}'$, there exists an $\mathbb{N}^+ \subseteq \mathbb{N}'$ and an $x \in M$ such that $(x_n)_{n \in \mathbb{N}^+} \underset{\tau}{\rightarrow} x$.

Theorem 4 ([6]). 1. If X is separated, then every compact sequence $(M,M_1,M_2, \ldots)$ is closed.

2. If $(M,M_1,M_2, \ldots)$ is compact, $(M',M_1',M_2', \ldots)$ is closed and $M_n' \subseteq M_n$ for every $n \in \mathbb{N}$, then $(M',M_1',M'_2, \ldots)$ is compact.

P r o o f. Simple.

A sequence $(M_1,M_2, \ldots)$ is said to be adherence compact if the sequence $(\overline{(M_1,M_2, \ldots)},M_1,M_2, \ldots)$ is compact. $(x_n)_{n \in \mathbb{N}'} \in \prod_{n \in \mathbb{N}'} E_n$, $\mathbb{N}' \in \mathfrak{N}$, is called adherence compact if $(M_1,M_2, \ldots)$ is adherence compact where $M_n = \{x_n\}$ for $n \in \mathbb{N}'$, $= \emptyset$ else.

IV. Continuity and related notions. Let $X = ((E,E_1,E_2, \ldots),\tau)$ and $X' = ((E',E_1',E_2', \ldots),\tau')$ be discrete convergence spaces. By F and F_n, $n \in \mathbb{N}$, we denote the set of all mappings $f:E \to E'$ or of all mappings $f_n:E_n \to E_n'$, respectively, and we define a mapping $\sigma:F \to P(\bigcup_{\mathbb{N}' \in \mathfrak{N}} \prod_{n \in \mathbb{N}'} F_n)$ by

$$(f_n)_{n \in \mathbb{N}'} \in \sigma(f),\ \mathbb{N}' \in \mathfrak{N},\ \Leftrightarrow\ \forall\, \mathbb{N}^+ \subseteq \mathbb{N}' : (x_n)_{n \in \mathbb{N}^+} \underset{\tau}{\rightarrow} x \text{ implies } (f_n(x_n))_{n \in \mathbb{N}^+} \underset{\tau'}{\rightarrow} f(x).$$

$C = ((F,F_1,F_2, \ldots),\sigma)$ is a discrete convergence space. If X is a discrete approximation, C is separated or Urysohn if and only if X' has the corresponding property ([2], Lemma 1).

A sequence $(f,f_1,f_2, \ldots) \in F \times \prod_{n \in \mathbb{N}} F_n$ is said to be continuous if $(f_n)_{n \in \mathbb{N}} \in \sigma(f)$, this means if $(x_n)_{n \in \mathbb{N}'} \underset{\tau}{\rightarrow} x$, $\mathbb{N}' \in \mathfrak{N}$, always implies $(f_n(x_n))_{n \in \mathbb{N}'} \underset{\tau'}{\rightarrow} f(x)$.

Theorem 5 ([5], Theorem 1). For the continuity of $(f,f_1,f_2, \ldots)$ the following equivalent conditions are necessary and if X' is Urysohn also sufficient:

1. $f[\overline{(M_1,M_2, \ldots)}] \subseteq \overline{(f_1[M_1],f_2[M_2], \ldots)}$ for arbitrary $M_n \subseteq E_n$, $n \in \mathbb{N}$.
2. $\overline{(f_1^{-1}[M_1'],f_2^{-1}[M_2'], \ldots)} \subseteq f^{-1}[\overline{(M_1',M_2', \ldots)}]$ for arbitrary $M_n' \subseteq E_n'$, $n \in \mathbb{N}$.
3. $f^{-1}[\underline{(M_1',M_2', \ldots)}] \subseteq \underline{(f_1^{-1}[M_1'],f_2^{-1}[M_2'], \ldots)}$ for arbitrary $M_n' \subseteq E_n'$,

$n \in N$.

4. $(M_1', M_2', \ldots)$ is a discrete neighbourhood of $f(x)$ $\Rightarrow$ $(f_1^{-1}[M_1'], f_2^{-1}[M_2'], \ldots)$ is a discrete neighbourhood of x.
5. $(M', M_1', M_2', \ldots)$ is closed (open) $\Rightarrow$ $(f^{-1}[M'], f_1^{-1}[M_1'], f_2^{-1}[M_2'], \ldots)$ is closed (open).

A sequence $(f, f_1, f_2, \ldots)$ is said to be conditionally continuous if for every $(x_n)_{n \in N'} \underset{\tau}{\Rightarrow} x$, $N' \in \mathfrak{N}$, with discretely converging $(f_n(x_n))_{n \in N'}$, we have $(f_n(x_n))_{n \in N'} \underset{\tau'}{\Rightarrow} f(x)$.

<u>Theorem 6</u> ([2], Lemma 2; cp. with [12], page 293, (1)). Consider the following assertions:

1. $(f, f_1, f_2, \ldots)$ is continuous.
2. $(f, f_1, f_2, \ldots)$ is conditionally continuous. If $(x_n)_{n \in N'} \underset{\tau}{\Rightarrow} x$, $N' \in \mathfrak{N}$, the sequence $(f_n(x_n))_{n \in N'}$ is adherence compact.

We have $1 \Rightarrow 2$ and, if X' is Urysohn, $2 \Rightarrow 1$.

A sequence $(f, f_1, f_2, \ldots)$ is said to be inversely continuous if for every $N' \in \mathfrak{N}$, every sequence $(x_n)_{n \in N'} \in \prod_{n \in N'} E_n$ and every $x \in E$, $(f_n(x_n))_{n \in N'} \underset{\tau'}{\Rightarrow} f(x)$ implies $(x_n)_{n \in N'} \underset{\tau}{\Rightarrow} x$. If f and every f_n, $n \in N$, are bijective, then $(f, f_1, f_2, \ldots)$ is inversely continuous if and only if $(f^{-1}, f_1^{-1}, f_2^{-1}, \ldots)$ is continuous. A sequence $(f_1, f_2, \ldots)$ is said to be regular if adherence compactness of $(f_n(x_n))_{n \in N'}$, $N' \in \mathfrak{N}$, implies that $(x_n)_{n \in N'}$ is adherence compact, too.

<u>Theorem 7</u> ([2], Theorem 1). Let X be Urysohn, X' be separated, f be injective, $(f, f_1, f_2, \ldots)$ be conditionally continuous and $(f_1, f_2, \ldots)$ be regular. Then $(f, f_1, f_2, \ldots)$ is inversely continuous.

Theorem 7 is an analog to a certain generalization of the well-known classical theorem that a continuous one-to-one mapping of a compact topological space onto a separated topological space is homeomorphic.

<u>Theorem 8</u> ([3], Lemma 5). Let X and X' be metric discrete convergence spaces and let δ_n, δ_n' ($n \in N$) denote the metrics on E_n or E_n', respectively. Let f be surjective and $(f, f_1, f_2, \ldots)$ be continuous. Moreover let us assume the following condition:

(**) For every $\varepsilon > 0$ there exists a $\delta > 0$ such that $\delta_n'(f_n(x_n), f_n(y_n)) \leq \delta$ implies $\delta_n(x_n, y_n) < \varepsilon$ for arbitrary $x_n, y_n \in E_n$ and $n \in N$.

Then $(f_1, f_2, \ldots)$ is regular.

If the mappings f_n, $n \in N$, are bijective, (**) means that $\{f_n^{-1} \mid n \in N\}$ is uniformly equicontinuous.

V. Numerical problems. Given an element $x' \in E'$, the equation $f(x) = x'$ is said to be uniquely approximation-solvable with respect to $(f_n)_{n\in\mathbb{N}}$ if there exist a unique solution x of $f(x) = x'$, a sequence $(x'_n)_{n\in\mathbb{N}} \overrightarrow{\tau'} x'$, for every $(x'_n)_{n\in\mathbb{N}} \overrightarrow{\tau'} x'$ a sequence $(x_n)_{n\in\mathbb{N}}$ of solutions x_n of $f_n(x_n) = x'_n$ and for every such sequence one has $(x_n)_{n\in\mathbb{N}} \overrightarrow{\tau} x$.

By means of Theorem 7 we easily get

Theorem 9 ([2], Theorem 2; cp. with [12], page 295, (4)). Let X be Urysohn, X' be separated and a discrete approximation, f be injective and every f_n, $n \in \mathbb{N}$, be surjective. Moreover, assume $(f,f_1,f_2, \ldots)$ is conditionally continuous and $(f_1,f_2, \ldots)$ is regular. Then for every $x' \in E'$ the equation $f(x) = x'$ is uniquely approximation-solvable with respect to $(f_n)_{n\in\mathbb{N}}$.

Theorem 9 is important for questions of numerical analysis. It gives conditions under which the problem stated in the introduction is solvable. With regard to error estimates, we can use

Theorem 10 (cp. with [12], section 5.2). Let X and X' be metric discrete convergence spaces and let δ_n, δ'_n ($n \in \mathbb{N}$) denote the metrics on E_n or E'_n, respectively. Condition (**) of Theorem 8 is necessary and sufficient for the following property: There exists a mapping $\varkappa: [0,\infty) \to [0,\infty]$ with $\varkappa(0) = 0$ which is continuous at 0 and monotonously increasing such that

$$\delta_n(x_n,y_n) \leq \varkappa(\delta'_n(f_n(x_n),f_n(y_n)))$$

for arbitrary x_n, $y_n \in E_n$ and $n \in \mathbb{N}$.

P r o o f. That (**) implies the mentioned property can be seen without difficulty by taking

$$\varkappa(\varrho) = \sup\{\delta_n(x_n,y_n) \mid x_n,\ y_n \in E_n \text{ with } \delta'_n(f_n(x_n),f_n(y_n)) \leq \varrho,\ n \in \mathbb{N}\}.$$

The inverse is obvious.

As a consequence of Theorem 10 we immediately get the following

Corollary. Let X and X' be as in Theorem 10. Let a sequence $(x'_n)_{n\in\mathbb{N}} \overrightarrow{\tau'} x'$ be given and let us assume there exist solutions x, x_n ($n \in \mathbb{N}$) of $f(x) = x'$ and $f_n(x_n) = x'_n$, respectively. Let $(f,f_1,f_2, \ldots)$ be consistent, which means that there exists a sequence $(y_n)_{n\in\mathbb{N}} \overrightarrow{\tau} x$ with $(f_n(y_n))_{n\in\mathbb{N}} \overrightarrow{\tau'} f(x) = x'$. Let (**) be true. Then

$$\delta_n(x_n,p_n(x)) \leq \varkappa(\delta'_n(x'_n,f_n(y_n))) + \delta_n(y_n,p_n(x))$$

for every $n \in \mathbb{N}$.

The last inequality gives an error estimate for the discrete convergence of the solutions x_n of $f_n(x_n) = x'_n$ to the solution x of $f(x) = x'$. If X and X' are linear normed discrete convergence spaces, the assumptions are often such (see for instance [13], page 50) that it is possible to find a positive homogeneous $\varkappa$ and to take $y_n = p_n(x)$, $n \in \mathbb{N}$, which leads to

$$\| x_n - p_n(x) \|_n \leq \varkappa(1) \| x'_n - f_n(p_n(x)) \|'_n$$

where $\| . \|_n$, $\| . \|'_n$ $(n \in \mathbb{N})$ denote the norms generating δ_n or δ'_n, respectively.

<u>VI. Applications</u>. Theorem 9 can be applied to several special cases, for instance to ordinary and partial differential equations, to integral equations and to eigenvalue problems. The methods being used naturally depend on the given problem. Our following considerations concern three such methods.

1. Generalized projection method. Let E be a metric space with metric δ. For every $n \in \mathbb{N}$, let E_n be a subspace of E and p_n be a mapping of E onto E_n such that for every $x \in E$, $(\delta(x,p_n(x)))_{n \in \mathbb{N}} \to 0$. Let X be the corresponding metric discrete convergence space and τ its discrete convergence. X is separated. For $(x_n)_{n \in \mathbb{N}'} \in \prod_{n \in \mathbb{N}'} E_n$, $\mathbb{N}' \in \mathfrak{N}$, and $x \in E$ we have $(x_n)_{n \in \mathbb{N}'} \xrightarrow[\tau]{} x$ if and only if $(\delta(x,x_n))_{n \in \mathbb{N}'} \to 0$. The pseudometric on E generated by the metrics of E_n, $n \in \mathbb{N}$, as in section 2 is identical with the original metric δ of E.

Besides X, let another metric discrete convergence space X' of the same special Type as X be given and with respect to X' let us use notations δ' and p'_n analogously to δ and p_n, respectively. Let [E'] be the metric discrete convergence space $((E^+,E_1^+,E_2^+, \ldots),\tau^+)$ where $E^+ = E_1^+ = E_2^+ = \ldots = E'$ and where the corresponding mappings $p_n^+ : E^+ \to E_n^+$, $n \in \mathbb{N}$, are the identities. $(id,p'_1,p'_2, \ldots)$ will be considered as a mapping sequence associated to [E'] and X'. Hence continuity of $(id,p'_1,p'_2, \ldots)$ means that for arbitrary $\mathbb{N}' \in \mathfrak{N}$ and arbitrary points $x'_n \in E'$, $n \in \mathbb{N}'$, and $x' \in E'$, $(\delta'(x',x_n))_{n \in \mathbb{N}'} \to 0$ implies $(p'_n(x_n))_{n \in \mathbb{N}'} \xrightarrow[\tau']{} x'$. Let $f: E \to E'$. Then in a natural way for every $n \in \mathbb{N}$ there is given a mapping $f_n : E_n \to E'_n$ by $f_n = p'_n \circ f | E_n$, to which the following two theorems refer.

<u>Theorem 11</u> ([3], Lemma 6). If $(id,p'_1,p'_2, \ldots)$ and f are continuous, then $(f,f_1,f_2, \ldots)$ is continuous, too.

In the underlying case, the fundamental Theorem 9 can be specialized to

<u>Theorem 12</u> ([3], Theorem 3). Let $(id,p'_1,p'_2, \ldots)$ and f be continuous and let f and every f_n, $n \in \mathbb{N}$, be surjective. Assume, (**) is fulfilled. Then for every $x' \in E'$ the equation $f(x) = x'$ is uniquely approximation-solvable with respect to $(f_n)_{n \in \mathbb{N}}$.

The method underlying the above considerations is a generalization of the well-known classical projection method which includes as special cases the Galerkin method (hence the method of least squares and the reduction method) as well as the collocation method ([10], page 9).

2. Quadrature method ([13], § 5). Let us consider the integral equation $x(t) = \int_D K(t,s)x(s)ds + y(t)$, where D is a closed bounded region in $\mathbb{R}^n$ ($\mathbb{R}$ being the set of all reals) and the functions $x, y: D \to \mathbb{R}$ and $K: D \times D \to \mathbb{R}$ are continuous. If for every $n \in \mathbb{N}$, there is given a quadrature formula $\int_D z(s)ds = \sum_{i=1}^{n} a_{in}z(s_{in}) + \varphi_n(z)$ with $a_{in} \in \mathbb{R}$ and $s_{in} \in D$ for every $i \in \{1, \ldots, n\}$ and with $(\varphi_n(z))_{n \in \mathbb{N}} \to 0$ for every continuous $z: D \to \mathbb{R}$, then we can easily subordinate the problem to find a solution to our general procedure. For this we have to introduce suitable discrete convergence spaces X, X' and mappings f_n, $n \in \mathbb{N}$. Let $X = ((E,E_1,E_2, \ldots),\tau)$ be a linear normed discrete convergence space, where E, E_n and p_n, $n \in \mathbb{N}$, are as follows: E is the linear space of all continuous functions on D. E_n is the set $\mathbb{R}^n$ which is equipped with the maximum norm $\|.\|_n$ given by $\|x_n\|_n = \max_i |x_{in}|$ where $x_n = (x_{1n}, \ldots, x_{nn})$. $p_n: E \to E_n$ is defined by $p_n(x) = (x(s_{1n}), \ldots, x(s_{nn}))$. Let X' coincide with X. The integral equation is identical with $f(x) = y$, where $f: E \to E'$ is given by $f(x)(t) = x(t) - \int_D K(t,s)x(s)ds$. The quadrature formulas suggest to take as $f_n: E_n \to E_n'$ the mapping defined by the equation $f_n(x_n) = x_n - (\sum_{i=1}^{n} a_{in}K(s_{1n},s_{in})x_{in}, \ldots, \sum_{i=1}^{n} a_{in}K(s_{nn},s_{in})x_{in})$.

3. Difference method ([13], § 6, 7). To illustrate the method which can be used for the (approximative) solution of initial and boundary value problems for usual and partial differential equations, let us restrict ourselves to the boundary value problem $x'' + a(t)x' + b(t)x = y(t)$ on [0,1] with $x(0) = x(1) = 0$ where a, b, y are continuous. Discretization by means of formulas for the numeric differentiation, for arbitrary $n \in \mathbb{N}$ gives

$$\frac{x_{i+1\,n} - 2x_{in} + x_{i-1\,n}}{h^2} + a(ih)\,\frac{x_{i+1\,n} - x_{i-1\,n}}{2h} + b(ih)x_{in} = y(ih), \quad i = 1, \ldots, n-1,$$

with $x_{0n} = x_{n+1\,n} = 0$ and $h = (n+1)^{-1}$. This suggests to take as $X = ((E,E_1,E_2, \ldots),\tau)$ and $X' = ((E',E_1',E_2', \ldots),\tau')$ linear normed discrete convergence spaces, where E, E_n, p_n and E', E_n', p_n' ($n \in \mathbb{N}$) are given as follows: E is the linear space of all functions x on [0,1] with $x(0) = x(1) = 0$ which are twice continuously differentiable. E_n is the space of all (n+2)-tuples $x_n = (x_{0n}, \ldots, x_{n+1\,n})$ of reals with $x_{0n} = x_{n+1\,n} = 0$ which is equipped with the norm $\|.\|_n$ given by

$$\|x_n\|_n = \max_i |x_{in}| + \max_i \left|\frac{x_{i+1\,n} - x_{i-1\,n}}{2h}\right| + \max_i \left|\frac{x_{i+1\,n} - 2x_{in} + x_{i-1\,n}}{h^2}\right| .$$

p_n is defined by $p_n(x) = (x(0), x(h), \ldots, x(1))$. E' is the linear space of all continuous functions on $[0,1]$, E_n' is the space $\mathbb{R}^n$ equipped with the maximum norm and p_n' is given by $p_n'(y) = (y(h), \ldots, y(nh))$.

References

[1] F. E. Browder, Approximation-solvability of nonlinear functional equations in normed linear spaces, Arch. Rat. Mech. Anal. 26 (1967), 33 - 42.

[2] S. Gähler, Contributions to the theory of discrete convergence, Proc. Intern. School: Convergence Structures and Applications II (Schwerin 1983), 49 - 56, Abh. Akad. Wiss. DDR, Jahrgang 1984.

[3] ——— , Metric discrete convergence spaces and their application in numerical analysis, Proc. Coll. on Topology (Eger 1983), to appear.

[4] ——— , Discrete convergence and regularity of mapping sequences, Proc. Conf.: Convergence and Generalized Functions (Katowice 1983), to appear.

[5] ——— , Remarks on Continuity in the Theory of Discrete Convergence, Proc. Conf.: Topology and Measure IV (Trassenheide 1983), to appear.

[6] W. Gähler, Contributions to a seminar on discrete convergence (not published).

[7] R. D. Grigorieff, Zur Theorie approximationsregulärer Operatoren I, II, Math. Nachr. 55 (1973), 233 - 249, 251 - 263.

[8] W. V. Petryshyn, Remarks on the approximation-solvability of nonlinear functional equations, Arch. Rat. Mech. Anal. 26 (1967), 43 - 49.

[9] ——— , On the approximation-solvability of nonlinear equations, Math. Ann. 177 (1968), 156 - 164.

[10] S. Prößdorf, B. Silbermann, Projektionsverfahren und die näherungsweise Lösung singulärer Gleichungen, Teubner, Leipzig 1977.

[11] F. Stummel, Diskrete Konvergenz linearer Operatoren, I, Math. Ann. 190 (1970), 52 - 92; II, Math. Z. 120 (1971), 231 - 264; III, Proc. Conf. Lin. Oper. and Approx. (Oberwolfach 1971), Intern. Series Num. Math. 20 (1972), 196 - 216.

[12] ——— , Discrete convergence of mappings, Proc. Conf. Num. Anal. (Dublin 1972), 285 - 310, Acad. Press, New York-London 1973.

[13] G. Vainikko, Funktionalanalysis der Diskretisierungsmethoden, Teubner, Leipzig 1976.

AXIOMS OF STRUCTURES AND FUNCTOR POWER SERIES

Werner Gähler [1)]

This paper is devoted to relations between convergence structures and algebraic structures. Such relations are given by means of a new notion, that of functor power series. As set-theoretic foundation none of the usual axiomatics can be used without further assumptions, neither that of Zermelo-Fraenkel nor that of von Neumann-Bernays-Gödel or its modification of Morse-Kelley. We need a notion of superclass and will choose that of conglomerate according to Herrlich and Strecker [5] and also some of its variants to be obtained of smaller enlargements of Zermelo-Fraenkel set theory. For establishing the relations between convergence structures and algebraic structures the notion of equation is used. This notion will be introduced in full generality. It contains as special case the well-known notion of equation from algebra.

1. On conglomerates

The idea of a conglomerate is the following one: As axiomatics is used that of Zermelo-Fraenkel where additionally the axiom of universe is assumed to be true. However, one speaks of conglomerates instead of sets and calls sets only the members of the universe. Hence, the axiomatics for conglomerates consists of the following axioms I - X for sake of simplicity partially verbally described:

<u>I (Extensionality).</u> $X = Y$ if and only if $x \in X \Leftrightarrow x \in Y$.

<u>II (Foundation).</u> Each non-empty conglomerate X has a member x such that x and X have no element in common.

<u>III (Comprehension Scheme).</u> For each set-theoretic formula A and each conglomerate X there exists a conglomerate Y consisting of all elements of X with property A (where X and Y do not occur in A).

<u>IV (Pairing).</u> For all conglomerates X and Y there is a conglomerate $\{X, Y\}$ consisting of X and Y only.

<u>V (Power).</u> For each conglomerate X there is a conglomerate PX having all subconglomerates of X as members.

<u>VI (Union).</u> For each conglomerate X there is a conglomerate $\cup X$ consisting of all elements of the members of X.

<u>VII (Infinity).</u> There exist the empty conglomerate $\emptyset$ and a conglomerate X such that $\emptyset \in X$ and $x \in X$ implies $x \cup \{x\} \in X$.

<u>VIII (Replacement Scheme).</u> If A is a set-theoretic formula and X is a conglomerate such that for each $x \in X$ there is precisely one conglomerate y with property A then there exists a conglomerate Y such that for each $x \in X$ there is a $y \in Y$ with property A (where Y does not occur in A).

1) Akademie der Wissenschaften der DDR, Institut für Mathematik, DDR-1086 Berlin, Mohrenstraße 39.

IX (Universe). There is a conglomerate $\mathcal{U}$ with the following properties: (U_1) $x \in X \in \mathcal{U} \Rightarrow x \in \mathcal{U}$. (U_2) $X, Y \in \mathcal{U} \Rightarrow \{X,Y\} \in \mathcal{U}$. (U_3) $X \in \mathcal{U} \Rightarrow PX \in \mathcal{U}$. (U_4) $X \in \mathcal{U} \Rightarrow \bigcup X \in \mathcal{U}$. (U_5) $N \in \mathcal{U}$. (U_6) For each map $f: X \to Y$ with $X \in \mathcal{U}$ and $f[X] \subseteq \mathcal{U}$ we have $f[X] \in \mathcal{U}$.

X (Choice). For every X there is a map f with $f(x) \in x$ for all $x \in X \setminus \{\emptyset\}$.

In this set theory for conglomerates denoted by ZFU ordinals and cardinals are defined as usual in Zermelo-Fraenkel set theory. If an ordinal is a member of $\mathcal{U}$ we speak of an ordinal number. Analogously cardinal numbers are cardinals lying in $\mathcal{U}$.

In ZFU we define a notion of level for conglomerates. Let $\mathcal{U}_0 = \mathcal{U}$ and for each ordinal $\alpha > 0$ let $\mathcal{U}_\alpha = P(\bigcup_{\beta<\alpha} \mathcal{U}_\beta)$. One gets that each conglomerate is a member of at least one $\mathcal{U}_\alpha$. For each conglomerate X the smallest α with $X \in \mathcal{U}_\alpha$ is called the level of X and is denoted by $\mathrm{lv}X$. According to $\mathrm{lv}X$ being equal to 0, 1 or α we speak of a set, a proper class or an α-conglomerate, respectively. X is a class if it is a set or a proper class. If $\mathrm{lv}X > 0$ then

(1.1) $$\mathrm{lv}\{X\} = \mathrm{lv}PX = \mathrm{lv}X + 1$$

and

(1.2) $$\mathrm{lv}X = \sup\{\mathrm{lv}x + 1 \mid x \in X\}.$$

In ZFU we use the following notion of ordered pair. For each set x let $qx = \{x\}$ and $rx = P\{x\}$. Furthermore, for each α-conglomerate x with $\alpha > 0$ let $qx = \{q\xi \mid \xi \in x\}$ and $rx = \{r\xi \mid \xi \in x\}$. Then for arbitrary conglomerates X and Y, $(X,Y) = \{qx \mid x \in X\} \cup \{ry \mid y \in Y\}$ is called the ordered pair of X and Y. Of course, (X,Y) has the usual properties of an ordered pair. Moreover,

(1.3) $$\mathrm{lv}(X,Y) = \max\{\mathrm{lv}X, \mathrm{lv}Y\}.$$

For all X and Y, $\iota_1: X \to (X,Y)$, $x \mapsto qx$, and $\iota_2: Y \to (X,Y)$, $y \mapsto ry$, are injections, $\iota_1[X] \cup \iota_2[Y] = (X,Y)$ and $\iota_1[X] \cap \iota_2[Y] = \emptyset$.

2. Smaller enlargements of Zermelo-Fraenkel set theory

As is reflected in the use of denotation "set" ZFU is in some sense an enlargement of Zermelo-Fraenkel set theory. There are also interesting "smaller enlargements" given by the following axioms I' - VI':

I' = I (Extensionality).

II' = II (Foundation).

III' = III (Comprehension Scheme).

IV' (Axiom of Universe, extended version of IX). There is a conglomerate $\mathcal{U}$ with the following properties: $(U_1') = (U_1)$. $(U_2') = (U_2)$ where $\{X,Y\} = \{Z \in \mathcal{U} \mid Z = X \text{ or } Z = Y\}$. (U_3') For each $X \in \mathcal{U}$ all subconglomerates $Y \subseteq X$ as well as $PX = \{Y \in \mathcal{U} \mid Y \subseteq X\}$ are members of $\mathcal{U}$. $(U_4') = (U_4)$ where $\bigcup X = \{Y \in \mathcal{U} \mid \exists x \in X\ (y \in x)\}$. (U_5') = VII where additionally $X \in \mathcal{U}$. $(U_6') = (U_6)$.

V' (Axiom of Exhaustion). There is a positive ordinal number ν with the following properties: For each ordinal number $\alpha < \nu$ there exists a conglomerate $\mathcal{U}_\alpha$ such that $\mathcal{U}_0 = \mathcal{U}$ and in the case of $\alpha > 0$, $X \in \mathcal{U}_\alpha$ iff for each $x \in X$ there is an ordinal number $\beta < \alpha$ with $x \in \mathcal{U}_\beta$. Each conglomerate is contained in at least one $\mathcal{U}_\alpha$.

VI' = X (Choice).

Ordinal numbers in particular those occuring in axiom V' are to be understood with respect to the universe $\mathcal{U}$, in which the Zermelo-Fraenkel set theory holds. Also cardinal numbers are to be understood with respect to $\mathcal{U}$. The set theory with axioms I' - VI' will be denoted by ZFν where ν is the positive ordinal number occuring in axiom V'. If ν is a successor then in axiom III' X can be replaced by the fixed maximal conglomerate $\mathcal{U}_{\nu-1}$. If ν is a limit then in the last part of axiom V' the phrase "is contained in" can be replaced by " is a member of". In ZFν we also define a notion of level for conglomerates. If for a conglomerate X there is no α with $X \in \mathcal{U}_\alpha$ then the level lvX of X is defined to be ν. For other X by the level lvX of X we mean the smallest α with $X \in \mathcal{U}_\alpha$. Analogously as in ZFU lvX > 0 implies (1.1) and (1.2). The notions set, class, proper class, α-conglomerate and ordered pair are defined as in ZFU. In ZFν for ordered pairs also (1.3) holds. ZF1 is the Morse-Kelley set theory. Obviously in ZFν the axiom VI of union is fulfilled. If ν is a limit then in ZFν also the axioms IV of pairing and V of power hold. However, in ZFν the axiom scheme VIII of replacement is never fulfilled.

3. ZFU or ZFν with $\nu \geq 2$

In the sequel all investigations are related to ZFU or ZFν where $\nu \geq 2$. The following figure illustrates the conglomerates of ZFU and ZFν, respectively, ordered by inclusion. If $X \subsetneqq Y$ then Y is lying higher than X.

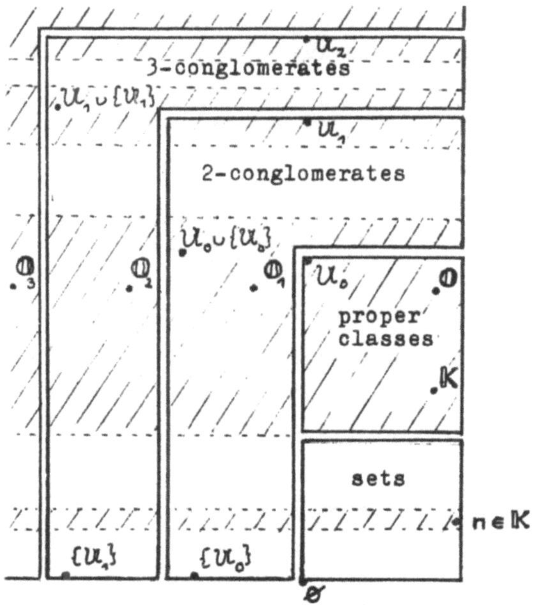

For sake of simplicity in this figure in case of $ZF\nu$ $\nu > 3$ is assumed. $\mathbb{O}$ is the class of all ordinal numbers and $\mathbb{K}$ the class of all cardinal numbers. In the figure $\mathbb{O}_n$ is the successor of $\mathbb{O}_{n-1}$ where $\mathbb{O}_0 = \mathbb{O}$. The parts indicated by hatching collect equipotent conglomerates.

According to [5] by a quasicategory we mean the notion defined analogously to that of a category, however with conglomerates. Functors between quasicategories are defined as in case of categories. In this paper by functor we always mean a covariant one. Let CAT denote the quasicategory of all categories with all functors between categories as morphisms. Moreover, let FUNC denote the quasicategory of all functors between categories with all natural transformations between those functors as morphisms. CAT and FUNC are 2-conglomerates. The categories SET of all sets and CARD of all cardinal numbers are proper classes. obCARD = $\mathbb{K}$ and morphisms of CARD are the pairs (n,m) of cardinal numbers with $n \leq m$.

4. Sequences and power series of set functors

By a (cardinal) sequence of set functors we mean a functor T: CARD → FUNC with the following properties:

(1) For each $n \in \mathbb{K}$, Tn is a functor Φ_n: SET → SET.

(2) For all $n,m \in \mathbb{K}$ with $n \leq m$, $T(n,m)$ is a natural transformation $\Gamma_{nm} = (\Gamma_{nm}X)_{X\in\mathcal{U}} : \Phi_n \to \Phi_m$ such that each map $\Gamma_{nm}X: \Phi_n X \to \Phi_m X$ is injective. That Γ_{nm} is a natural transformation means that for each map $f: X \to Y$ between sets we have $\Phi_m f \circ \Gamma_{nm}X = \Gamma_{nm}Y \circ \Phi_n f$. $T = ((\Phi_n)_{n\in\mathbb{K}}, (\Gamma_{nm})_{n\leq m})$ is a 2-conglomerate. If T has a direct limit $(\Phi, (\Gamma_n)_{n\in\mathbb{K}})$ then T is said to converge to $(\Phi, (\Gamma_n)_{n\in\mathbb{K}})$. Such direct limit consists of a set functor Φ and a family of natural transformations $\Gamma_n: \Phi_n \to \Phi$ for which $\Gamma_n = \Gamma_m \Gamma_{nm}$ for all $n,m \in \mathbb{K}$ with $n \leq m$. $(\Phi, (\Gamma_n)_{n\in\mathbb{K}})$ being a direct limit means that for any pair $(\Phi', (\Gamma_n')_{n\in\mathbb{K}})$ consisting of a set functor Φ' and a family of natural transformations $\Gamma_n': \Phi_n \to \Phi'$ with $\Gamma_n' = \Gamma_m' \Gamma_{nm}$ for all $n,m \in \mathbb{K}$ with $n \leq m$ there is precisely one natural transformation $\Gamma: \Phi \to \Phi'$ with $\Gamma \Gamma_n = \Gamma_n'$ for all $n \in \mathbb{K}$.

Proposition 4.1. A sequence $T = ((\Phi_n)_{n\in\mathbb{K}}, (\Gamma_{nm})_{n\leq m})$ of set functors converges if and only if for each set X, $\{\operatorname{card} \Phi_n X \mid n \in \mathbb{K}\}$ is bounded in $\mathbb{K}$ or – equivalently – $\{\operatorname{card} \Phi_n X \mid n \in \mathbb{K}\}$ is a set. If T converges then for each set X, $\sup\{\operatorname{card} \Phi_n X \mid n \in \mathbb{K}\}$ can be used as ΦX.

Proposition 4.2. If T converges to $(\Phi, (\Gamma_n)_{n\in\mathbb{K}})$ then for each set X the maps $\Gamma_n X: \Phi_n X \to \Phi X$ are injective. Moreover, then for each set X there is a cardinal number l_X such that for all $n \geq l_X$ the maps $\Gamma_n X$ are bijective.

Power series of set functors will be defined as special sequences of set functors. Each power series of set functors is given by a pair of families $(a_n)_{n\in\mathbb{K}}$ and $(b_n)_{n\in\mathbb{K}}$ with the following properties:

(1) For each $n \in \mathbb{K}$, a_n is a set and b_n is an equivalence functor of $\sum_{k\le n} a_k \mathrm{id}^k$ 1).

(2) For all $n,m \in \mathbb{K}$ with $n \le m$ and each set X,

$$b_n X = b_m X \cap (\sum_{k\le n}(a_k \times X^k) \times \sum_{k\le n}(a_k \times X^k)).$$

Let Φ_n denote the quotient functor $\sum_{k\le n} a_k \mathrm{id}^k / b_n$ 2) for each $n \in \mathbb{K}$. Property (2) implies that for all $n,m \in \mathbb{K}$ with $n \le m$ and each set X the map $\mu_{nm}X\colon \Phi_n X \to \Phi_m X$, $(k,\alpha,g)^{\sim b_n} \mapsto (k,\alpha,g)^{\sim b_m}$ is injective where $(k,\alpha,g)^{\sim b_n}$ and $(k,\alpha,g)^{\sim b_m}$ denote the equivalence classes of the triples (k,α,g) with respect to $b_n X$ and $b_m X$, respectively. We even have $(k,\alpha,g)^{\sim b_n} \subseteq (k,\alpha,g)^{\sim b_m}$ for these equivalence classes. As is easily seen $\mu_{nm} = (\mu_{nm}X)_{X\in\mathcal{U}}$ is a natural transformation of Φ_n to Φ_m for all $n,m \in \mathbb{K}$ with $n \le m$. Hence, $((\Phi_n)_{n\in\mathbb{K}}, (\mu_{nm})_{n\le m})$ is a sequence of set functors called the <u>power series</u> generated by $(a_n)_{n\in\mathbb{K}}$ and $(b_n)_{n\in\mathbb{K}}$. Because $(\mu_{nm})_{n\le m}$ is uniquely determined by $(\Phi_n)_{n\in\mathbb{K}}$ we also consider $(\Phi_n)_{n\in\mathbb{K}}$, that is

$$(\sum_{k\le n} a_k \mathrm{id}^k / b_n)_{n\in\mathbb{K}}$$

as the <u>power series</u> generated by $(a_n)_{n\in\mathbb{K}}$ and $(b_n)_{n\in\mathbb{K}}$. According to Proposition 4.1 a power series $(\sum_{k\le n} a_k \mathrm{id} / b_n)_{n\in\mathbb{K}}$ converges if and only if for each set X, $\{\mathrm{card} \sum_{k\le n}(a_k \times X^k)/b_n X \mid n \in \mathbb{K}\}$ is bounded. If a power series converges to $(\Phi, (\mu_n)_{n\in\mathbb{K}})$ we also say that this power series <u>converges</u> to Φ with respect to $(\mu_n)_{n\in\mathbb{K}}$.

By the following theorem we get that for each set functor Φ there is at least one power series converging to Φ.

<u>Theorem 4.3.</u> Let Φ be a set functor and let $a_n = \Phi n$ for all $n \in \mathbb{K}$. Then for each $n \in \mathbb{K}$ an equivalence functor b_n of $\sum_{k\le n} a_k \mathrm{id}^k$ and a natural transformation $\mu_n\colon \sum_{k\le n} a_k \mathrm{id}^k / b_n \to \Phi$ are defined letting

$$b_n X = \{((k,\alpha,g),(k',\alpha',g')) \mid (\Phi g)(\alpha) = (\Phi g')(\alpha')\} \ 3)$$

and $\mu_n X\colon (k,\alpha,g)^{\sim b_n} \mapsto (\Phi g)(\alpha)$ for all sets X. Then the power series <u>generated by</u> $(a_n)_{n\in\mathbb{K}}$ and $(b_n)_{n\in\mathbb{K}}$ converges to Φ with respect to $(\mu_n)_{n\in\mathbb{K}}$.

1) $\sum_{k\le n} a_k \mathrm{id}^k$ is the set functor which assigns to each set X the disjoint union $(\sum_{k\le n} a_k \mathrm{id}^k)X = \sum_{k\le n}(a_k \times X^k)$ $(:= \bigcup_{k\le n}\{k\}\times a_k \times X^k)$ and to each map $f\colon X \to Y$ between sets the map $(\sum_{k\le n} a_k \mathrm{id}^k)f\colon (k,\alpha,g) \mapsto (k,\alpha, fg)$. Of course, in the sums k runs over all cardinal numbers $\le n$. b_n being an equivalence functor of $\sum_{k\le n} a_k \mathrm{id}^k$ means that for each set X, $b_n X$ is an equivalence relation of $\sum_{k\le n}(a_k \times X^k)$ and for each map $f\colon X \to Y$ between sets, $b_n f$ is the map $((k,\alpha,g),(k',\alpha',g')) \mapsto ((k,\alpha,fg),(k',\alpha',fg'))$ of $b_n X$ into $b_n Y$. For arbitrary set functors Φ and Ψ, Ψ is called an <u>equivalence functor</u> of Φ if for each set X, ΨX is an equivalence relation of ΦX and for each map $f\colon X \to Y$ between sets, $\Psi f(\mathfrak{m},\mathfrak{n}) = (\Phi f(\mathfrak{m}), \Phi f(\mathfrak{n}))$ for all $(\mathfrak{m},\mathfrak{n}) \in \Psi X$.

2) For each set functor Φ and each equivalence functor Ψ of Φ the <u>quotient functor</u> Φ/Ψ is defined to be the set functor which assigns to each set X the quotient set $\Phi X/\Psi X$ and to each map $f\colon X \to Y$ between sets the corresponding quotient map of $\Phi f\colon \Phi X \to \Phi Y$.

3) Of course, $k,k' \le n$, $\alpha \in \Phi k$, $\alpha' \in \Phi k'$, $g \in X^k$ and $g' \in X^{k'}$, hence $(\Phi g)(\alpha)$, $(\Phi g')(\alpha') \in \Phi X$.

We call the power series appearing in this theorem the canonical power series of Φ.

Sometimes it is useful to replace the canonical power series $(\sum_{k\leq n} a_k \mathrm{id}^k / b_n)_{n\in\mathbb{K}}$ of Φ by a power series which is in some sense minimal and which can be constructed as follows: For each $n\in\mathbb{K}$ let $a_n' = a_n \setminus \{\alpha\in a_n \mid$ there are $m<n$, $\beta\in a_m$ and $h\colon m\to n$ with $\alpha = (\Phi h)(\beta)\}$ and let $\bar{a}_n$ be the quotient set of $a_n' \times n^n$ with respect to the equivalence relation defined by $(\alpha,h)\sim(\beta,k) \Leftrightarrow (\Phi h)(\alpha) = (\Phi k)(\beta)$. We denote the corresponding equivalence class of a pair (α,h) by $(\alpha,h)^{\sim n}$. $(\alpha,h)^{\sim n} \mapsto (\Phi h)(\alpha)$ is an injection of $\bar{a}_n$ into a_n', hence we have card $\bar{a}_n \leq$ card a_n'. For each $n\in\mathbb{K}$ an equivalence functor $\bar{b}_n$ of $\sum_{k\leq n}\bar{a}_k \mathrm{id}^k$ and a natural transformation $\bar{\mu}_n$: $\sum_{k\leq n}\bar{a}_k \mathrm{id}^k/\bar{b}_n \to \Phi$ are defined letting

$\bar{b}_n X = \{((k,(\alpha,h)^{\sim k},g),(k',(\alpha',h')^{\sim k'},g')) \mid (\Phi gh(\alpha) = (\Phi g'h')(\alpha')\}$

and $\bar{\mu}_n X\colon (k,(\alpha,h)^{\sim k},g)^{\sim \bar{b}_n} \mapsto (\Phi gh)(\alpha)$ for all sets X. $(\sum_{k\leq n}\bar{a}_k \mathrm{id}^k/\bar{b}_n)_{n\in\mathbb{K}}$ is a power series converging to Φ with respect to $(\bar{\mu}_n)_{n\in\mathbb{K}}$. It is called the reduced canonical power series of Φ.

5. Polynomials of set functors

Polynomials of set functors can be considered as special power series of set functors. By a polynomial of set functors we mean a quotient functor $\sum_{k\leq n} a_k \mathrm{id}^k/b_n$ where $n\in\mathbb{K}$, a_k, $k\leq n$, are sets and b_n is an equivalence functor of $\sum_{k\leq n} a_k \mathrm{id}^k$. The maximal k with $a_k \neq \emptyset$ is called the degree of the polynomial. There is a polynomial with no degree defined in this way, namely the trivial set functor $X\mapsto\emptyset$, $f\mapsto\emptyset$. We define its degree to be -1. Each polynomial $\sum_{k\leq n} a_k \mathrm{id}^k/b_n$ will be identified with the power series $(\sum_{k\leq m} a_k' \mathrm{id}^k/b_m')_{m\in\mathbb{K}}$ where

$$a_m' = \begin{cases} a_m & \text{if } m\leq n \\ \emptyset & \text{if } m>n \end{cases}$$

and for each set X

$$b_m' X = \begin{cases} b_n X \cap (\sum_{k\leq m}(a_k\times X^k)\times\sum_{k\leq m}(a_k\times X^k)) & \text{if } m\leq n \\ b_n X & \text{if } m>n. \end{cases}$$

Then (as power series) each polynomial of set functors converges to itself with respect to an obviously defined family of natural transformations μ_n. Isomorphic polynomials of set functors may differ in their degrees. For instance, the identity set functor id and the set functor $2\mathrm{id}^2/b$ where $bX = \{((\alpha,g),(\alpha',g')) \mid \alpha,\alpha'\in 2,\ g,g'\in X^2,\ g(\alpha) = g'(\alpha')\}$ for all sets X are isomorphic but have different degrees. The canonical power series of a non-trivial polynomial is never a polynomial. However, the reduced canonical power series of a polynomial always is, and its degree is less than or equal to that of the given polynomial.

A set functor Φ is said to have rank n if it is isomorphic to a polynomial of set functors with degree n and there is no polynomial with smaller degree isomorphic to Φ. There are set functors without rank, for instance the power set functor P and the filter functor $\mathcal{F}$. The rank of a non-trivial set functor is the smallest cardinal number n such that with respect to the canonical power series for each set X card$(\sum_{k\leq m} a_k \mathrm{id}^k/b_m)X$

is constant for all $m \geq n$. A set functor has rank n if and only if its reduced canonical power series is a polynomial of degree n.

6. Extended algebraic structures

For each set functor Φ let ΦCON denote the category of all Φ-convergence spaces, that is, of all pairs (X,s) consisting of a set X and a subset s of $\Phi X \times X$. s is called a Φ-convergence structure of X. Morphisms of ΦCON are the Φ-continuous maps $f: (X,s) \to (Y,t)$ between Φ-convergence spaces, which are triples $(f,(X,s),(Y,t))$ with (X,s) and (Y,t) Φ-convergence spaces and f a map of X into Y such that $(\mathfrak{M},x) \in s$ implies $(\Phi f(\mathfrak{M}),f(x)) \in t$. By ΦCON_{map} and ΦCON_{parmap} we denote the full subcategory of ΦCON of all Φ-convergence spaces (X,s) with s a map or a partial map, respectively. Of course, s being a map or a partial map means that s is a map with ΦX or a subset of ΦX as domain, respectively.

In what follows let Φ_n be a polynomial $\sum_{k\leq n} a_k id^k/b_n$ of set functors, let $\Psi_n = \sum_{k\leq n} a_k id^k$ and let $\Omega_n = \sum_{k\leq n} a_k$ $(:= \bigcup_{k\leq n} \{k\} \times a_k)$. Each Ψ_n-convergence structure s of a set X is called an extended Ω_n-algebraic structure consisting of the extended algebraic operations $s_{k\alpha} \subseteq X^k \times X$, $(k,\alpha) \in \Omega_n$, defined by

$$(g,x) \in s_{k\alpha} \Leftrightarrow ((k,\alpha,g),x) \in s.$$

A Ψ_n-convergence structure s is a map if and only if all relations $s_{k\alpha}$ are maps of X^k into X (algebraic operations). Then s is an Ω_n-algebraic structure. Analogously we have that s is a partial map if and only if all relations $s_{k\alpha}$ are partial maps between X^k and X (partial algebraic operations). Then s is a partial Ω_n-algebraic structure.

Each Φ_n-convergence structure of a set X can be identified with a Ψ_n-convergence structure fulfilling some equations, as will be pointed out in the sequel. The notion of equation we need is an extension of that of algebra. Here we only consider some special equations. A more detailed treatment on equations will be given in section 9.

For each pair $(k,\alpha) \in \Omega_n$ and each map $f: k \to l$, $l \in \mathbb{K}$, we define a functor $G_{k\alpha f}: \Psi_n CON \to id^l CON$ as follows: For every Ψ_n-convergence space (X,s), $G_{k\alpha f}(X,s) = (X, \{(g,x) \in X^l \times X \mid ((k,\alpha,g \circ f),x) \in s\}$ and for every Ψ_n-continuous map $g: (X,s) \to (Y,t)$, $G_{k\alpha f}(g,(X,s),(Y,t)) = (g, G_{k\alpha f}(X,s), G_{k\alpha f}(Y,t))$. Let

(6.1) $E_n = \{\{G_{k\alpha f}, G_{k'\alpha' f'}\} \mid (k,\alpha),(k',\alpha') \in \Omega_n,\ f: k \to k + k',$
$f': k' \to k + k'$ and $((k,\alpha,f),(k',\alpha',f')) \in b_n(k + k')\}$.

A Ψ_n-convergence structure s of a set X and also (X,s) is said to fulfill the equation $\{G_{k\alpha f}, G_{k'\alpha' f'}\}$ if $G_{k\alpha f}(X,s) = G_{k'\alpha' f'}(X,s)$. If s fulfills all equations of E_n then s is called an extended (Ω_n,E_n)-algebraic structure. If additionally s is a map or a partial map, we speak of an (Ω_n,E_n)-algebraic structure or a partial (Ω_n,E_n)-algebraic structure, respectively. For defining (Ω_n,E_n)-algebraic structures the functors $G_{k\alpha f}$ can be replaced by their restrictions $G_{k\alpha f}': \Psi_n CON_{map} \to id^l CON_{map}$. Analogously in the case of partial (Ω_n,E_n)-algebraic structures the restrictions $G_{k\alpha f}''$: $\Psi_n CON_{parmap} \to id^1 CON_{parmap}$ can be used. X equipped with an extended (Ω_n,E_n)-algebraic structure of X is an extended (Ω_n,E_n)-algebra. Let

$$EXTALG_{\Omega_n,E_n}$$

denote the full subcategory of $\Psi_n CON$ of all extended (Ω_n,E_n)-algebras. Analogously the categories

$$ALG_{\Omega_n,E_n} \text{ and } PARALG_{\Omega_n,E_n}$$

of all (Ω_n,E_n)-algebras and all partial (Ω_n,E_n)-algebras are defined as subcategories of $\Psi_n CON$.

Proposition 6.1. For each Φ_n-convergence structure s of a set X,

$$s' = \{((k,\alpha,g),x) \mid ((k,\alpha,g)^{\sim b_n},x) \in s\}$$

is an extended (Ω_n,E_n)-algebraic structure of X and $(X,s) \mapsto (X,s')$, $(g,(X,s),(Y,t)) \mapsto (g,(X,s'),(Y,t'))$ is an isomorphism $\Phi_n CON \cong EXTALG_{\Omega_n,E_n}$. Restricting this isomorphism to $\Phi_n CON_{map}$ and $\Phi_n CON_{parmap}$ gives isomorphisms $\Phi_n CON_{map} \cong ALG_{\Omega_n,E_n}$ and $\Phi_n CON_{parmap} \cong PARALG_{\Omega_n,E_n}$.

7. Convergence structures as limits of extended algebraic structures

In this section some main results of previous sections are applied to establishing relations between convergence structures and algebraic ones. Let X be a set. By

$$CON_X$$

we mean the category of all convergence structures of X defined as follows: Objects of CON_X are all pairs (Φ,s) consisting of a set functor and a Φ-convergence structure s of X. Morphisms g: $(\Phi,s) \to (\Psi,t)$ are the triples $(g,(\Phi,s),(\Psi,t))$ with $(\Phi,s),(\Psi,t) \in obCON_X$ and g a map of ΦX into ΨX for which $(\mathfrak{m},x) \in s$ implies $(g(\mathfrak{m}),x) \in t$.
A functor S: CARD $\to CON_X$ is called a (cardinal) sequence of convergence structures of X provided for all $n,m \in \mathbb{K}$ with $n \le m$, $S(n,m)$ is an injection. If a sequence $S = ((\Phi_n,s_n)_{n\in\mathbb{K}},(\nu_{nm})_{n\le m})$ of convergence structures of X has a direct limit $((\Phi,s),(\nu_n)_{n\in\mathbb{K}})$ we say that S converges to (Φ,s) with respect to $(\nu_n)_{n\in\mathbb{K}}$.

Proposition 7.1. Let $T = ((\Phi_n)_{n\in\mathbb{K}},(\rho_{nm})_{n\le m})$ be a sequence of set functors converging to $(\Phi,(\rho_n)_{n\in\mathbb{K}})$ and let (X,s) be a Φ-convergence space. Then $S = ((\Phi_n,s_n)_{n\in\mathbb{K}},(\rho_{nm}X)_{n\le m})$ with

$$(7.1) \qquad s_n = \{(\mathfrak{m},x) \mid (\rho_n X(\mathfrak{m}),x) \in s\}$$

is a sequence of convergence structures of X converging to (Φ,s) with respect to $(\rho_n X)_{n\in\mathbb{K}}$.

If especially $(\Phi_n)_{n\in\mathbb{K}}$ is a power series of set functors converging to $(\Phi,(\rho_n)_{n\in\mathbb{K}})$ and hence $(\rho_{nm})_{n\le m}$ is uniquely determined then we also speak of convergence of the sequence of Φ_n-convergence structures s_n $(n\in\mathbb{K})$ to the Φ-convergence structure s with respect to $(\rho_n X)_{n\in\mathbb{K}}$.

In the sequel let Φ be an arbitrary set functor and let (X,s) be an arbitrary Φ-convergence space. By means of Theorem 4.3 it follows that there exists at least one power series $(\sum_{k\leq n} a_k id^k/b_n)_{n\in\mathbb{K}}$ converging to Φ with respect to a family $(\Gamma_n)_{n\in\mathbb{K}}$ of natural transformations. Let one of these power series and a corresponding family $(\Gamma_n)_{n\in\mathbb{K}}$ be fixed. $(\sum_{k\leq n} a_k id^k/b_n)_{n\in\mathbb{K}}$ could be the canonical power series or the reduced canonical power series of Φ. For each $n\in\mathbb{K}$ let $\Phi_n = \sum_{k\leq n} a_k id^k/b_n$ and $\Omega_n = \sum_{k\leq n} a_k$ and let E_n be defined by (6.1) and s_n by (7.1). According to Proposition 6.1,

$$s_n' = \{((k,\alpha,g),x) \mid ((k,\alpha,g)^{\sim b_n},x) \in s_n\}$$

is an extended (Ω_n,E_n)-algebraic structure, which we identify with the Φ_n-convergence structure s_n. Taking into account Proposition 7.1, we get the following result which says that each Φ-convergence structure is a "limit" of a cardinal sequence of extended algebraic structures.

<u>Theorem 7.2.</u> The sequence of extended (Ω_n,E_n)-algebraic structures s_n' $(n\in\mathbb{K})$ converges to the Φ-convergence structure s with respect to $(\Gamma_n X)_{n\in\mathbb{K}}$.

Of course, if s is a map or a partial map, then each s_n' is an (Ω_n,E_n)-algebraic structure or a partial (Ω_n,E_n)-algebraic structure, respectively.

From Theorem 7.2 follows for instance, that to each topology and more general, to each filter convergence structure there converges a cardinal sequence of extended algebraic structures.

8. On a modification

The results of sections 6 and 7 have partially an interesting modification as will be shown in the following.

For each set functor Φ let ΦSP denote the category of all <u>Φ-spaces</u>, that is, of all pairs (X,s) consisting of a set X and a subset s of ΦX. s is called a <u>Φ-structure</u> of X. Morphisms $f\colon (X,s) \to (Y,t)$ of ΦSP are the triples $(f,(X,s),(Y,t))$ where (X,s) and (Y,t) are Φ-spaces and f is a map of X into Y for which $\Phi f[s] \subseteq t$.

Let Φ_n be a polynomial $\sum_{k\leq n} a_k id^k/b_n$ of set functors and let $\Psi_n = \sum_{k\leq n} a_k id^k$ and $\Omega_n = \sum_{k\leq n} a_k$. Each Ψ_n-structure s of a set X is also called an <u>Ω_n-structure</u>. For each pair $(k,\alpha)\in\Omega_n$ and each map $f\colon k \to l$, $l\in\mathbb{K}$, a functor $G_{k\alpha f}\colon \Psi_n SP \to id^l SP$ is defined letting $G_{k\alpha f}(X,s) = (X,\{g\in X^l \mid (k,\alpha,g\circ f)\in s\})$ for each Ψ_n-space (X,s) and $G_{k\alpha f}(g,(X,s),(Y,t)) = (g,G_{k\alpha f}(X,s), G_{k\alpha f}(Y,t))$ for each Ψ_nSP-morphism $g\colon (X,s) \to (Y,t)$. Let a collection E_n of equations be given analogously to (6.1). In an obvious manner <u>(Ω_n,E_n)-structures</u> are introduced as Ω_n-structures <u>fulfilling all equations</u> of E_n. A set X equipped with an (Ω_n,E_n)-structure of X is an <u>(Ω_n,E_n)-space</u>. Analogously as the first isomorphism in Proposition 6.1 an isomorphism

$$\Phi_n SP \cong SP_{\Omega_n,E_n}$$

between Φ_nSP and the full subcategory of Ψ_nSP of all (Ω_n,E_n)-spaces can be established.

Let X be a set. By

$$STR_X$$

we denote the category of all functor structures of X. STR_X has as objects all pairs (Φ,s) consisting of a set functor Φ and a Φ-structure s of X and as morphisms $g: (\Phi,s) \to (\Psi,t)$ all triples $(g,(\Phi,s),(\Psi,t))$ where $(\Phi,s),(\Psi,t) \in obSTR_X$ and g is a map of ΦX into ΨX with $g[s] \subseteq t$. A functor $S: CARD \to STR_X$ is called a (cardinal) sequence of functor structures of X provided for all $n,m \in K$ with $n \leq m$, $S(n,m)$ is an injection. In an obvious manner we can modify Proposition 7.1 to a statement for functor structures where the notion of convergence of a sequence of functor structures is defined analogously to that of convergence of a sequence of convergence structures.

In the sequel let Φ be an arbitrary set functor and let $(\sum_{k \leq n} a_k id^k/b_n)_{n \in K}$ be a power series converging to Φ with respect to a family $(\mu_n)_{n \in K}$ of natural transformations. Moreover, let (X,s) be a Φ-space. For each $n \in K$ let $\Phi_n = \sum_{k \leq n} a_k id^k/b_n$ and $\Omega_n = \sum_{k \leq n} a_k$ and let E_n be defined analogously to (6.1). Each Φ_n-structure $s_n = \{m \mid \mu_n X(m) \in s\}$ can be identified with the (Ω_n,E_n)-structure $s_n' = \{(k,\alpha,g) \mid (k,\alpha,g)^{\sim b_n} \in s_n\}$. Analogously to Theorem 7.2 we have the following result.

Theorem 8.1. The sequence of (Ω_n,E_n)-structures s_n' $(n \in K)$ converges to the Φ-structure s with respect to $(\mu_n X)_{n \in K}$.

For instance, to each uniform structure and more general, to each uniform filter convergence structure (which is an $\mathcal{F} id^2$-structure) a cardinal sequence of (Ω_n,E_n)-structures for some Ω_n and E_n $(n \in K)$ converges.

9. Equations

By a structure kind over SET we mean a concrete transportable category over SET. Of course, a concrete category over SET is a pair $(\mathcal{C},F)$ consisting of a category $\mathcal{C}$ and a faithful functor $F: \mathcal{C} \to SET$, the corresponding forgetful functor. Transportable means, that for each object A of $\mathcal{C}$ and each isomorphism $f: FA \to Y$ in SET there is precisely one isomorphism $g: A \to B$ in $\mathcal{C}$ with $Fg = f$. Frequently, for sake of simplicity we only write $\mathcal{C}$ for a concrete category $(\mathcal{C},F)$ over SET. Let

STRUCT

denote the quasicategory of all structure kinds over SET. Morphisms of STRUCT are the concrete functors $G: \mathcal{C} \to \mathcal{C}'$ between structure kinds over SET where here concrete means that $GF' = F$ holds for the forgetful functors F of $\mathcal{C}$ and F' of $\mathcal{C}'$.

Let $\mathcal{C}$ be a structure kind over SET with forgetful functor F. A subkind of $\mathcal{C}$ is a concrete category consisting of a full isomorphism-closed subcategory $\mathcal{K}$ of $\mathcal{C}$ and the restriction $F|\mathcal{K}$ of F to $\mathcal{K}$. Of course, $(\mathcal{K},F|\mathcal{K})$ is a structure kind over SET. We consider each subkind $\mathcal{K}$ of a structure kind $\mathcal{C}$ defining an axiom for the objects of $\mathcal{C}$ which is fulfilled for an object of $\mathcal{C}$ if and only if it is an object of $\mathcal{K}$. There is the important case of axiom determined by an equation. Let $\mathcal{C}$ be a structure kind over

SET. A $\mathcal{C}$-equation is an unordered pair $\{G,H\}$ of concrete functors G,H: $\mathcal{C} \to \mathcal{C}'$ where $\mathcal{C}'$ is any structure kind over SET. An object A of $\mathcal{C}$ is said to fulfill the equation $\{G,H\}$ if $GA = HA$. Let $\mathcal{C}_{G,H}$ denote the full subcategory of $\mathcal{C}$ of all objects of $\mathcal{C}$ fulfilling the equation $\{G,H\}$. $\mathcal{C}_{G,H}$ (together with $F|\mathcal{C}_{G,H}$) is the subkind of $\mathcal{C}$ determined by the equation $\{G,H\}$.

By the following proposition equationally determined subkinds can be characterized as equalizers.

Proposition 9.1. Let $\mathcal{C}$ be a structure kind over SET and $\{G,H\}$ a $\mathcal{C}$-equation. Then $\mathcal{C}_{G,H}$ together with the identical embedding of $\mathcal{C}_{G,H}$ into $\mathcal{C}$ is an equalizer of G,H in STRUCT (and also in CAT).

There is the following interior characterization of equationally determined subkinds suggested by J. Adámek.

Theorem 9.2. Let $\mathcal{C}$ be a structure kind over Set. The equationally determined subkinds of $\mathcal{C}$ are given precisely by those full subcategories of $\mathcal{C}$ which are closed under retracts [1].

Proposition 9.3. Let $\mathcal{C}$, $\mathcal{k}$ and $\mathcal{k}'$ be structure kinds over SET. If $\mathcal{k}$ is an equationally determined subkind of $\mathcal{C}$ and $\mathcal{k}'$ is an equationally determined subkind of $\mathcal{k}$ then $\mathcal{k}'$ is an equationally determined subkind of $\mathcal{C}$.

For the following let $\mathcal{C}$ be a structure kind over SET and let $E = \{\{G_i,H_i: \mathcal{C} \to \mathcal{C}_i\} \mid i \in I\}$ be a collection of $\mathcal{C}$-equations indexed by a set I. By $\mathcal{C}'$ we denote the product of the concrete categories $\mathcal{C}_i$ $(i \in I)$ defined as follows: Objects of the category $\mathcal{C}'$ are the families $(A_i)_{i\in I}$ of objects A_i of $\mathcal{C}_i$ with F_iA_i independent of i where F_i is the forgetful functor of $\mathcal{C}_i$. Morphisms $(f_i)_{i\in I}: (A_i)_{i\in I} \to (B_i)_{i\in I}$ of $\mathcal{C}'$ are the families of $\mathcal{C}_i$-morphisms $f_i: A_i \to B_i$ with F_if_i independent of i. The forgetful functor F' of $\mathcal{C}'$ is given by $F'(A_i)_{i\in I} = F_iA_i$ and $F'(f_i)_{i\in I} = F_if_i$. $\mathcal{C}'$ is a structure kind over SET. Let $G: \mathcal{C} \to \mathcal{C}'$ be the concrete functor such that $GA = (G_iA)_{i\in I}$ for each object A of $\mathcal{C}$ and $Gf = (Gf_i)_{i\in I}$ for each $\mathcal{C}$-morphism $f: A \to B$. Let H be defined analogously by the functors H_i $(i \in I)$. Then we have the following result which says that it is no generalization to determine subkinds of $\mathcal{C}$ by a collection of $\mathcal{C}$-equations indexed by a set instead of one equation.

Proposition 9.4. $\mathcal{C}_{G,H}$ is the full subcategory of $\mathcal{C}$ of all objects of $\mathcal{C}$ which fulfill all equations of E.

The last part of this section is devoted to a special case in which both concrete functors appearing in an equation can be identified with natural subobjects where the notion of natural subobject is a generalization of that of a natural transformation between set-valued functors. Let $S: \mathcal{C} \to$ SET be a functor. By a natural subobject of S we mean a family

1) A subcategory $\mathcal{k}$ of $\mathcal{C}$ is closed under retracts if each object B of $\mathcal{C}$ which is a retract of an object A of $\mathcal{k}$ is an object of $\mathcal{k}$, too. To be a retract means that there are $\mathcal{C}$-morphisms $e: B \to A$ and $m: A \to B$ with $me = 1_B$.

$\mu = (\mu A)_{A \in ob\mathcal{C}}$ of subsets μA of SA such that

(9.1) $\quad Sf[\mu A] \subseteq \mu B \quad$ for each $\mathcal{C}$-morphism $f: A \to B$.

If in particular S is a functor pair $\langle S_1, S_2 \rangle: \mathcal{C} \to$ SET [1] and for each object A of $\mathcal{C}$ μA is a map of S_1A into S_2A then (9.1) means that μ is a natural transformation of S_1 to S_2.

In the sequel let Φ and Ψ be set functors, let $\mathcal{C}$ be a full isomorphism-closed subcategory of ΦSP and let $F: \mathcal{C} \to$ SET be the restriction of the forgetful functor of ΦSP to $\mathcal{C}$. $\mathcal{C}$ together with F is a structure kind over SET. We are interested in natural subobjects μ of $S = \Psi F: \mathcal{C} \to$ SET. Each natural subobject μ of ΨF can be assigned the concrete functor G_μ: $\mathcal{C} \to \Psi$SP for which $G_\mu(X,S) = (X, \mu(X,s))$ for each object (X,s) of $\mathcal{C}$ and $G_\mu(f,(X,s),(Y,t)) = (f, G_\mu(X,s), G_\mu(Y,t))$ for each $\mathcal{C}$-morphism $f: (X,s) \to (Y,t)$.

Proposition 9.5. $\mu \mapsto G_\mu$ is a bijection between the collection of all natural subobjects of ΨF and the collection of all concrete functors of $\mathcal{C}$ to ΨSP.

An equation $\{G,H\}$ consisting of concrete functors $G,H: \mathcal{C} \to \Psi$SP can also be characterized by the unordered pair $\{\mu, \nu\}$ of the corresponding natural subobjects μ and ν of ΨF, for which $G_\mu = G$ and $G_\nu = H$. Indeed, for each object A of $\mathcal{C}$, GA = HA iff $\mu A = \nu A$. If Ψ is a product functor $\Psi' \times$ id and G and H are ranging in the subcategory $\Psi'CON_{map}$ of ΨSP then μ and ν are natural transformations of $\Psi'F$ to F.

Let $E = \{\{G_i, H_i: \mathcal{C} \to \Psi_i SP\} \mid i \in I\}$ be a collection of equations indexed by a set I where the corresponding concrete functors are ranging in categories Ψ_iSP with $\Psi = \sum_{i \in I} \Psi_i$ [2]. The product $\mathcal{C}'$ of the concrete categories Ψ_iSP is concretely isomorphic to ΨSP. T: $\mathcal{C}' \to \Psi$SP given by $T((X, s_i)_{i \in I}) = (X, \bigcup_{i \in I} \{i\} \times s_i)$ and $T((f_i)_{i \in I}, (X, s_i)_{i \in I}, (Y, t_i)_{i \in I}) = ((f_i)_{i \in I}, T((X, s_i)_{i \in I}, T((Y, t_i)_{i \in I}))$ is a concrete isomorphism. Hence, taking into account Proposition 9.4 we get that instead of the collection E we can use the equation $\{G,H\}$ of the concrete functors $G,H: \mathcal{C} \to \Psi$SP which are defined as compositions of the corresponding functors of Proposition 9.4 with T.

In the following we give some examples of equatinally determined subkinds of structure kinds.

(1) Let Φ' be a set functor and let $\Phi = \Phi' \times$ id. $\Phi'CON_{parmap}$ and $\Phi'CON_{map}$ are subkinds of $\mathcal{C} = \Phi'CON$ which can be determined by an equation of concrete functors $G,H: \mathcal{C} \to \Psi$SP with $\Psi = id^2$ and $id^2 + \Phi'$, respectively. For any monad $\boldsymbol{\Phi} = (\Phi', \eta, \rho)$ determined with Φ' the category of all $\boldsymbol{\Phi}$-algebras (see [7] or [8]) is a subkind of $\mathcal{C} = \Phi'CON$ which can be determined by an equation of concrete functors $G,H: \mathcal{C} \to (id^2 + 3\Phi')$SP.

1) S_1 and S_2 are functors of $\mathcal{C}$ to SET and for for each $A \in ob\mathcal{C}$, $\langle S_1, S_2 \rangle A = S_1A \times S_2A$ and for each $\mathcal{C}$-morphism $f: A \to B$, $\langle S_1, S_2 \rangle f: (x,y) \mapsto (S_1f(x), S_2f(y))$ $(x \in S_1A,\ y \in S_2A)$.

2) For each set X, $\Psi X = \bigcup_{i \in I} \{i\} \times \Psi_i X$ and for each map $f: X \to Y$ between sets, $\Psi f: (i, m) \mapsto (i, \Psi_i f(m))$ $(i \in I,\ m \in \Psi_i X)$.

(2) Let $\Phi = \mathcal{F} \times id$. By PTOP, $PTOP_{sep}$ and $PTOP_{sep\ comp}$ we denote the category of all, of all separated and of all separated and compact pseudo-topological spaces, respectively. Moreover, by TOP, TOP_{sep} and $TOP_{sep\ comp}$ we denote the corresponding categories of topological spaces. These categories are subkinds of $\mathcal{C} = \mathcal{F}CON$ which can be determined by an equation of concrete functors $G,H: \mathcal{C} \to \Psi SP$ where respectively $\Psi = 3(\mathcal{F} \times id)$, $id^2 + 3(\mathcal{F} \times id)$, $id^2 + \mathcal{F} + 3(\mathcal{F} \times id)$, $4(\mathcal{F} \times id)$, $id^2 + 4(\mathcal{F} \times id)$ and $id^2 + \mathcal{F} + 4(\mathcal{F} \times id)$.

(3) Let $\Phi = \mathcal{F}id^2$. The categories PUN and UN of all pseudo-uniform and all uniform spaces are subkinds of $\mathcal{C} = \mathcal{F}id^2SP$ which can be determined by an equation of concrete functors $G,H: \mathcal{C} \to \Psi SP$ with $\Psi = 5\mathcal{F}id^2$ and $6\mathcal{F}id^2$, respectively.

10. Extension to the case of arbitrary base category

It is possible to extend our theory using set-valued functors instead of set functors. Here we will make some remarks how to do it. What we need at first is an analogous notion to that of CARD with respect to an arbitrary base category.

Let $\mathcal{D}$ be a category which has hom-sets, that is for all $X,Y \in ob\mathcal{D}$, $[X,Y]_{\mathcal{D}}$ is a set. By a cardinal choice of $\mathcal{D}$ we mean a category $CARD_{\mathcal{D}}$ with the following properties:

(1) $obCARD_{\mathcal{D}} \subseteq ob\mathcal{D}$ and for each $X \in ob\mathcal{D}$ there is precisely one object $card_{\mathcal{D}}X$ of $CARD_{\mathcal{D}}$ $\mathcal{D}$-isomorphic to X.

(2) For all $n,m \in obCARD_{\mathcal{D}}$, $[n,m]_{CARD_{\mathcal{D}}}$ consists of (n,m) only, or is empty according to wether there exists a $\mathcal{D}$-monomorphism of n to m or not.

(3) $obCARD_{\mathcal{D}}$ is directed by the preorder $\preccurlyeq$ defined by

$$n \preccurlyeq m \iff \text{there is a } \mathcal{D}\text{-monomorphism of } n \text{ to } m.$$

(4) For each $n \in obCARD_{\mathcal{D}}$, $\{k \in obCARD_{\mathcal{D}} \mid k \preccurlyeq n\}$ is a set.

$obCARD_{\mathcal{D}}$ is also denoted by $\mathbb{K}_{\mathcal{D}}$. The elements of $\mathbb{K}_{\mathcal{D}}$ are called $\mathcal{D}$-cardinal numbers with respect to $CARD_{\mathcal{D}}$. The map $X \mapsto card_{\mathcal{D}}X$ ($X \in ob\mathcal{D}$) is the corresponding cardinality map. For this map we have $card_{\mathcal{D}} \circ card_{\mathcal{D}} = card_{\mathcal{D}}$ and

$card_{\mathcal{D}}X \preccurlyeq card_{\mathcal{D}}Y \iff$ there is a $\mathcal{D}$-monomorphism of X to Y.

Two cardinal choices $CARD_{\mathcal{D}}$ and $CARD'_{\mathcal{D}}$ are isomorphic. An isomorphism is given by $n \mapsto card'_{\mathcal{D}}n$ ($n \in obCARD_{\mathcal{D}}$) where $card'_{\mathcal{D}}$ is the cardinality map of $CARD'_{\mathcal{D}}$.

In the following we give some examples of cardinal choices.

(1) For $\mathcal{D}$ = SET, CARD is a cardinal choice.

(2) Let $\mathcal{D}$ be a directed class with initial sets $\{k \mid k \preccurlyeq n\}$ made to a category as usual. Then $\mathcal{D} = CARD_{\mathcal{D}}$.

(3) Let I be a set and let $\mathcal{D} = SET^I$. Objects of SET^I are the families $(X_i)_{i \in I}$ of sets indexed by I and morphisms $f: (X_i)_{i \in I} \to (Y_i)_{i \in I}$ are the families $(f_i)_{i \in I}$ of maps $f_i: X_i \to Y_i$. Here a cardinal choice is given by taking as $\mathcal{D}$-cardinal numbers all families $(n_i)_{i \in I}$ of usual cardinal numbers. In this case $\preccurlyeq$ is a partial order.

(4) Let $\mathcal{D}$ = TOP. Then a cardinal choice is given by taking as $\mathcal{D}$-cardinal numbers all topological spaces $(n,\mathcal{O})$ where n is a usual cardinal number and $\mathcal{O}$ is always one among isomorphic topologies on n. In this case $\preccurlyeq$ fails to be a partial order.

In the following we assume that a category $\mathcal{D}$ with hom-sets and a cardinal choice $CARD_{\mathcal{D}}$ of $\mathcal{D}$ are given.
Under some assumptions we are able to develop an arithmetic of $\mathcal{D}$-cardinal numbers using the following definitions. Let $(n_i)_{i\in I}$ be a family of $\mathcal{D}$-cardinal numbers with I a set. If this family has a product X, $f_i: X \to n_i$ $(i \in I)$ in $\mathcal{D}$ then $card_{\mathcal{D}} X$ is the product $\prod_{i\in I} n_i$ of the $\mathcal{D}$-cardinal numbers n_i. If this family has a coproduct Y, $g_i: n_i \to Y$ $(i\in I)$ then $card_{\mathcal{D}} Y$ is the sum $\sum_{i\in I} n_i$ of the $\mathcal{D}$-cardinal numbers n_i. Both notions do not depend on a specification of products and coproducts in $\mathcal{D}$. Let n and m be $\mathcal{D}$-cardinal numbers. If (n,m) has an exponentiation Z, $v: Z\times m \to n$ [1] then $card_{\mathcal{D}} Z$ is defined to be n^m. Of course, also n^m does not depend on a specification of exponentiations.

The notion of sequence of set functors can be extended as follows: By a sequence of set-valued functors with base category $\mathcal{D}$ we mean a functor $T: CARD_{\mathcal{D}} \to FUNC$ with the following properties:

(1) For each $n \in \mathbb{K}_{\mathcal{D}}$, Tn is a functor $\Phi_n: \mathcal{D} \to SET$.

(2) For all $n,m \in \mathbb{K}_{\mathcal{D}}$ with $n \preccurlyeq m$, T(n,m) is a natural transformation $\Gamma_{nm} = (\Gamma_{nm}X)_{X\in ob\mathcal{D}}: \Phi_n \to \Phi_m$ such that each map $\Gamma_{nm}X: \Phi_n X \to \Phi_m X$ is injective.

Convergence of sequences of set-valued functors is defined as in case of set functors.
Proposition 4.1 and the first part of Proposition 4.2 can be extended taking as T a sequence of set-valued functors with base category $\mathcal{D}$ and as X objects of $\mathcal{D}$. card is not to change. The second part of Proposition 4.2 can be extended also in an obvious manner, however, only if $\mathbb{K}_{\mathcal{D}}$ is a proper class.
The notion of power series of set-valued functors with base category $\mathcal{D}$

$$\left(\sum_{k\preccurlyeq n} a_k id_{\mathcal{D}}^k / b_n\right)_{n\in\mathbb{K}_{\mathcal{D}}}$$

is defined as in case of set functors where the following holds:

(1) For each $n \in \mathbb{K}_{\mathcal{D}}$, a_n is a set, $id_{\mathcal{D}}^n$ is the hom-functor $[n,\cdot]_{\mathcal{D}}$ and b_n is an equivalence functor of $\sum_{k\preccurlyeq n} a_k id_{\mathcal{D}}^k: \mathcal{D} \to SET$ [2].

(2) For all $n,m \in \mathbb{K}_{\mathcal{D}}$ with $n \preccurlyeq m$ and each $X \in ob\mathcal{D}$

$$b_n X = b_m X \cap \left(\sum_{k\preccurlyeq n}(a_k \times [k,X]_{\mathcal{D}}) \times \sum_{k\preccurlyeq n}(a_k \times [k,X]_{\mathcal{D}})\right).$$

Also the notion of canonical power series of a set-valued functor Φ converging to Φ and the notion of a polynomial of set-valued functors

---

1) We mean that Z is an object of $\mathcal{D}$ and v is a $\mathcal{D}$-morphism of a product of Z and m to n and for each morphism f of a product $A\times m$ to n, $A \in ob\mathcal{D}$, there is precisely one morphism $\bar{f}: A \to Z$ with $v(\bar{f}\times 1_m) = f$.

2) Equivalence functors are defined as in case of set functors. For each $X \in ob\mathcal{D}$, $(\sum_{k\preccurlyeq n} a_k id_{\mathcal{D}}^k / b_n)X = \sum_{k\preccurlyeq n}(a_k \times [k,X]_{\mathcal{D}})$ and for each $\mathcal{D}$-morphism $f: X \to Y$, $(\sum_{k\preccurlyeq n} a_k id_{\mathcal{D}}^k / b_n)f: (k,\alpha,g) \mapsto (k,\alpha,fg)$.

can be defined as in case of set functors. However, since $\mathbb{K}_{\mathcal{D}}$ is not well-ordered by $\preccurlyeq$ in general, there need not be a reduced canonical power series of a set-valued functor Φ converging to it. If a set-valued functor Φ is isomorphic to a polynomial $\sum_{k \preccurlyeq n} a_k \mathrm{id}_{\mathcal{D}}^k / b_n$ then we say that Φ has rank. But there need not be a minimal n with this property. Hence, if Φ has rank we cannot fix a $\mathcal{D}$-cardinal number as rank in general.

In case of $\mathcal{D} = \mathrm{SET}^I$ each hom-functor $\mathrm{id}_{\mathcal{D}}^n$, $n = (n_i)_{i \in I}$, coincides with the product functor $\prod_{i \in I} p_i^{n_i}$ of the compositions $p_i^{n_i} = \mathrm{id}^{n_i} p_i$ where p_i is the ith projection functor of $\mathcal{D}$ to SET, for which $p_i X = X_i$ and $p_i f = f_i$. Hence, in this case power series have "several variables":

$$\Big(\sum_{k=(k_i)_{i \in I} \preccurlyeq n} a_k \prod_{i \in I} p_i^{k_i} / b_n\Big)_{n \in \mathbb{K}_{\mathcal{D}}}.$$

For each set-valued functor $\Phi: \mathcal{D} \to \mathrm{SET}$ the category ΦSP of all Φ-spaces (X,s) $(X \in \mathrm{ob}\mathcal{D},\ s \subseteq \Phi X)$ is defined analogously as in case of set functors. If Φ is a sum functor $\sum_{j \in J} (\Phi_j \times \mathrm{id}_{\mathcal{D}}^j)$ 1) with J a set of $\mathcal{D}$-cardinal numbers then ΦSP is also denoted by

$$\sum_{j \in J} (\Phi_j, \mathrm{id}_{\mathcal{D}}^j)\mathrm{CON}$$

and we speak of $\sum_{j \in J} (\Phi_j, \mathrm{id}_{\mathcal{D}}^j)$-convergence structures and $\sum_{j \in J} (\Phi_j, \mathrm{id}_{\mathcal{D}}^j)$-convergence spaces instead of Φ-structures and Φ-spaces, respectively.

In the sequel we assume that $\mathcal{D}$ has coproducts of pairs of objects and therefore the sum n + n' of $\mathcal{D}$-cardinal numbers n and n' is defined as above. Let a set J of $\mathcal{D}$-cardinal numbers and a $\mathcal{D}$-cardinal number n be fixed. For each $j \in J$ let $\Phi_{jn} = \sum_{k \preccurlyeq n} a_{jk} \mathrm{id}_{\mathcal{D}}^k / b_{jn}: \mathcal{D} \to \mathrm{SET}$ be a polynomial and let $\Psi_{jn} = \sum_{k \preccurlyeq n} a_{jk} \mathrm{id}_{\mathcal{D}}^k$. Moreover, let $\Omega_n = \{ (j,k,\alpha) \mid j \in J,\ k \preccurlyeq n,\ \alpha \in a_{jk} \}$. Each $\sum_{j \in J} (\Psi_{jn}, \mathrm{id}_{\mathcal{D}}^j)$-convergence structure is called an extended (J,Ω_n)-algebraic structure. For each $(j',k,\alpha) \in \Omega_n$ and each $\mathcal{D}$-morphism $f: k \to l$, $l \in \mathbb{K}_{\mathcal{D}}$, we define a functor $G_{j'k\alpha f}: \sum_{j \in J} (\Psi_{jn}, \mathrm{id}_{\mathcal{D}}^j)\mathrm{CON} \to \sum_{j \in J} (\mathrm{id}_{\mathcal{D}}^l, \mathrm{id}_{\mathcal{D}}^j)\mathrm{CON}$ as follows: For each $\sum_{j \in J} (\Psi_{jn}, \mathrm{id}_{\mathcal{D}}^j)$-convergence space (X,s), $G_{j'k\alpha f}(X,s) = (X, \{(j',g,h) \mid g \in [l,X]_{\mathcal{D}},\ h \in [j,X]_{\mathcal{D}},\ (j,(k,\alpha,gf),h) \in s\})$ and for each $\sum_{j \in J} (\Psi_{jn}, \mathrm{id}_{\mathcal{D}}^j)$CON-morphism $g: (X,s) \to (Y,t)$, $G_{j'k\alpha f}(g,(X,s),(Y,t)) = (g, G_{j'k\alpha f}(X,s), G_{j'k\alpha f}(Y,t))$. Let

$$E_n = \{ \{ G_{jk\alpha f}, G_{jk'\alpha' f'} \} \mid (j,k,\alpha),(j,k',\alpha') \in \Omega_n,\ f: k \to k + k' \text{ and } f': k' \to k + k' \text{ are } \mathcal{D}\text{-morphisms},\ ((k,\alpha,f),(k',\alpha',f')) \in b_{nj}(k + k') \}.$$

A $\sum_{j \in J} (\Psi_{jn}, \mathrm{id}_{\mathcal{D}}^j)$-convergence structure s of an object X of $\mathcal{D}$ is called an extended (J,Ω_n,E_n)-algebraic structure provided s fufills all equations $\{G_{jk\alpha f}, G_{jk'\alpha' f'}\} \in E_n$, that is for these unordered pairs we have $G_{jk\alpha f}(X,s) = G_{jk'\alpha' f'}(X,s)$. Analogously as $\mathrm{ALG}_{\Omega_n,E_n}$, $\mathrm{PARALG}_{\Omega_n,E_n}$ and $\mathrm{EXTALG}_{\Omega_n,E_n}$ the categories $\mathrm{ALG}_{J,\Omega_n,E_n}$, $\mathrm{PARALG}_{J,\Omega_n,E_n}$ and $\mathrm{EXTALG}_{J,\Omega_n,E_n}$ are defined. Both last categories are also interesting in case of $\mathcal{D}$ = SET. In an obvious manner Proposition 6.1 can be extended to $\sum_{j \in J} (\Phi_{jn}, \mathrm{id}_{\mathcal{D}}^j)$-convergence structures. In particular we have that there is an isomorphism

$$\sum_{j \in J} (\Phi_{jn}, \mathrm{id}_{\mathcal{D}}^j)\mathrm{CON} \cong \mathrm{EXTALG}_{J,\Omega_n,E_n}.$$

---

1) This functor assigns the disjoint sum $\sum_{j \in J} (\Phi_j X \times [j,X]_{\mathcal{D}})$ to each object X of $\mathcal{D}$.

As is easily seen all notions and results of sections 7 and 8 can be extended to our general case. Of course, we assume that $\mathcal{D}$ has coproducts of pairs of objects. Instead of CON_X we can use the category

$$CON_{J,X}$$

where J is a set of $\mathcal{D}$-cardinal numbers and X is an object of $\mathcal{D}$. This category has as objects all pairs $((\Phi_j)_{j\in J}, s)$ consisting of a family of functors $\Phi_j: \mathcal{D} \to SET$ and a $\sum_{j\in J}(\Phi_j, id_{\mathcal{D}}^j)$-convergence structure s of X. Morphisms $g: ((\Phi_j)_{j\in J}, s) \to ((\Psi_j)_{j\in J}, t)$ are the triples $(g, ((\Phi_j)_{j\in J}, s), ((\Psi_j)_{j\in J}, t))$ with $((\Phi_j)_{j\in J}, s), ((\Psi_j)_{j\in J}, t) \in obCON_{J,X}$ and g a family of maps $g_j: \Phi_j X \to \Psi_j X$ with $(j, \mathfrak{M}, h) \in s \Rightarrow (j, g_j(\mathfrak{M}), h) \in t$.
The extension of Theorem 7.2 says that each $\sum_{j\in J}(\Phi_j, id_{\mathcal{D}}^j)$-convergence structure is a limit of a sequence of extended (J, Ω_n, E_n)-algebraic structures $(n \in K_{\mathcal{D}})$.
With exception of the last mentioned examples section 9 can be extended completely in an obvious manner to the case of arbitrary base category. For the notion of structure kind over an arbitrary base category see also [3] and [4].

References

[1] J. Adámek, Theory of Mathematical Structures, D. Reidel Publishing Company, Dordrecht Boston Lancaster 1983.

[2] W. Gähler, A topological approach to structure theory, I, Math. Nachr. 100 (1981), 93-144.

[3] - , General convergence and applications, Proc. Topology and Measure IV, Trassenheide (GDR) 1983.

[4] - , Structure problems in fuzzy set theory, systems theory and optimization, Proc. 4 th Joint Polish-GDR Symposium on "Nonconventional Optimization Problems", Warsaw 1984.

[5] H. Herrlich and G. E. Strecker, Category Theory, Heldermann Verlag, Berlin 1979.

[6] M. Kühnrich, Superclasses in a finite extension of Zermelo set theory, Zeitschr. f. math. Logik und Grundlagen d. Math. 24 (1978), 539-552.

[7] E. G. Manes, Algebraic Theories, Springer Verlag, New York Heidelberg Berlin 1976.

[8] G. Richter, Kategorielle Algebra, Akademie-Verlag Berlin 1979.

SEQUENTIAL CONVERGENCES ON ABELIAN LATTICE-ORDERED GROUPS

Matúš Harminc[1)]

In this paper the notion of convergence of sequences on abelian lattice-ordered groups is introduced. We investigate order properties of the set L_G of all convergences on a given abelian lattice-ordered group G (§4 and §5). The set L_G, equipped with the natural partial order, is shown to be isomorphic to a certain system of subsemigroups of the semigroup $(G^+)^N$, where G^+ is the positive cone of G (§2). A smallest convergence with a given subset of positive zero sequences is constructed (§3) and the construction is used to show that L_G need not be a lattice.

1. NOTATION AND DEFINITION

Let us recall the concept of an abelian lattice-ordered group (cf. [1, 2]):

Let $(G, +, \leq)$ be a triple such that

G is an abelian group with respect to $+$,

G is a lattice with respect to $\leq$ and

$a \leq b$ implies $a + g \leq b + g$ whenever a, b, g are elements of G.

The following denotation will be used:

N - the set of all natural numbers;

Z - the set of all integer numbers;

Q - the set of all rational numbers;

Irr - the set of all irrational numbers;

G - the underlying set of an abelian lattice-ordered group;

G^N - the set of sequences with elements belonging to G;

0 - the neutral element of a group;

G^+ - the set of all positive elements of G;

$\leq$ denotes the order on G or the induced pointwise order on G^N;

u, v denote monotone mappings of N into N;

$S \circ u$ denotes a subsequence of the sequence S;

$\langle \dot{g} \rangle$ denotes the constant sequence;

$S(n)$ denotes the n-th term of the sequence S;

$R + S$ denotes the sequence whose n-th term is $R(n) + S(n)$.

The sequences $R \wedge S$, $R \vee S$, $-S$, $|S|$ are defined analogously.

---

1) Matematický ústav SAV, Karpatská 5, 040 01 Košice, Czechoslovakia

In the following definition the axioms (i) - (vi) define a FLUSH-convergence structure for G, i. e., a convergence group (cf. P. MIKUSIŃSKI [3], J. NOVÁK [4]); the remaining three axioms concern relations between the convergence and the order on G.

DEFINITION 1. Let $(G, +, \leq)$ be an abelian lattice-ordered group. A set $\mathcal{L} \subset G^N \times G$ is said to be a convergence on $(G, +, \leq)$, if the following nine axioms are satisfied:

(i) $(S, s) \in \mathcal{L}$ implies $(S \circ u, s) \in \mathcal{L}$ for every subsequence $S \circ u$ of S.

(ii) If for each u there exists v such that $(S \circ u \circ v, s) \in \mathcal{L}$, then $(S, s) \in \mathcal{L}$.

(iii) $(\langle \dot{s} \rangle, s) \in \mathcal{L}$ for each $s \in G$.

(iv) $(S, a) \in \mathcal{L}$ and $(S, b) \in \mathcal{L}$ imply $a = b$.

(v) $(S, s) \in \mathcal{L}$ implies $(-S, -s) \in \mathcal{L}$.

(vi) $(R, r) \in \mathcal{L}$ and $(S, s) \in \mathcal{L}$ imply $(R + S, r + s) \in \mathcal{L}$.

(vii) $(R, r) \in \mathcal{L}$ and $(S, s) \in \mathcal{L}$ imply $(R \wedge S, r \wedge s) \in \mathcal{L}$.

(viii) $(R, r) \in \mathcal{L}$ and $(S, s) \in \mathcal{L}$ imply $(R \vee S, r \vee s) \in \mathcal{L}$.

(ix) If $(R, g) \in \mathcal{L}$ and $(T, g) \in \mathcal{L}$ and $R \leq S \leq T$, then $(S, g) \in \mathcal{L}$.

2. POSITIVE ZERO SEQUENCES

The system of all convergences on $(G, +, \leq)$ is denoted by L_G. Let $\mathcal{L}$ be an element of L_G. We denote by $P(\mathcal{L})$ the set of all $S \in (G^+)^N$ such that $(S, 0) \in \mathcal{L}$.

THEOREM 1. Let $\mathcal{L} \in L_G$. Then $P(\mathcal{L})$ is a convex semigroup of $(G^+)^N$ and the following three conditions are fulfilled:

(I) $S \in P(\mathcal{L})$ implies $S \circ u \in P(\mathcal{L})$ for every subsequence $S \circ u$ of S.

(II) Let $S \in (G^+)^N$. If for each u there exists v such that $S \circ u \circ v \in P(\mathcal{L})$, then $S \in P(\mathcal{L})$.

(III) $\langle \dot{s} \rangle \in P(\mathcal{L})$ if and only if $s = 0$.

Conversely, for every convex semigroup P_1 of $(G^+)^N$ fulfilling the conditions (I) - (III) there exists $\mathcal{L}_1 \in L_G$ such that $P(\mathcal{L}_1) = P_1$; if $\mathcal{L}_2 \in L_G$ and $P(\mathcal{L}_2) = P_1$, then $\mathcal{L}_2 = \mathcal{L}_1$.

A SKETCH OF THE PROOF. The axioms (i) - (ix) imply immediately that $P(\mathcal{L})$ is a convex semigroup of $(G^+)^N$ fulfilling the conditions (I) - (III). For proving the converse part of the theorem we construct the needed convergence in the following way: Let P be a convex semigroup of $(G^+)^N$ such that the conditions (I) -

- (III) are fulfilled ($P(\mathcal{L})$ is replaced by P). Let $\mathcal{L}(P) = \{(S, s) \in G^N \times G$ such that $|S - s| \in P\}$ (or, equivalently, such that $S = A - B + s$ for some $A, B \in P$). Then $\mathcal{L}(P)$ is a convergence on $(G, +, \leq)$ and $P(\mathcal{L}(P)) = P$.

The system of all convex subsemigroups of $(G^+)^N$ for which the conditions (I) - (III) are fulfilled is denoted by P_G.

REMARK. If $P \in P_G$, then $P(\mathcal{L}(P)) = P$. If $\mathcal{L} \in L_G$, then $\mathcal{L}(P(\mathcal{L})) = \mathcal{L}$. The mappings $\mathcal{L} \mapsto P(\mathcal{L})$ and $P \mapsto \mathcal{L}(P)$ are mutually inverse order isomorphisms between the partially ordered sets $(L_G, \subset)$ and $(P_G, \subset)$. We use this fact for to investigate order properties of the set L_G.

3. CONSTRUCTION OF $[\langle \sigma X \rangle]^*$

A natural question arises, whether a given sequence can be a zero sequence for some convergence. More generally: What set of sequences can be a subset of zero sequences of some convergence? (It is easy to see that it suffices to consider positive zero sequences only.) The theorem below gives an answer.

Let X be a subset of $(G^+)^N$. We use the following symbols (cf. F. ZANOLIN [5, 6]):

$\sigma X = \{S \in (G^+)^N$: there exists $R \in X$ such that $S = R \circ u$ for some $u\}$;

$\langle X \rangle = \{S \in (G^+)^N$: there exist $n \in N$ and $S_1, S_2, \ldots, S_n$ elements of X such that $S = S_1 + S_2 + \ldots + S_n\}$;

$[X] = \{S \in (G^+)^N$: there exists $T \in X$ such that $\langle \dot{0} \rangle \leq S \leq T\}$;

$X^* = \{S \in (G^+)^N$: for each u there exists v such that $S \circ u \circ v \in X\}$.

Let $X \subset (G^+)^N$. We denote by $L_G(X)$ the set of all $\mathcal{L} \in L_G$ with $X \subset P(\mathcal{L})$.

THEOREM 2. Let $X \subset (G^+)^N$. If $[\langle \sigma X \rangle]$ contains a sequence $\langle \dot{s} \rangle$ such that $s \neq 0$, then $L_G(X) = \emptyset$. In the opposite case, $[\langle \sigma X \rangle]^* \in P_G$ and $\mathcal{L}([\langle \sigma X \rangle]^*)$ is the smallest element of $L_G(X)$.

A SKETCH OF THE PROOF. If $P \in P_G$, then $X \subset P$ implies $[\langle \sigma X \rangle]^* \subset P$, thus if there exists $\mathcal{L} \in L_G$ such that $X \subset P(\mathcal{L})$, then $[\langle \sigma X \rangle] \subset P(\mathcal{L})$. Now there cannot exist any

$s \neq 0$ for which $\langle \dot{s} \rangle \in [\langle \delta X \rangle]$. In the opposite case the semigroup $[\langle \delta X \rangle]^*$ determines the smallest convergence such that each sequence of X converges to zero in this convergence.

4. THE PARTIALLY ORDERED SET $(L_G, \subset)$

The discrete convergence $\{(S, s) \in G^N \times G : S(n) = s$ for all but finitely many $n \in N\}$ is obviously the smallest element of the partially ordered set $(L_G, \subset)$. Some other properties of this set are discussed in this section.

Let I be a system of indices.

LEMMA 1. Let $P_i \in P_G$ for each $i \in I$. Then $\inf\{P_i : i \in I\} = \bigcap_{i \in I} P_i$.

LEMMA 2. Let $P_i \in P_G$ for each $i \in I$. If there exists $P \in P_G$ such that $P_i \subset P$ for each $i \in I$, then $\sup\{P_i : i \in I\} = \langle \bigcup_{i \in I} P_i \rangle^*$.

THEOREM 3. (1) $(L_G, \subset)$ is a lower semilattice.

(2) If $\mathcal{L} \in L_G$, then there exists $\mathcal{K} \in L_G$ such that
(i) $\mathcal{L} \subset \mathcal{K}$ and (ii) $\mathcal{K}$ is a maximal element of L_G.

(3) Every closed interval of L_G is a complete distributive lattice.

COROLLARY. $(L_G, \subset)$ is a complete distributive lattice if and only if it has a greatest element.

5. THE EXAMPLE

The following example shows that $(L_G, \subset)$ need not be a lattice. We construct two elements of L_G with no common upper bound.

Let e be a fixed irrational number. Call a function f from $Q \cap (e, e + 1)$ to Z piecewise constant if there exists a finite partition $\{M_j : j = 1, 2, \ldots, n\}$ of the set $Q \cap (e, e + 1)$ such that for each $j \in \{1, 2, \ldots, n\}$

(i) there are $a_j, b_j \in Irr$ such that $e \leq a_j < b_j \leq e + 1$ and $M_j = Q \cap (a_j, b_j)$ and

(ii) $f(x) = f(y)$ whenever $x, y \in M_j$.

Let G be the set of all piecewise constant functions from

$Q \cap (e, e + 1)$ to Z. Then G with respect to the pointwise operation $+$ and with respect to the pointwise order $\leq$ is an abelian lattice-ordered group. Define two sequences of $(G^+)^N$ as follows:

For each $n \in N$ let

$S(n)$ be the characteristic function of the set

$$\bigcup_{k=0}^{m} (e + 2k.3^{-n} , e + (2k + 1).3^{-n}) ,$$

where $m = (3^n - 1).2^{-1}$ and let

$T(n)$ be the characteristic function of the set

$$\bigcup_{k=1}^{m} (e + (2k - 1).3^{-n} , e + 2k.3^{-n}) ,$$ where again

$m = (3^n - 1).2^{-1}$.

By means of Thm. 2 it can be shown that both $[\langle\delta\{S\}\rangle]^*$ and $[\langle\delta\{T\}\rangle]^*$ belong to P_G and that $\mathcal{L}([\langle\delta\{S\}\rangle]^*)$ and $\mathcal{L}([\langle\delta\{T\}\rangle]^*)$ have no common upper bound in L_G (because $S + T = \langle \dot{i} \rangle$) . Now we infer that $(L_G , \subset)$ is not a lattice.

6. MULTIVALUED CONVERGENCES

We can also define the notion of multivalued convergence on a given abelian lattice-ordered group in the following way:

DEFINITION 2. Let $(G, +, \leq)$ be an abelian lattice-ordered group. A set $\mathcal{L} \subset G^N \times G$ is said to be a multivalued convergence on $(G, +, \leq)$, if the conditions (i) - (iii) and (v) - (ix) (see §1) are fulfilled.

The system of all multivalued convergences on $(G, +, \leq)$ is denoted by M_G . For $\mathfrak{M} \in M_G$ we put $R(\mathfrak{M}) = \{S \in (G^+)^N : (S, 0) \in \mathfrak{M}\}$.

THEOREM 4. Let $\mathfrak{M} \in M_G$. Then $R(\mathfrak{M})$ is a convex subsemigroup of $(G^+)^N$, closed with respect to subsequences and fulfilling the following two conditions:

(S) $\langle \dot{o} \rangle \in R(\mathfrak{M})$.

(U) Let $S \in (G^+)^N$. If for each u there exists v such that $S \circ u \circ v \in R(\mathfrak{M})$, then $S \in R(\mathfrak{M})$.

Conversely, whenever R_1 is a convex semigroup of $(G^+)^N$ which is closed with respect to subsequences and fulfills the conditions (S) and (U), then there exists $\mathfrak{M}_1 \in M_G$ such that $R(\mathfrak{M}_1) = R_1$; moreover, if $\mathfrak{M}_2 \in M_G$ and $R(\mathfrak{M}_2) = R_1$, then $\mathfrak{M}_2 = \mathfrak{M}_1$.

The system of all convex subsemigroups of $(G^+)^N$ closed with respect to subsequences and fulfilling the conditions (S) and (U) is denoted by R_G.

In the same way as in Thm. 1 (§2) we can construct an order isomorphism between partially ordered sets $(M_G, \subset)$ and $(R_G, \subset)$. It is clear that $L_G \subset M_G$ and for $\mathcal{L} \in L_G$ and $\mathcal{M} \in M_G$, $\mathcal{M} \subset \mathcal{L}$ implies $\mathcal{M} \in L_G$. The discrete convergence is the smallest element of M_G and $G^N \times G$ is the greatest element of M_G.

The symbols used below are defined in §3. Let I be a system of indices.

LEMMA 3. Let $R_i \in R_G$ for each $i \in I$. Then
$\inf\{R_i : i \in I\} = \bigcap_{i \in I} R_i$ and
$\sup\{R_i : i \in I\} = \langle \bigcup_{i \in I} R_i \rangle^*$.

THEOREM 5. $(M_G, \subset)$ is a complete distributive lattice and $(L_G, \subset)$ is a lower subsemilattice of the lattice $(M_G, \subset)$.

REFERENCES

[1] Birkhoff, G.: Lattice theory, third ed.. Amer. Math. Soc., Providence, 1973.

[2] Fuchs, L.: Partially ordered algebraic systems. Pergamon Press, 1963.

[3] Mikusiński, P.: Problems posed at the conference. Proc. Conf. on Convergence, Szczyrk, 1979, Katowice, 1980, 110-112.

[4] Novák, J.: On convergence groups. Czechoslovak Math. J. 20 (1970), 357-374.

[5] Zanolin, F.: Solution of a problem of Josef Novák about convergence groups. Bolletino Un. Mat. Ital. (5) 14-A, (1977), 375-381.

[6] Zanolin, F.: A note on convergences in commutative groups. Rend. Inst. Mat. Univ. Trieste, 13 (1981), 79-86.

CONVERGING SEQUENCES AND EXAMPLES

Klaas Pieter HART[1)]

(Take it to the limit, Eagles [Ea])

Introduction.

The purpose of this contribution is to describe a general method for constructing examples.
The method was devised by Ostaszewski [Os]. Important modifications were made in [JuKuRu] and [vD].
The basic idea is to make sure that each member of a certain family of sets gets sufficiently many accumulation points. It is then possible to determine the behaviour of more subsets of the space.
Usually the construction proceeds by transfinite induction. We shall discuss three examples: two in which the topology is constructed outright from some subsets of the underlying set, and one in which we start with the reals and refine the usual topology to get what we want.
The examples treated here are reasonably easy, but they illustrate the basic ideas nicely. More elaborate constructions can be found in e.g. [vDWi], [Ha], [JuKuRu], [Pr], [Va1], [We2] and [We3].
All undefined terms can be found in [En] or [Ku]. We consider an ordinal to be the set of its predecessors, so $\alpha \in \beta$ and $\alpha < \beta$ mean the same. Furthermore cardinals are initial ordinals and $2^\omega = |\mathbb{R}|$.
Finally, to explain the title, the simplest device for adding a limit point is a converging sequence ....

Example 0. A countably compact, perfectly normal, hereditarily separable, noncompact space.

This is Ostaszewski's original example [Os]. To construct it we need some extra set theoretical assumptions, since there are models of set theory in which all countably compact perfectly normal spaces are compact [We1].
We shall assume the continuum hypothesis (CH) and
♣: there exists a sequence $\langle S_\lambda : \lambda \in \Lambda \rangle$ such that

- each S_λ is an ω-sequence cofinal in λ, and
- whenever $X \subseteq \omega_1$ is uncountable then $S_\lambda \subseteq X$ for at least one $\lambda \in \Lambda$..

Here $\Lambda = \{\alpha \in \omega_1 : \alpha \text{ is a limit}\}$. See [De] for more information.
Now, by CH, let $\{X_\lambda : \lambda \in \Lambda\}$ enumerate $\{A \subseteq \omega_1 : |A| = \omega\}$, such that always $X_\lambda \subseteq \lambda$.
Furthermore let $\langle S_\lambda : \lambda \in \Lambda \rangle$ be a ♣-sequence.

KMA PF UPJŠ
Jesenná 5; 041 54 Košice
Czechoslovakia

We define, inductively, for each $\lambda \in \Lambda$ a topology τ_λ on λ such that:

(i) if $\gamma \in \lambda$ then $\gamma \in \tau_\lambda$,

(ii) if $\lambda' < \lambda$ then $\langle\lambda',\tau_{\lambda'}\rangle$ is an open subspace of $\langle\lambda,\tau_\lambda\rangle$,

(iii) $\langle\lambda,\tau_\lambda\rangle$ is locally compact, hence metrizable,

(iv) if $\lambda' < \lambda$ then $[\lambda',\lambda) \subseteq \overline{S_{\lambda'}}$ (τ_λ-closure) and

(v) X_λ has an accumulation point in $\langle\lambda+\omega,\tau_{\lambda+\omega}\rangle$.

Then let τ be the topology on ω_1 generated by $\bigcup_{\lambda\in\Lambda}\tau_\lambda$. Let us verify the properties of $\langle\omega_1,\tau\rangle$.

Countable compactness: if $C \subseteq \omega_1$ is infinite, then $X_\lambda \subseteq C$ for some $\lambda \in \Lambda$, hence C has an accumulation point in $\lambda+\omega$ and hence in ω_1. This is where CH was needed: $\langle S_\lambda : \lambda\in\Lambda\rangle$ only tells us something about uncountable subsets of ω_1.

For the rest ♣ is sufficient:

Hereditary separability: if $X \subseteq \omega_1$ is uncountable find $\lambda \in \Lambda$ with $S_\lambda \subseteq X$.

Then $\overline{S_\lambda} = S_\lambda \cup [\lambda,\omega_1)$ by (iv).

It follows that $X \cap \lambda$ is dense in X.

Perfect normality:

Let F be closed and uncountable. Using (iv) again we find that $[\lambda,\omega_1) \subseteq F$ for some $\lambda \in \Lambda$. So $\omega_1 \setminus F$ is countable and $F = \bigcap\{\omega_1 \setminus \{\alpha\} : \alpha \in \omega_1 \setminus F\}$ is a G_δ.

If F is countable, then $F \subseteq \gamma$ for some γ, hence $F = \gamma \cap \bigcap\{\omega_1 \setminus \alpha : \alpha \in \gamma \setminus F\}$ is a G_δ.

Finally, let F and G be closed and disjoint. One of them, F say, is countable and hence compact. But then it is easy to separate F and G by disjoint open sets. We need countable compactness and hence CH here but if we define the τ_λ's carefully, normality will follow from ♣ alone. It is a nice exercise to work this out.

It remains to perform the construction:

To begin with let $\tau_\omega = P(\omega)$.

Let τ_λ be given, we construct $\tau_{\lambda+\omega}$.

By (i) S_λ is closed and discrete in $\langle\lambda,\tau_\lambda\rangle$.

If X_λ has no accumulation point in λ let $A = S_\lambda \cup X_\lambda$, otherwise let $A = S_\lambda$.

Now A is closed and discrete in λ, so by (iii) we can find a discrete collection of τ_λ-clopen sets $\{C_a : a \in A\}$ such that $C_a \cap A = \{a\}$ for each $a \in A$ and each C_a is compact (we can have $\bigcup_{a\in A} C_a = \lambda$, but this is not necessary).

Split A into ω disjoint parts: $A = \bigcup_{n\in\omega} A_n$; such that for all $n \in \omega$ $|A_n \cap S_\lambda| = \omega$ and, if necessary, $|A_n \cap X_\lambda| = \omega$. Write $A_n = \{a_{nm} : m \in \omega\}$.

For $n \in \omega$ we let

$$\{\{\lambda+n\} \cup \bigcup_{i\geq m} C_{a_{ni}} : m \in \omega\}$$

be a local base at $\lambda+n$. In addition we make $\langle\lambda,\tau_\lambda\rangle$ an open subspace of $\langle\lambda+\omega,\tau_{\lambda+\omega}\rangle$. It is straightforward to check (i), (ii) and (iii).

For (iv): by contruction $[\lambda,\lambda+\omega) \subseteq \overline{S_\lambda}$ and if $\lambda' < \lambda$ then $[\lambda,\lambda+\omega) \subseteq \overline{S_\lambda \cap [\lambda',\lambda)} \subseteq \overline{S_{\lambda'}}$ (all closures in $\tau_{\lambda+\omega}$).

For (v): if X_λ has no accumulation point in λ, then each $\lambda+n$ becomes one in $\lambda+\omega$.

If λ is a limit of limits, we let τ_λ be generated by $\bigcup_{\lambda'<\lambda}\tau_{\lambda'}$. Again (i), (ii) and (iii) are easy, (iv) follows by induction and (v) needs no checking. □

In this example the special sets were the countably infinite subsets of ω_1 plus some special sequences. The first collection determined countable paracompactness, the second collection determined the behaviour of all uncountable closed sets up to some countable set. But this was sufficient for our purposes.

Example 1. A first-countable countably compact non-normal space.

This example is due to Vaughan [Va2]. It was motivated by the observation that first-countable countably compact spaces are regular (countable compactness subsumes Hausdorff). The question of normality arises naturally.

We fix some notation:

$A = \Lambda \cup \{\alpha\in\omega_1 : \alpha \text{ is even}\}$ and $B = \Lambda \cup \{\alpha \in \omega_1 : \alpha \text{ is odd}\}$.

We shall define a topology on $X = (\omega_1 \times \omega) \cup A\times\{\omega\} \cup B\times\{\omega+1\}$.

To do this we put for $\alpha \in \omega_1$:

$H_\alpha = \{\langle\beta,\omega\rangle : \beta \in A \wedge \beta \le \alpha\}$, $K_\alpha = \{\langle\beta,\omega+1\rangle : \beta \in B \wedge \beta\le\alpha\}$ and $X_\alpha = ([0,\alpha]\times\omega) \cup H_\alpha \cup K_\alpha$.

We shall define for each $\alpha\in\omega_1$ a topology τ_α on X_α such that

(i) $\beta \in \alpha \rightarrow \langle X_\beta,\tau_\beta\rangle$ is an open subspace of $\langle X_\alpha,\tau_\alpha\rangle$,

(ii) $\langle X_\alpha,\tau_\alpha\rangle$ is compact,

(iii) $[0,\alpha]\times\omega$ - with its usual product topology - is an open subspace of $\langle X_\alpha,\tau_\alpha\rangle$,

(iv) H_α and K_α are closed (and disjoint) in $\langle X_\alpha,\tau_\alpha\rangle$,

(v) if α is even then $\langle\alpha,n\rangle \rightarrow \langle\alpha,\omega\rangle$, and
if α is odd then $\langle\alpha,n\rangle \rightarrow \langle\alpha,\omega+1\rangle$.

Again we let τ be generated by $\bigcup_{\alpha\in\omega_1}\tau_\alpha$.

We verify the properties of $\langle X,\tau\rangle$:

First countable: each $\langle X_\alpha,\tau_\alpha\rangle$ is countable and compact hence first-countable, each X_α is also open in X, hence X is first-countable.

Countably compact: if $C\subseteq X$ is countable then $C\subseteq X_\alpha$ for some α, hence $\overline{C}$ is compact. It follows that every infinite subset of X has an accumulation point.

Not normal: note that $A\times\{\omega\} = \bigcup_{\alpha\in\omega_1}H_\alpha$ and $B\times\{\omega+1\} = \bigcup_{\alpha\in\omega_1}K_\alpha$ and that by construction these sets are closed and disjoint.

Let $U \supseteq A\times\{\omega\}$ and $V \supseteq B\times\{\omega+1\}$ be open. We show that $U\cap V \ne \emptyset$.

By (v) find for each odd α $n_\alpha \in\omega$ such that $\{\alpha\}\times(n_\alpha,\omega) \subseteq V$. Next find n such that $I = \{\alpha : n_\alpha = n\}$ is uncountable (infinite is sufficient).

Let α be a limit point of I. Then, by (iii), we find $\{\alpha\}\times(n,\omega) \subseteq \overline{V}$, hence by (v) (limits are even) $\langle\alpha,\omega\rangle \in \overline{V}$. But $\langle\alpha,\omega\rangle\in U$ so $U\cap V\ne\emptyset$.

We now construct the τ_α:

For each α let $X_\alpha^- = \bigcup_{\beta\in\alpha}X_\beta$. Then $X_\alpha^- = \emptyset$ if $\alpha = 0$ and $X_\alpha^- = X_\beta$ if $\alpha = \beta+1$.

Let α be 0 or a successor.

Then (i) and (ii) force us to make X_α^- clopen in $\langle X_\alpha,\tau_\alpha\rangle$. By (iii) each $\langle\alpha,n\rangle$ must be isolated and by (v) if α is even $\{\{\langle\alpha,\omega\rangle\}\cup\{\langle\alpha,n\rangle : n\ge m\} : m\in\omega\}$ must be a local base at $\langle\alpha,\omega\rangle$; if α is odd we must take $\{\{\langle\alpha,\omega+1\rangle\}\cup\{\langle\alpha,n\rangle : n\ge m\} : m\in\omega\}$ as a local base at $\langle\alpha,\omega+1\rangle$. Being a sum of two compact spaces, $\langle X_\alpha,\tau_\alpha\rangle$ is compact and (iv) is automatic.

Let α be a limit.

First we give X_α^- the topology determined by $\bigcup_{\beta\in\alpha}\tau_\beta$ and make it open in $\langle X_\alpha,\tau_\alpha\rangle$.

Second (iii) forces us to let $\{(\beta,\alpha]\times\{n\} : \beta\in\alpha\}$ be a local base at $\langle\alpha,n\rangle$ for $n\in\omega$.

Finally we deal with $\langle\alpha,\omega\rangle$ and $\langle\alpha,\omega+1\rangle$.

So far $X_\alpha^- \cup \{\langle\alpha,n\rangle : n\in\omega\}$ is a countable locally compact space, hence it is metrizable.

So we can find a partition $\{C_n: n \in \omega\}$ of it into compact open sets, such that each C_n meets at most one of $H_\alpha \cap X_\alpha^-$, $K_\alpha \cap X_\alpha^-$ and $\{\langle\alpha,m\rangle: m \in \omega\}$. Each of these three sets is not compact, hence each of them intersects infinitely many C_n. For definiteness say $C_n \cap K_\alpha \neq \emptyset$ iff n is odd.

We define local bases at $\langle\alpha,\omega\rangle$ and $\langle\alpha,\omega+1\rangle$ as follows:

at $\langle\alpha,\omega\rangle$: $\{\{\langle\alpha,\omega\rangle\} \cup \bigcup_{i\geq n} C_{2i}: n \in \omega\}$ and

at $\langle\alpha,\omega+1\rangle$: $\{\{\langle\alpha,\omega+1\rangle\} \cup \bigcup_{i\geq n} C_{2i+1}: n \in \omega\}$.

It is now easy to check (i), (ii) (X_α is a two-point compactification of $X_\alpha^- \cup \{\langle\alpha,m\rangle: n \in \omega\}$), (iii), (iv) (choice of the C_n) and (v) (each C_n contains only finitely many $\langle\alpha,m\rangle$). □

In this example the special sets were the countable sets and the H_α and K_α. We achieved countable compactness by giving each countable set a compact closure, hence this example is not separable.

This is in contrast with example 0, in which many countable sets have a non compact closure, e.g. all S_λ.

Example 2. A separable, normal, first-countable, locally compact space which is not Lindelöf (and hence not paracompact).

This example is different from the other two in that we don't construct our topology from scratch, but we start with a topology and hence a collection of converging sequences and use these to determine the new topology.

The idea to use Ostaszewski's technique in this context goes back to [JuKuRu]. The present example is due to van Douwen [vD].

The key observation is that if $F \subseteq \mathbb{R}$ is closed in the Euclidean topology then $|F| \leq \omega$ or $|F| = 2^\omega$ see e.g. [En; 4.5.5].

We shall construct a topology τ on $\mathbb{R}$, finer than the usual topology such that:

- $\langle \mathbb{R},\tau\rangle$ is locally compact and locally countable, hence it is first-countable and not Lindelöf: no cover by countable open sets has a countable subcover.

- $\mathbb{Q}$ is dense in $\langle \mathbb{R},\tau\rangle$
- If $A,B \subseteq \mathbb{R}$ are countable and disjoint and if $|\overline{A} \cap \overline{B}| = 2^\omega$ then $\overline{A}^\tau \cap \overline{B}^\tau \neq \emptyset$.[2)]

  This implies normality:

  Let $F,G \subseteq \mathbb{R}$ be τ-closed and disjoint.

  Find countable $A \subseteq F$ and $B \subseteq G$ respectively such that $\overline{F} = \overline{A}$ and $\overline{G} = \overline{B}$. Then $\overline{A}^\tau \cap \overline{B}^\tau = \emptyset$, hence $|\overline{A} \cap \overline{B}| < 2^\omega$ and hence $|\overline{F} \cap \overline{G}| \leq \omega$.

  Let $\mathcal{U}$ be a countable collection of open intervals such that $\bigcup\mathcal{U} = \mathbb{R}\setminus(\overline{F} \cap \overline{G})$ and $\forall U \in \mathcal{U}$: $\overline{U} \cap A = \emptyset$ or $\overline{U} \cap B = \emptyset$.

  Furthermore for each $x \in \overline{F} \cap \overline{G}$ find a τ-clopen neighbourhood U_x such that $U_x \cap F = \emptyset$ or $U_x \cap G = \emptyset$. Using $\mathcal{U} \cup \{U_x: x \in \overline{F} \cap \overline{G}\}$ it is now easy to find disjoint τ-open sets around F and G, see [En; 1.5.14].

We turn to the construction.

To begin with let $\{x_\alpha: \alpha \in 2^\omega\}$ enumerate $\mathbb{R}$ so that $\mathbb{Q} = \{x_n: n \in \omega\}$. Furthermore let

---

2) If $A \subseteq \mathbb{R}$ then $\overline{A}$ denotes its euclidean closure and $\overline{A}^\tau$ its closure in the new topology.

$\{\langle A_\alpha, B_\alpha\rangle : \alpha\in 2^\omega\}$ enumerate the set of all pairs $\langle A,B\rangle$ of disjoint countable subsets of A with $|\overline{A}\cap\overline{B}| = 2^\omega$.

To faciliate a streamlined construction we make some definitions:

Let $\mathcal{B} = \{B_i : i\in\omega\}$ be a countable base for $\mathbb{R}$ with $B_0 = \mathbb{R}$. For $x\in\mathbb{R}$ and $i\in\omega$ we put

$$E(x,i) = \bigcap\{B_j : j\le i \wedge x\in B_j\}.$$

Then $\{E(x,i) : i\in\omega\}$ is a local base at x, and for all $i\in\omega$: $E(x,i+1)\subseteq E(x,i)$ and $y\in E(x,i) \rightarrow E(y,i)\subseteq E(x,i)$. Next we define $\phi : 2^\omega \rightarrow 2^\omega\backslash\omega$ by

$$\phi(\gamma) = \min\{\alpha\in 2^\omega\backslash(\omega\cup\{\phi(\beta) : \beta\in\gamma\}) : A_\gamma\cup B_\gamma \subseteq \{x_\beta : \beta\in\alpha\} \wedge x_\alpha\in\overline{A}_\gamma\cap\overline{B}_\gamma\}.$$

Then ϕ is well-defined, since no countable set is cofinal in 2^ω, $|\omega\cup\{\phi(\beta) : \beta\in\gamma\}| < 2^\omega$ and $|\overline{A}_\gamma\cap\overline{B}_\gamma| = 2^\omega$.

Now we are ready to define our new topology:

For $\alpha\in 2^\omega\backslash\omega$ choose a sequence $\langle S_\alpha(i) : i\in\omega\rangle$ in $\{x_\beta : \beta\in\alpha\}$ such that

(i) for all $i\in\omega$ $S_\alpha(i)\in E(x_\alpha, i)$,

(ii) if $\alpha = \phi(\gamma)$ for some (unique) γ then for all $i\in\omega$: $S_\alpha(3i)\in\mathbb{Q}$, $S_\alpha(3i+1)\in A_\gamma$ and $S_\alpha(3i+2)\in B_\gamma$,

(iii) if $\alpha\ne\phi(\gamma)$ for all γ then for all $i\in\omega$ $S_\alpha(i)\in\mathbb{Q}$.

Next we define by induction (each $S_\alpha(i)$ precedes x_α in the enumeration !) local bases $\{L(x_\alpha, j) : j\in\omega\}$ at each x_α as follows:

- $\alpha\in\omega \rightarrow L(x_\alpha, j) = \{x_\alpha\}$ $(j\in\omega)$,
- $\alpha\in 2^\omega\backslash\omega \rightarrow L(x_\alpha, j) = \{x_\alpha\}\cup\bigcup_{i\ge j} L(S_\alpha(i), i)$ $(j\in\alpha)$.

We verify inductively the following facts:

(i) $L(x_\alpha, j)\subseteq\{x_\beta : \beta\le\alpha\}$,

(ii) $y\in L(x_\alpha, j) \rightarrow \exists k\in\omega : L(y,k)\subseteq L(x_\alpha, j)$,

(iii) $L(x_\alpha, j)$ is countable and compact,

(iv) $L(x_\alpha, j)\cap\mathbb{Q}\ne\emptyset$.

These facts are trivially true if $\alpha\in\omega$, so assume $\alpha\ge\omega$:

For (i) : $L(x_\alpha, j) \subseteq \{x_\alpha\}\cup\bigcup_{i\ge j}\{x_\beta : x_\beta \text{ precedes or equals } S_\alpha(i)\}$
$\subseteq \{x_\beta : \beta\le\alpha\}$.

For (ii) : if $y = x_\alpha$ take $k = j$; if $y\in L(s_\alpha(i), i)$ for some i then we find k by our inductive hypothesis.

For (iii) : - each $L(S_\alpha(i), i)$ is countable hence so is $L(x_\alpha, j)$;
- each $L(S_\alpha(i), i)$ is compact and $L(x_\alpha, j)$ contains all but finitely many of them.

For (iv) : For every $i\in\omega$ $L(S_\alpha(i), i)\cap\mathbb{Q}\ne\emptyset$.

Of course we let $\{L(x_\alpha, j) : \alpha\in 2^\omega, j\in\omega\}$ generate a topology τ on $\mathbb{R}$ after we have established (i) and (ii), it is then justified to talk about compactness of $L(x_\alpha, j)$. Then (iii) says that τ is a locally compact and locally countable topology on $\mathbb{R}$ and (iv) says that $\mathbb{Q}$ is τ-dense. Finally let $A,B\subseteq\mathbb{R}$ be countable and disjoint such that $|\overline{A}\cap\overline{B}| = 2^\omega$. Then $\langle A,B\rangle = \langle A_\gamma, B_\gamma\rangle$ for some $\gamma\in 2^\omega$. But then $x_{\phi(\gamma)}\in\overline{A}^\tau\cap\overline{B}^\tau$. So τ also satisfies the last condition.

To see that τ is finer than the euclidean topology note that $L(x,j)\subseteq E(x,j)$ for all and j □ .

In this example the special subsets were countable and they were selected because of their behaviour in the euclidean topology. Then τ was constructed in such a way that some of this behaviour was preserved, in this case to make τ normal.

Remark.
In examples 0. and 2. we could have defined the topology using converging sequences only. We chose not to do this for two reasons: some verifications in example 0. would become more cumbersome and in both cases we would loose first-countability and local compactness. The reader is encouraged to check these things him- or herself.

References:

[De] K.J. Devlin, Variations on ◇. J. Symb. logic 44 (1979) 51-58.

[vD] E.K. van Douwen, A technique for constructing honest locally compact submetrizable spaces, to appear.

[vDWi] ----------- and H.H. Wicke, A real weird topology on the reals, Houston J. Math. 3 (1977) 141-152.

[Ea] Eagles, One of these nights, Asylum 7E-1039 WEA records (1975).

[En] R. Engelking, General Topology, PWN Warszawa (1977).

[Ha] K.P. Hart, Almost-n-fully normal spaces, Fund. Math. 123 (to appear).

[JuKuRu] I. Juhasz, K. Kunen and M.E. Rudin, Two more herediratilly separable non-Lindelöf spaces. Can. J. Math. 28 (1976) 998-1005.

[Ku] K. Kunen, Set Theory, North Holland Amsterdam (1980).

[Pr] T. Przymusinski, Normality and paracompactness in finite and countable Cartesian products, Fund. Math. 105 (1980) 89-104.

[Os] A.J. Ostaszewski, On countably compact, perfectly normal spaces, J. London Math. Soc. 14 (1976) 505-516.

[Va1] J.E. Vaughan, Products of perfectly normal, sequentially compact spaces, J. London Math. Soc. 14 (1976) 517-520.

[Va2] ------------, A countably compact, first-countable non normal T_2-space, Proc. Amer. Math. Soc. 75 (1979) 329-342.

[We1] W. Weiss, Countably compact spaces and Martin's Axiom, Can. J. Math. 30 (1978) 243-249.

[We2] --------, A countably paracompact non normal space, Proc. Amer. Math. Soc. 79 (1980) 487-490.

[We3] --------, Small Dowker spaces, Pac. J. Math. 94 (1981) 485-492.

SEQUENTIAL STRUCTURES: CATEGORICAL ASPECTS

Horst Herrlich[1)]

INTRODUCTION

Small concrete categories A can always be embedded, via simple categorical constructions, into better behaved concrete categories. In particular, with any such A, there can be associated the following topological categories:

(1) Max A, the largest final completion of A

(2) Univ A, the universal final completion of A

(3) Min A, the smallest final completion or Mac Neille completion of A.

The general features of these categories are well understood. In particular, Max A is always cartesian closed.

Here the constructions of Max A and Univ A are applied to 3 concrete categories, each having only a single object:

(a) Seq (a "convergent sequence"), yielding categories for sequential convergence structures,

(b) USeq (a "pair of adjacent sequences"), yielding categories for uniformly sequential structures,

(c) CSeq (a "Cauchy sequence"), yielding categories of sequential Cauchy structures.

The resulting categories are described axiomatically and compared with the familiar categories of sequential structures, which have been obtained rather by intuition than by a systematic procedure.

[1)] Fachbereich Mathematik und Informatik, Universität Bremen, Postfach 33o44o, 28 Bremen 33, Fed. Rep. Germany

§ 1 FINAL COMPLETIONS OF C-CATEGORIES

In this section we give an explicit description of Max $\underline{A}$ and Univ $\underline{A}$ for the kind of categories $\underline{A}$, which are of interest in the investigation of sequential structures.
All categories are supposed to be <u>c-categories</u> (cf. [9]), i.e. categories, whose objects are structured sets (X,τ), whose morphisms are certain functions $f: (X,\tau) \to (X',\tau')$, subject to the following conditions:

(1) if (X,τ) is an object, than $id_X: (X,\tau) \to (X,\tau)$ is a morphism

(2) if $f: (X,\tau) \to (X',\tau')$ and $g: (X',\tau') \to (X'',\tau'')$ are morphisms, than $g\circ f: (X,\tau) \to (X'',\tau'')$ is a morphism

(3) if (X,τ) and (X',τ') are objects and $f: X \to X'$ is a constant map, than $f: (X,\tau) \to (X',\tau')$ is a morphism,

and whose composition law is the usual composition law for maps.

From now on let $\underline{A}$ be a c-category with a single object (A,τ):

<u>Construction of Max $\underline{A}$:</u>

The objects of Max $\underline{A}$ are pairs (X,F), where X is a set and F is a collection of maps $f: A \to X$, subject to the following conditions:

(1) if $f \in F$ and $g: (A,\tau) \to (A,\tau)$ is an $\underline{A}$-morphism, then $f\circ g \in F$

(2) if $f: A \to X$ is a constant map, then $f \in F$.

A map $g: (X,F) \to (X',F')$ between Max $\underline{A}$-objects is a morphism in Max $\underline{A}$ iff

$$f \in F \quad \text{implies} \quad g\circ f \in F'.$$

The full subcategory of Max $\underline{A}$, consisting of the single object (A,Mor $\underline{A}$), is isomorphic to $\underline{A}$. Via this isomorphism, $\underline{A}$ will be regarded as a subcategory of Max $\underline{A}$, and Max $\underline{A}$ as an extension of $\underline{A}$.

<u>Properties of Max $\underline{A}$:</u>

Max $\underline{A}$ is usually very large, in fact it is the largest extension of $\underline{A}$, in which $\underline{A}$ is finally dense (cf.[7],[8]).

No matter how wild $\underline{A}$ may be, Max $\underline{A}$ is always an extremely well-behaved category (cf.[9]):

(1) Max $\underline{A}$ is a topological (= finally complete and small-fibred) category; in particular Max $\underline{A}$ is complete, cocomplete, well-powered and cowell-powered;

(2) Max $\underline{A}$ is a quasitopos in the sense of Penon [13]; in particular Max $\underline{A}$ is cartesian closed;

(3) Max $\underline{A}$ is a topological universe in the sense of Nel [12], i.e. satisfies (1) and (2) above;

(4) in Max $\underline{A}$ arbitrary products of quotients are quotients.

Although, from a purely categorical point of view Max $\underline{A}$ is exceedingly nice, from a concrete point of view smaller categories are often more interesting. Among these Univ $\underline{A}$ is of particular interest. It can be characterized as the finally complete reflection of $\underline{A}$ (with respect to finality-preserving concrete functors), alternatively as the largest extension of $\underline{A}$, in which $\underline{A}$ is finally dense and for which the embedding of $\underline{A}$ preserves final structures (cf.[7],[8]).

Construction of Univ $\underline{A}$:

A collection G of $\underline{A}$-morphisms is called final, provided a map $f: (A,\tau) \to (A,\tau)$ is an $\underline{A}$-morphism if all compositions $f \circ g: (A,\tau) \to (A,\tau)$, with $g \in G$, are $\underline{A}$-morphisms. Univ $\underline{A}$ is the full subcategory of Max $\underline{A}$, consisting of those objects (X,F) of Max $\underline{A}$, which satisfy the following condition:

if $f: A \to X$ is a map such that
$\{g \in \text{Mor } \underline{A} \mid f \circ g \in F\}$ is final in $\underline{A}$, then $f \in F$

§ 2 SEQUENTIAL STRUCTURES: TERMINOLOGY AND NOTATION

$\mathbb{N}$ denotes the set of natural numbers o,1,... .
$\mathbb{N}^* = \mathbb{N} \cup \{\infty\}$
$\mathbb{N}_2 = \mathbb{N} \times \{o,1\}$

Elements of $X^{\mathbb{N}}$, i.e. sequences in X, will be denoted by ξ or η
Elements of $X^{\mathbb{N}^*}$ will be denoted by (ξ,x) or (η,y)
Elements of $X^{\mathbb{N}_2}$, i.e. pairs of sequences in X, will be denoted by

(ξ,η) or (ξ',η')

$\langle x\rangle$ denotes the constant sequence with value x

$\xi\wedge\eta$ denotes the sequence, whose (2n)-th term is $\xi(n)$ and whose (2n-1)-th term is $\eta(n)$

ξ is called <u>eventually constant</u> iff the following condition is satisfied: $\exists x\ \exists n\ \forall m\geq n\ \xi(m) = x$; in this case x is called the <u>final value</u> of ξ

(ξ,η) and (ξ',η') <u>eventually coincide</u> iff the following condition is satisfied: $\exists n\ \exists k \in \mathbb{Z}\ \forall\ m\geq n\quad (\xi(m),\eta(m)) = (\xi'(m+k),\eta'(m+k))$

(ξ,η) and (ξ',η') are <u>symmetrically equivalent</u> iff the following condition is satisfied: $\forall n\ \{\xi(n),\eta(n)\} = \{\xi'(n),\eta'(n)\}$

Mon_s denotes the set of all <u>strictly monotone maps</u> $\alpha\colon \mathbb{N} \to \mathbb{N}$

Mon_i denotes the set of all <u>monotone maps</u> $\alpha\colon \mathbb{N} \to \mathbb{N}$ with <u>infinite range</u>

Mon_f denotes the set of all <u>monotone maps</u> $\alpha\colon \mathbb{N} \to \mathbb{N}$ with <u>finite range</u>

Perm denotes the set of all <u>permutations</u> $\alpha\colon \mathbb{N} \to \mathbb{N}$

Fin denotes the set of all <u>finite-to-one maps</u> $\alpha\colon \mathbb{N} \to \mathbb{N}$

<u>Subsequences</u> of ξ are usually understood to be sequences of the form $\xi\circ\alpha$ with $\alpha \in Mon_s$. Similarly, sequences of the form $\xi\circ\alpha$ with $\alpha \in$ Fin may be regarded as <u>generalized</u> <u>subsequences</u> of X.

L denotes a <u>sequential convergence structure</u> on X, i.e. a subset of $X^{\mathbb{N}*}$ (elements (ξ,x) of L are interpreted as "the sequence ξ L-converges to x" and written more suggestively as $\xi \xrightarrow{L} x$)

maps $f\colon (X,L) \to (X',L')$ are called <u>sequentially continuous</u> iff $\xi \xrightarrow{L} x$ implies $f\circ\xi \xrightarrow{L'} f(x)$

U denotes a <u>uniformly sequential structure</u> on X, i.e. a subset of $X^{\mathbb{N}}{}^2$ (elements (ξ,η) of U are interpreted as "U-adjacent sequences" or "U-near sequences")

maps $f\colon (X,U) \to (X',U')$ are called <u>uniformly sequentially continuous</u> iff $(\xi,\eta) \in U$ implies $(f\circ\xi,\ f\circ\eta) \in U'$

C denotes a <u>sequential Cauchy structure</u> on X, i.e. a subset of $X^{\mathbb{N}}$ (elements of C are interpreted as "C-Cauchy sequences")

maps $f\colon (X,C) \to (X',C')$ are called <u>sequentially Cauchy continuous</u> iff $\xi \in C$ implies $f\circ\xi \in C'$

For further terminology and background information compare [4].

§ 3 SEQUENTIAL CONVERGENCE STRUCTURES

3.1 AXIOMS

Consider the following axioms for sequential convergence structures L on X:

(L_1) $\langle x\rangle \xrightarrow{L} x$ for each $x \in X$

(L_1-Event) if ξ is eventually constant with final value x and there exists $\eta \xrightarrow{L} y$ with image $\xi \subset$ image η, then $\xi \xrightarrow{L} x$

(L_2) $\xi \xrightarrow{L} x$ and $\alpha \in Mon_s$ imply $\xi\circ\alpha \xrightarrow{L} x$

(L_2-Fin) $\xi \xrightarrow{L} x$ and $\alpha \in Fin$ imply $\xi\circ\alpha \xrightarrow{L} x$

(L_2-Perm) $\xi \xrightarrow{L} x$ and $\alpha \in Perm$ imply $\xi\circ\alpha \xrightarrow{L} x$

(L_2-Mon_i) $\xi \xrightarrow{L} x$ and $\alpha \in Mon_i$ imply $\xi\circ\alpha \xrightarrow{L} x$

(L_2-Mon_f) $\xi \xrightarrow{L} x$ and $\alpha \in Mon_f$ and y is the final value of $\xi\circ\alpha$ imply $\xi\circ\alpha \xrightarrow{L} y$

(L_2-Mix) $\xi \xrightarrow{L} x$ implies $\xi \wedge \langle x\rangle \xrightarrow{L} x$

(L_3) if $(\xi,x) \in X^{\mathbb{N}^*}$ satisfies the condition $\forall\ \alpha \in Mon_s\ \ \exists\ \beta \in Mon_s\ \ \xi\circ\alpha\circ\beta \xrightarrow{L} x$, then $\xi \xrightarrow{L} x$

<u>Definition</u> ([1],[2],[3],[4]): A sequential convergence structure L on X is called

(1) an <u>*L*-structure</u> on X iff it satisfies (L_1) and (L_2)

(2) an <u>*L**-structure</u> on X iff it satisfies (L_1),(L_2) and (L_3).
<u>*L*</u> (resp. <u>*L**</u>) denotes the category, whose objects are pairs (X,L) with L being an *L*-structure (resp. *L**-structure) on X, and whose morphisms are the sequentially continuous maps.

3.2 FINAL COMPLETIONS OF <u>Seq</u>

<u>Seq</u> denotes the full subcategory of <u>Top</u>, consisting of the single object $(\mathbb{N}^*,\tau)$ with $\tau = \{A \subset \mathbb{N}^* \mid \infty \in A \Rightarrow \mathbb{N}^*\setminus A \text{ finite}\}$.
In particular, a map $f\colon (\mathbb{N}^*,\tau) \to (\mathbb{N}^*,\tau)$ is a morphism in <u>Seq</u> iff either $f(\infty) = \infty$ and $f^{-1}(n)$ is finite for each $n \in \mathbb{N}$ or f is eventually constant.

Theorem 1: Equivalent are:

(1) (X,L) is an object of Max (Seq)

(2) (X,L) satisfies (L_1) and the following condition: if $(\xi,x) \in L$ and f is a Seq-morphism, then $(\xi,x)\circ f \in L$

(3) (X,L) satisfies the conditions (L_1), $(L_1$-Event), $(L_2$-Fin) and $(L_2$-Mix)

(4) (X,L) satisfies the conditions (L_1), $(L_2$-Perm), $(L_2\text{-Mon}_i)$, $(L_2\text{-Mon}_f)$ and $(L_2$-Mix)

Corollary: Max(Seq) is a full subcategory of $\mathcal{L}$.

Proposition: For a collection $\mathcal{F}$ of morphisms in Seq the following conditions are equivalent:

(1) $\mathcal{F}$ is final in Seq

(2) $\mathcal{F}$ is final in Top

(3) for each infinite subset M of $\mathbb{N}$ there exists some f in $\mathcal{F}$ such that $f^{-1}[M]$ is infinite and does not contain ∞.

Theorem 2: Equivalent are

(1) (X,L) is an object of Univ(Seq)

(2) L is an $\mathcal{L}^*$-structure.

Corollary: $\mathcal{L}^*$ = Univ(Seq), i.e. $\mathcal{L}^*$ is the universal final completion of Seq.

§ 4 UNIFORMLY SEQUENTIAL STRUCTURES

4.1 AXIOMS

Consider the following axioms for uniformly sequential structures U on X:

(U_1) U is an equivalence relation on $X^{\mathbb{N}}$

(U_1-Ref) $(\xi,\xi) \in U$ for each $\xi \in X^{\mathbb{N}}$

(U_1-Ref$_1$) $(\langle x\rangle,\langle x\rangle) \in U$ for each $x \in X$

(U_1-Ref$_2$) $(\xi,\eta) \in U$ implies $(\xi,\xi) \in U$

(U_1-Ref$_3$) image $\xi \subset$ image (η,ζ) and $(\eta,\zeta) \in U$ imply $(\xi,\xi) \in U$

(U_1-Sym) $(\xi,\eta) \in U$ implies $(\eta,\xi) \in U$

(U_1-Sym$_1$) if (ξ,η) and (ξ',η') are symmetrically equivalent and $(\xi,\eta) \in U$, then $(\xi',\eta') \in U$

(U_1-Trans) $(\xi,\eta) \in U$ and $(\eta,\zeta) \in U$ imply $(\xi,\zeta) \in U$

(U_2) $(\xi,\eta) \in U$ and $\alpha \in \mathrm{Mon}_s$ imply $(\xi\circ\alpha,\eta\circ\alpha) \in U$

(U_2-Event) if (ξ,η) and (ξ',η') eventually coincide and image $(\xi',\eta') \subset$ image (ξ,η) and $(\xi,\eta) \in U$, then $(\xi',\eta') \in U$

(U_2-Mix) if $(\xi,\eta) \in U$ and $\alpha \in \mathrm{Fin}$ and image $\zeta \subset$ image (ξ,η), then $(\xi\circ\alpha \wedge \zeta,\ \eta\circ\alpha \wedge \zeta) \in U$

(U_3) if $(\xi,\eta) \in X^{\mathbb{N}2}$ satisfies the condition
$\forall\ \alpha \in \mathrm{Mon}_s\ \ \exists\ \beta \in \mathrm{Mon}_s\ \ (\xi\circ\alpha\circ\beta,\ \eta\circ\alpha\circ\beta) \in U$,
then $(\xi,\eta) \in U$.

<u>Definition</u> ([4],[5],[6]): A uniformly sequential structure U on X is called

(1) a <u>*U*-structure</u> on X iff it satisfies (U_1) and (U_2)

(2) a <u>*U**-structure</u> on X iff it satisfies (U_1),(U_2) and (U_3).

<u>*U*</u> (resp. <u>*U**</u>) denotes the category, whose objects are pairs (X,U) with U being a *U*-structure (resp. *U**-structure) on X, and whose morphisms are the uniformly sequentially continuous maps.

4.2 FINAL COMPLETIONS OF <u>USeq</u>

<u>USeq</u> denotes the full subcategory of <u>Unif</u>, consisting of the single object $(\mathbb{N}_2,\mu)$, where μ consists of all covers $\mathcal{C}$ of $\mathbb{N}_2$ satisfying the following condition:

$$\exists\, n\ \ \forall\, m \geq n\ \ \exists\, C \in \mathcal{C}\ \ \{m\}\times\{o,1\} \subset C.$$

In particular, a map $f\colon (\mathbb{N}_2,\mu) \to (\mathbb{N}_2,\mu)$ is a morphism in <u>USeq</u> iff it satisfies the following condition:

$$\forall n\ \exists m\ \forall m'\geq m\ \ \exists n'\geq n\ \ f(m',o)\neq f(m',1) \Rightarrow \{f(m',o),f(m',1)\}=\{(n',o),(n',1)\}$$

Theorem 3: Equivalent are:

(1) (X,U) is an object of Max (USeq)

(2) (X,U) satisfies (U_1-Ref_1) and the following condition:

if $(\xi,\eta) \in U$ and f is a USeq-morphism, then $(\xi,\eta)\circ f \in U$

(3) (X,U) satisfies the conditions (U_1-Ref_1),(U_1-Ref_3),(U_1-Sym_1), (U_2),(U_2-Event) and (U_2-Mix).

Remark: None of the categories U and Max (USeq) is a subcategory of the other: $\mathcal{U}$-objects need not satisfy any of the conditions (U_1-Sym_1), (U_2-Event) or (U_2-Mix); vice versa Max (USeq)-objects need not satisfy (U_1-Ref) or (U_1-Trans). It seems that the condition (U_1-Trans) - and hence (U_1) - is unnaturally strong. As the following example demonstrates, not even all nearness-induced uniformly sequential structures satisfy it.

Example: Consider the nearness space (X,μ), where $X = \mathbb{N}\times\{o,1,2\}$, and where μ consists of all covers $\mathcal{C}$ of X satisfying the following condition:

$\exists n \quad \forall m\geq n \quad \exists C \in \mathcal{C} \; \exists D \in \mathcal{C} \quad \{m\}\times\{o,1\} \subset C$ and $\{m\}\times\{1,2\} \subset D$.

Let U be the uniformly sequential structure, induced by μ, i.e. $U = \{(\xi,\eta) \in X^{\mathbb{N}2} \mid \forall\, \mathcal{C} \in \mu \; \exists n \; \forall m\geq n \; \exists\, C \in \mathcal{C} \; \{\xi(m),\eta(m)\} \subset C\}$.

Then (X,U) satisfies all the axioms, mentioned in 4.1, with the exception of (U_1-Trans). If $\xi(n) = (n,o)$, $\eta(n) = (n,1)$ and $\zeta(n) = (n,2)$, then $(\xi,\eta) \in U$ and $(\eta,\zeta) \in U$, but $(\xi,\zeta) \notin U$.

Proposition: For a collection $\mathcal{F}$ of morphisms in USeq the following conditions are equivalent:

(1) $\mathcal{F}$ is final in USeq

(2) for each infinite subset M of $\mathbb{N}$ there exists some f in $\mathcal{F}$ such that the set

$$\{m \in M \mid \exists n \in \mathbb{N} \; \{f(n,o),f(n,1)\}=\{(m,o),(m,1)\}\}$$

is infinite.

Theorem 4: Equivalent are:

(1) (X,U) is an object of Univ (USeq)

(2) (X,U) satisfies the conditions (U_1-Ref_1), (U_1-Ref_2), (U_1-Sym), (U_2) and (U_3).

Corollary: $\underline{U}$* is a full subcategory of Univ (USeq). An object of Univ (USeq) belongs to $\underline{U}$* iff it satisfies (U_1-Ref) and (U_1-Trans).

§ 5 SEQUENTIAL CAUCHY STRUCTURES

5.1 AXIOMS

Consider the following axioms for sequential Cauchy structures C on X:

(C_1) $\langle x\rangle \in C$ for each $x \in X$

(C_2) $\xi \in C$ and $\alpha \in Mon_s$ imply $\xi\circ\alpha \in C$

(C_2-Fin) $\xi \in C$ and $\alpha \in Fin$ imply $\xi\circ\alpha \in C$

(C_2-Perm) $\xi \in C$ and $\alpha \in Perm$ imply $\xi\circ\alpha \in C$

(C_2-Mon_i) $\xi \in C$ and $\alpha \in Mon_i$ imply $\xi\circ\alpha \in C$

(C_2-Mon_f) $\xi \in C$ and $\alpha \in Mon_f$ imply $\xi\circ\alpha \in C$

(C_3) $\xi \in C$ and $\eta \in C$ and $\xi\circ\alpha = \eta\circ\beta$ for some $\alpha \in Mon_s$ and $\beta \in Mon_s$ imply $\xi\wedge\eta \in C$

(C_4) if $\xi \in X^{\mathbb{N}}$ satisfies the following two conditions:

(1) $\forall\alpha \in Mon_s\ \exists\beta \in Mon_s\quad \xi\circ\alpha\circ\beta \in C$

(2) $\forall\alpha \in Mon_s\ \forall\beta \in Mon_s\quad (\xi\circ\alpha\in C \text{ and } \xi\circ\beta\in C) \Rightarrow \xi\circ\alpha\wedge\xi\circ\beta\in C$

then $\xi \in C$.

Definition ([4],[1o],[11]): A sequential Cauchy structure C on X is called

(1) a C-structure on X iff it satisfies (C_1),(C_2) and (C_3)

(2) a C*-structure on X iff it satisfies (C_1),(C_2),(C_3) and (C_4).

$\underline{C}$ (resp. $\underline{C}$*) denotes the category, whose objects are pairs (X,C) with C being a C-structure (resp. C*-structure) on X, and whose morphisms are the sequentially Cauchy continuous maps.

5.2 FINAL COMPLETIONS OF CSeq

CSeq denotes the full subcategory of Unif, consisting of the single object $(\mathbb{N},\mu)$, where μ consists of all covers of $\mathbb{N}$ containing a member C such that $\mathbb{N}\setminus C$ is finite.

Definition: A subset M of $\mathbb{N}$ is called balanced iff M and $\mathbb{N}\setminus M$ are infinite; otherwise M is called unbalanced.

Proposition: For a map $f\colon \mathbb{N} \to \mathbb{N}$ the following conditions are equivalent:

(1) $f\colon (\mathbb{N},\mu) \to (\mathbb{N},\mu)$ is a morphism in CSeq

(2) f is finite-to-one or eventually constant

(3) $f^{-1}(n)$ is unbalanced for each $n \in \mathbb{N}$

(4) $f^{-1}[M]$ is unbalanced for each unbalanced set M.

Theorem 5: Equivalent are:

(1) (X,C) is an object of Max (CSeq)

(2) (X,C) satisfies the conditions (C_1), $(C_2\text{-Fin})$ and $(C_2\text{-Mon}_f)$

(3) (X,C) satisfies the conditions (C_1), $(C_2\text{-Perm})$, $(C_2\text{-Mon}_i)$ and $(C_2\text{-Mon}_f)$.

Proposition: For a collection $\mathcal{F}$ of morphisms in CSeq the following conditions are equivalent:

(1) $\mathcal{F}$ is final in CSeq

(2) for each balanced set M there exists some f in $\mathcal{F}$ such that $f^{-1}[M]$ is balanced.

Theorem 6: Equivalent are:

(1) (X,C) is an object of Univ (CSeq)

(2) (X,C) is an object of Max (CSeq), which satisfies the following condition:

a sequence ξ belongs to C provided for each balanced set M there exists some α in Fin with $f\circ\alpha\in C$ and $\alpha^{-1}[M]$ balanced.

Proposition: Univ (CSeq) is a full subcategory of $\underline{C}^*$ and $\underline{C}^*$ is a full subcategory of Max (CSeq).

Remark: The following example shows that Univ (CSeq) is very thin.

Example: Let C be the collection of all Cauchy sequences in the closed unit interval I. Then (I,C) is a $\underline{C}^*$-object, which does not belong to Univ (CSeq). In fact, if $\xi: \mathbb{N} \to I \cap \mathbb{Q}$ is a bijection, then $\xi \notin L$, but for each balanced set M there exists $\alpha \in Mon_s$ with $\xi \circ \alpha \in L$ and $\alpha^{-1}[M]$ balanced.

REFERENCES

[1] P. Alexandroff and P. Urysohn, Une condition nécessaire et suffisante pour qu'une classe (L) soit une classe (D). C.R.Acad.Sci. Paris 177 (1923) 1274-1276.

[2] M. Fréchet: Sur quelques points du calcul fonctionel. Rend. Circ. Mat. Palermo 22 (19o6) 1-74.

[3] M. Fréchet: Les espaces abstraits. Paris 1926.

[4] R. Frič and V. Koutník: Sequential structures. Convergence Structures and Applications to Analysis, Abh. Akad. Wiss. DDR, Abt. Math.-Naturwiss.-Technik, 1979, Nr. 4N. Akademie Verlag, Berlin 198o, 37-56.

[5] A. Goetz: A notion of uniformity for L-spaces of Fréchet. Gen. Topol. and its Relat. to Mod. Analysis and Algebra, Proc. First Prague Topol. Symp, 1961, Praha 1962, 177-178.

[6] A. Goetz: A notion of uniformity for L-spaces of Fréchet. Colloq. Math. 9 (1962) 223-231.

[7] H. Herrlich: Initial completions. Mathem. Z. 15o (1976) 1o1-11o.

[8] H. Herrlich: Initial and final completions. Categorical Topology, Lecture Notes in Mathematics 719, Springer Verlag 1979, 137-149.

[9] H. Herrlich: Universal topology. Proc. Conf. Categ. Topology, Toledo 1983, Heldermann Verlag (to appear).

[1o] J.M. Irwin and D.C. Kent: Sequential Cauchy spaces. Proc. Spec. Session Convergence Structures, University of Nevada 1976, 6o-82.

[11] J.M. Irwin and D.C. Kent: Sequential Cauchy spaces. Math. Slovaca 29 (1979) 117-13o.

[12] L.D. Nel: Topological universes and smooth Gelfand-Naimark duality. Math. appl. of category theory, Proc. AMS Spec. Session Denver 1983, Contemporary Mathematics (to appear).

[13] J. Penon: Sur les quasi-topos, Cahier Topol. Géom. Diff. 18 (1977) 181-218.

SEQUENTIAL UNIFORM SPACES

Miroslav Husek [1]

This talk is a survey of what is known about uniform spaces which can be described by sequences,i.e. a counterpart in uniform spaces of sequential topological spaces.In fact,it will be a survey of basic properties only because,up to now,there was no intensive study of these structures.For that reason,one may look at this paper also as at an advertisement of sequential uniform spaces; I am really convinced that they may be useful in various situations.The readers will see that some properties correspond to the analogous properties of sequential topological spaces but some of them are quite different (like e.g. productivity), which might be used in those situations where the topological approach does not work.At the end,some suggestions of possible investigation are given.

In comparison with the situation described above,there are relatively many papers concerning uniform sequential (convergence) spaces. Before we come to the main topic,let me tell a few words about the concepts mentioned just now.Uniform convergence spaces in a general meaning are convergence spaces (i.e.,their structure is described by means of nets or filters),where the defining axioms are uniform in some sense. Usually,general topological or uniform convergence spaces form a class containing all topological or uniform spaces; therefore,they may be regarded as an outside look at classical structures.Various examples of uniform convergence spaces and their investigation may be found e.g. in the papers of A.Goetz,C.H.Cook and H.R.Fischer,D.C.Kent,H.H.Keller, H.C.Biesterfeldt Jr.,B.Sjöberg,R.Frič,V.Koutník,R.J.Gazik,E.E.Reed, W.Gähler and others (this list is certainly not complete).Among the quoted papers one can find both general convergence structures and their restrictions to sequential ones.

Now,what about the exchange of the words "uniform" and "convergence",i.e. the convergence uniform spaces? These should be uniform spaces described by means of nets or filters,i.e. an inside look at the classical structures.Of course,we must define in a precise way what we mean by "described by nets or filters".As in topological spaces,one may expect that all the uniform spaces can be (in a natural way) described by nets or filters; then,of course,it is reasonable to consider only spaces described by means of some special nets or filters.The structures described by sequences are certainly the most important example of such a situation (another example are structures described by well-ordered nets).One can find sequential uniform spaces (and sequential proximity

[1]) Mathematical Institute of Charles University,Sokolovská 83,
186 00 Prague,Czechoslovakia

spaces) already in papers of A.Goetz, V.Z.Poljakov, R.Frič and V.Koutník but they investigate these structures in a bigger class of generalized uniform or proximity spaces. Our sequential uniform spaces are uniform modifications of their structures similarly as functionally sequential topological spaces are just completely regular modifications of sequential topological spaces (by a functionally sequential space I mean a completely regular space on which continuous and sequentially continuous real-valued functions coincide; they were investigated, under different names, e.g. by S.Mazur, N.Noble, V.Koutník).

Imitating the known situations, one must find some net-spaces that inductively generate (see the definition below) all the structures and then the sequential structures are those inductively generated by the sequence-spaces. In topological spaces, the underlying set of a net-space is the net itself together with some other (say limit) point endowed with the topology having the unique accumulation point, namely the limit point, with the "tails" of the net as its neighborhoods. Before we shall proceed in an analogous way in uniform spaces, we recall several basic concepts in order to avoid misunderstanding (for other concepts used here see [Č], [E], [I]).

By *uniform space* we mean a pair (X,u), where u is a filter of covers on the set X (filter with respect to star-refinement); when convenient, we shall use the corresponding uniform neighborhoods of diagonal instead of uniform covers. The family $\{f_i:(X_i,u_i)\to(X,u)\mid i\in I\}$ *inductively generates* the space (X,u) or the uniformity u if u is the finest uniformity rendering f_i uniformly continuous (u is finer than v if $u\supset v$); then f on (X,u) into a uniform space is uniformly continuous iff all compositions $f\circ f_i$ are uniformly continuous. The inner description of u by means of u_i's and f_i's is not simple: u consists of all covers initiating a normal sequence of covers in the set $\{U\mid U$ is a cover of $X, f^{-1}(U)\in u_i$ for all $i\in I\}$ (a sequence $\{U_n\}$ of covers of X is said to be normal if U_{n+1} is a star-refinement of U_n for any n). The dual notion to that of inductive generation is called *projective generation*. The special cases of inductive generation are *sum* and *quotient*. The former is obtained by taking for X the sum ΣX_i (a disjoint union of X_i's) and for f_i the canonical embedding of X_i into ΣX_i. The sum-uniformity has for its base the covers the restrictions of which to X_i's belong to u_i. The quotient of a space (Y,v) along a surjective mapping f on Y onto X is the finest uniformity u on X such that $f:(Y,v)\to(X,u)$ is uniformly continuous. Notice that if $f:(Y,v)\to(X,u)$ is inductively generating and not surjective, then (X,u) is the sum of the quotient $(fY,u\mid fY)$ and of the space $X-fY$, endowed with the uniformly discrete uniformity (any cover is uniform). It is easy to show that $\{f_i:(X_i,u_i)\to(X,u)\}$ is inductively generating iff $(\cup f_iX_i, u\mid\cup f_iX_i)$ is the quotient of $\Sigma(X_i,f_i)$ along the map induced by $\{f_i\}$ and the restriction of u on $X-\cup f_iX_i$ is uniformly discrete. Thus the expression

"C is closed under inductive generation", which means that if $\{f_i:(X_i,u_i)\to(X,u)\}$ is inductively generating and $(X_i,u_i)\in C$ for all i then $(X,u)\in C$ as well, is equivalent to "C contains all uniformly discrete spaces and is closed under forming sums and quotients"; such classes are called *coreflective* since for any space (X,u) there is a finer uniform space (X,u_C) belonging to C and such that every uniformly continuous mapping on a member of C into (X,u) is uniformly continuous into (X,u_C) as well ((X,u_C) is called the *coreflection* or the *modification* of (X,u) in C). Clearly, $(X,u)\in C$ iff $u=u_C$.

Now, we are prepared to define the desired net-spaces in the class of uniform spaces. One point of view is to take all uniform neighborhoods of the diagonal of a uniform space (X,u), to think of the diagonal as a point and to take all the nets in $X\times X$ converging to the diagonal in the usual (topological) way. Formally: two nets $\{x_a|a\in A\}$, $\{y_a|a\in A\}$ in X over the same index set A are said to be *adjacent* in a uniform space (X,u) if for every uniform neighborhood U of Δ_X (or every uniform cover $\mathcal{U}$ from u) there is $a\in A$ such that for every $b\in A$ with $b\geq a$ we have $(x_b,y_b)\in U$ (or $x_b\in \mathrm{St}_{\mathcal{U}}y_b$, resp.). The equivalence relation of adjacency replaces the notion of convergence used in topological spaces (notice that if $\{y_a\}$ is a constant net with a value y, then $\{x_a\}$, $\{y_a\}$ are adjacent iff $\{x_a\}$ converges to y in the topology induced by the uniformity). The corresponding net-space has for its underlying set $A\times(0,1)$ (i.e. the disjoint union of two copies of the index set A) and the uniformity has for its base covers $\{(x)\,|\,x\in A\times(0,1)\}\cup\{((a,0),(a,1))\,|\,a\geq b\}$ for all $b\in A$:

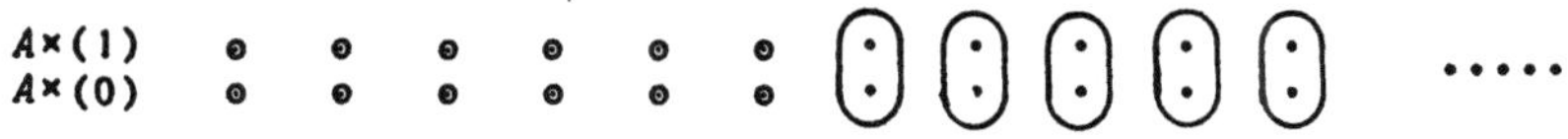

The corresponding map of this space into X is $f(a,0)=x_a$, $f(a,1)=y_a$. We will not prove here the fact that all adjacent pairs of nets determine the uniformity (or equivalently, that every uniform space is inductively generated by all uniformly continuous maps from the above net-spaces into the given uniform space). If u is a uniformity on X and C is the class of all adjacent pairs of nets with respect to (X,u), then u is determined backwards by C as follows: $U\subset X\times X$ is said to be a semiuniform neighborhood of Δ_X with respect to C if for every $(\{x_a\},\{y_a\})\in C$ the net $\{(x_a,y_a)\}$ lies eventually in U, and U is a uniform neighborhood of Δ_X in (X,u) if it initiates a normal sequence of semiuniform neighborhoods of Δ_X with respect to C. To describe completely the equivalences determined by uniformities one must use a condition similar to the Uryson axiom for sequential topological spaces, see e.g. [Go_2].

One can see from the above procedure that one may restrict the class C determining a uniformity u to those pairs of nets having for ordered domains the ordered set u (the order in u is "refinement" or "star-refinement") or its base. Thus metrizable uniform spaces are determined by their equivalence class of adjacent sequences, or equivalently, they are inductively generated by maps from the sequence-space $N\times(0,1)$ described above.

One more observation we shall need is the fact that a mapping on a uniform space (X,u) into a uniform space (Y,v) is uniformly continuous iff the nets $\{fx_a\},\{fy_a\}$ are adjacent in (Y,v) provided that the nets $\{x_a\},\{y_a\}$ are adjacent in (X,u).(Notice the analogy with the topological case.) Historically,the description of uniform structures by adjacent nets was used in [EŠ],and their corresponding net-spaces appeared in [I].

DEFINITION. A uniform space or a uniformity is said to be *sequential* if it is inductively generated by mappings from the sequence-space $N\times(0,1)$.

We already know that metrizable uniform spaces are sequential. Since the class of sequential uniform spaces is coreflective,hence closed under forming sums and quotients,one obtains many nonmetrizable sequential uniform spaces e.g.by taking infinite sums of uniformly nondiscrete metrizable spaces (such a sum is never metrizable as a uniform space) or a convenient uniform quotient of a metrizable space (by [HP],such a quotient may be far from being metrizable).

Using the adjacency-relation,we could define sequential uniform spaces as those that are determined by some equivalence relation between sequences.This approach is convenient for instance for describing the *sequential uniform modification* (coreflection) of a given uniform space (X,u): it is the space (X,u_s) where u_s is determined in the sense described above by all pairs of sequences which are adjacent in (X,u) (thus u is a sequential uniformity iff $u=u_s$).

The reader acquainted with sequential topological spaces can see a close analogy between the topological and uniform approaches.We shall introduce another such analogical situation but after that,several quite opposite properties of the two cases will be shown.

DEFINITION. A mapping $f:(X,u)\to(Y,v)$ of uniform spaces is said to be *sequentially uniformly continuous* if $\{fx_n\}$ and $\{fy_n\}$ are adjacent sequences in (Y,v) for any two adjacent sequences $\{x_n\},\{y_n\}$ in (X,u).

PROPOSITION 1. *A uniform space X is sequential iff every sequentially continuous mapping on X into any uniform (or metrizable uniform) space is uniformly continuous.*

The proof follows for instance from the fact that $\{f_i:(X_i,u_i)\to(X,u)\}$ is inductively generating iff a mapping $g:(X,u)\to(Y,v)$ is uniformly continuous if and only if all the compositions $g\circ f_i$ are uniformly continuous.(Realize that Proposition 1 is clearly true if we take for X the sequence space $N\times(0,1)$.) We may restrict the ranges of maps from Proposition 1 to metrizable uniform spaces because these spaces (projectively) generate all uniform spaces (every uniform space can be embedded into a product of pseudometrizable spaces).Notice that such a restriction of ranges is not possible for sequential topological spaces (one would obtain the functionally sequential spaces).The characterization given in Proposition 1 is useful in some situations as one can see from the following assertion,where uX is the finest uniform space inducing the completely regular modification of X (uX is projectively generated by all continuous mappings on X into metric spaces); uX is called the fine uniform space generated by X.

PROPOSITION 2. *If X is a sequential topological space,then the fine uniform space uX generated by X is a sequential uniform space.*

Proof. Take a sequentially uniformly continuous mapping f on uX.Since, as we have shown above,any converging sequence is adjacent to its limit point,f is sequentially continuous on X,hence continuous on X.By the definition of uX,f is uniformly continuous on uX,which,by Proposition 1, completes the proof.

The Proposition 2 provides a big supply of new sequential uniform spaces,but the class of sequential uniform spaces is still larger (not all metric uniform spaces are fine,e.g.the sequence-space is not fine). The converse of Proposition 2 is not true - see Example 3.

The next table compares the basic properties of sequential spaces in topological structures and in uniform structures.The symbol + (or -) means that the class in the corresponding column is (or is not,resp.) closed under the construction in the corresponding row;the symbol +- means dependence on models of set-theory: in some of them "+" is valid, in the others "-" is valid.The name of a column means in what class the sequential spaces are considered: $\mathcal{Cl}$ are closure spaces in the sense of [Č],*Top* topological spaces,Top_{cr} completely regular spaces,*SUnif* semi-uniform spaces in the sense of [Č],*Unif* uniform spaces.

	$\mathcal{Cl}$	*Top*	Top_{cr}	*SUnif*	*Unif*
sums	+	+	+	+	+
quotients	+	+	+	+	+
subspaces	+	- [F]	-[Ko_1,Ex.4]	+	- Ex.1
open subspaces	+	+	+	+	- Ex.1
closed subspaces	+	+	- Ex.3	+	- Ex.3
finite products	-[Č,35C.14]	- [F]	-[Ko_1,Ex.5]	- Ex.2	+ Prop.3
countable products	-	-	-	-	+ Prop.4
products	-	-	-	-	+-Prop.5

All the symbols "+" follow almost directly from the definitions except the productivity properties in the last column and,perhaps,the open heredity of functionally sequential spaces.This last property can be proved as follows: Suppose that X is a functionally sequential space and Y is its open subspace. Assume there is a sequentially continuous real-valued function f on Y and $x\in\bar{A}$ for $x\in Y, A\subset Y$ such that $fx\notin\overline{fA}$; we may suppose that $fx=1$ and $fa=0$ for $a\in A$. One can find a continuous function g on X such that $gx=1$ and $gz=0$ for $z\in X-G$, where G is a neighborhood of x with $\bar{G}\subset Y$. Then the function defined as 0 on $X-Y$ and $gy\cdot fy$ on Y is sequentially continuous on X, hence continuous but it is 0 on A and 1 at x, which is a contradiction.

An example showing that the class of sequential uniform spaces is not hereditary is similar to the analogous example from *Top*, [F] and Top_{ch}, [Ko_1] (moreover, it shows that the corresponding property is not hereditary with respect to G_δ- and F_σ-subspaces).

<u>EXAMPLE 1</u>. Let $X=N\times N\cup N\cup(\infty)$ with the topology determined by the following convergent sequences:

$$(k)\times N\to k \text{ for } k\in N, \text{and } N\to\infty.$$

The uniformity uX is, by Proposition 2, a sequential uniformity. Basic uniform covers of uX are of the form

$$\{(x)\,|\,x\in X\}\cup\{\{(k,n)\,|\,n>fk\}\cup(k)\,|\,k\in N\}\cup(\{(k,n)\,|\,k>m,n>fk\}\cup(\infty))$$

for $m\in N, f\in {}^N N$ (a function on N into N). In *Top*, the corresponding subspace which is not sequential is $N\times N\cup(\infty)$. In *Unif* it is better to look at the subspace $N\times N$. The only nontrivial adjacent sequences $\{x_n\},\{y_n\}$ (i.e., $x_n\neq y_n$ for all n) in $N\times N$ are infinite subsets of the same column $(k)\times N$ for some $k\in N$; indeed, if one of the sequence, say $\{x_n\}$, has an infinite projection on the first coordinate space then, finding an $f\in {}^N N$ with the values greater than the corresponding x_n's, there is a basic cover U such that the stars of the corresponding x_n's with respect to U are singletons and, hence, equal to y_n's for all but finitely many indices, which contradicts the assumption. Consequently, the sequential uniform modification of $N\times N$ has basic covers $\{(x)\,|\,x\in N\times N\}\cup\{\{(k,n)\,|\,n>fk\}\,|\,k\in N\}$ for $f\in {}^N N$. Since $N\times N$ is open in X, our example shows also that the class of sequential uniform spaces is not open-hereditary. The example showing that, unlike in *Top*, the classes of sequential uniform spaces and of functionally sequential spaces are not closed-hereditary will be given later (Example 3).

Now, we shall look at products. As it is shown in our table, the classes of sequential topological structures are not finitely productive. The same is true for sequential semiuniform spaces:

<u>EXAMPLE 2</u>. Let X be the set ω_1 of countable ordinals, C_1 be the set of sequences $\{(\alpha_n,\alpha_n+k)\,|\,n\in\omega\}$ in $X\times X$ for $k\in\omega$ and for increasing sequences $\{\alpha_n\}$ in X, and C_2 be the set of sequences $\{(\beta,\beta+n)\,|\,n\in\omega\}$ in $X\times X$ for $\beta\in X$. The semiuniformity u_1 generated by C_1 has for its base the sets $U\cup U^{-1}\cup\Delta_X$, where $U=\cup\{\{(\alpha,\alpha+n)\,|\,\alpha\in X\}-F_n\,|\,n\in\omega\}$ and F_n are finite sets. The semiuniformity u_2 generated by C_2 has for its base the sets $V\cup V^{-1}\cup\Delta_X$, where $V=\cup\{\{(\beta,\beta+n)\,|\,n\in\omega\}-F_\beta\,|\,\beta\in X\}$ and F_β are finite sets. The uniformity

$u=\inf(u_1,u_2)$ has for its base the sets $(U\cap V)\cup(U^{-1}\cap V^{-1})\cup\Delta_X$ for the above sets U,V. One can easily show that u is not generated by pairs of sequences (u-adjacent sequences must eventually coincide). Since (X,u) can be embedded into the product $(X,u_1)\times(X,u_2)$ (onto the diagonal), the product cannot be sequential.

The situation in *Unif* concerning products is quite different from the other classes. The following results about products of sequential uniform spaces were proved in [HR], partially also in [H_3]. To keep this survey self-contained, we shall give here (a little bit modified) proofs with all the details modulo the following facts from the theory of uniform spaces:

(I) In the class of uniform spaces, every product of quotient mappings is quotient again, [HR].

(II) In the class of uniform spaces, the product $X\times Y$ is the finest space coarser than $DX\times Y$ and $X\times DY$, where DP means the set P endowed with the uniformly discrete uniformity. (Thus $X\times Y$ is inductively generated by identity maps from $DX\times Y$ and $X\times DY$.)

PROPOSITION 3. *Finite products of sequential uniform spaces are sequential.*

Proof. It suffices to show that $X\times Y$ is sequential provided both X,Y are sequential uniform spaces. Because of (II), we may assume that X is uniformly discrete. By definition, Y is a quotient of a sum ΣM_i of metric spaces. Hence, by (I), $X\times Y$ is a quotient of $X\times\Sigma M_i$. Since X is uniformly discrete, $X\times\Sigma M_i$ equals to $\Sigma(X\times M_i)$ and, thus, $X\times Y$ is a quotient of the sum $\Sigma(X\times M_i)$ of metric spaces.

The next step is to show validity of Proposition 3 for countable products (the seventh row in our table). The passing from finite to countable productivity uses special properties of uniform spaces again.

PROPOSITION 4. *Countable products of sequential uniform spaces are sequential.*

Proof. Take sequential uniform spaces X_n for $n\in\omega$ and a sequentially uniformly continuous mapping f on ΠX_n into a metric space (M,d). The product ΠDX_n is finer than ΠX_n and it is metrizable so that f is uniformly continuous on ΠDX_n. For any $\varepsilon>0$ there is a finite set $F\subset\omega$ such that $d(fx,fy)<\varepsilon$ whenever $x,y\in\Pi X_n$, $pr_F x=pr_F y$. Take some point $a\in\Pi X_n$ and for $x\in\Pi X_n$ define $gx=fx_a$, where $pr_n x_a$ equals to $pr_n x$ or to $pr_n a$ depending on whether $n\in F$ or $n\notin F$ (roughly speaking, g is determined by the restriction of f to the canonical embedding of $\prod_F X_n$ onto $(pr_{\omega-F}a)\times\prod_F X_n\to\Pi X_n$). Clearly, g is sequentially uniformly continuous, hence uniformly continuous by Proposition 3, and $d(fx,gx)<\varepsilon$ for all $x\in\Pi X_n$. Consequently, f is uniformly approximated by uniformly continuous mappings, which implies that f is uniformly continuous, too. By Proposition 1, ΠX_n is sequential.

It would not be nice to continue in proving productivity of ω_1,ω_2 etc. of coordinate spaces. Fortunately, there exists a method invented by S. Mazur in [M], how to prove the productivity results for all the remaining possible cases. We shall see that Proposition 4 will be needed in our proof, and to prove Proposition 4, we needed Proposition 3 (so, we did not waste time up to now and also, since the proofs of both preceding Propositions are simple, it may be convenient to have them formulated and proved separately).

S. Mazur investigated products of functionally sequential topological spaces (see also [N] for generalizations). Denote by δ the first cardinal such that there exists a sequentially continuous mapping ${}^{\delta}2 \to R$ which is not continuous. It was proved in [M] that δ is weakly inaccessible and by D.V. Čudnovskii (see [Ču], [AČu]) that there are many weakly inaccessible cardinals before δ and that δ coincides with both the real-valued and two-valued (Ulam) measurable cardinals m_R, m_2 under some additional assumptions on set-theory (e.g. under Martin axiom or under the weaker condition that any intersection of less than 2^{ω} open dense subsets of R is nonempty). We can transfer this situation to uniform spaces.

<u>DEFINITION</u>. Denote by u the first cardinal such that there exists a sequentially uniformly continuous mapping ${}^{u}2 \to R$ which is not continuous.

Trivially, u is uncountable, and it is not difficult to show that $\delta \le u \le m_R$, [HR], so that the above results for δ apply to u as well.

<u>PROPOSITION 5</u>. *The product of less than u of sequential uniform spaces is sequential.*

Proof. Take a cardinal κ with $\kappa < u$, and for $\alpha \in \kappa$ take a sequential uniform space X_α. Let $f: \Pi X_\alpha \to (M,d)$ be a sequentially uniformly continuous mapping into a metric space. We want to show that f is uniformly continuous. By Proposition 4, it suffices to find a countable $J \subset K$ such that f depends on J, i.e. that $f = g \circ pr_J$ for some $g: \Pi_J X_\alpha \to M$ (because g is sequentially uniformly continuous). Define $J_n = \{\alpha \in \kappa \mid d(fx^\alpha, fy^\alpha) > 1/n$ for some x^α, y^α with $pr_\beta x^\alpha = pr_\beta y^\alpha$ for all $\beta \ne \alpha\}$. Then each J_n is finite (otherwise the corresponding x^α's and y^α's would form adjacent sequences with non-adjacent f-images). Thus the set $J = \cup\{J_n \mid n \in \omega - (0)\}$ is countable and we claim that f depends on J. Assume that $x, y \in \Pi X_\alpha$ with $pr_J x = pr_J y$ and define $h: {}^{\kappa}2 \to R$ as follows:

$$hr = d(f(r \cdot x), f(r_J \cdot x)) ,$$

where $r \cdot x$ is the point of ΠX_α with

$$pr_\alpha(r \cdot x) = \begin{cases} pr_\alpha x & \text{if } pr_\alpha r = 1 \\ pr_\alpha y & \text{if } pr_\alpha r = 0 \end{cases} ,$$

and r_J is the point of ${}^{\kappa}2$ with

$$pr_\alpha(r_J) = \begin{cases} pr_\alpha r & \text{if } \alpha \in J \\ 0 & \text{if } \alpha \notin J \end{cases}$$

The mapping h is a result of composition of the uniformly continuous maps

${}^{\kappa}2 \to \Pi X_\alpha$ (namely $\{r \to r\cdot x\}$,$\{r \to r_J\cdot x\}$ that can be regarded as products of mappings on 2 into X_α for $\alpha\in\kappa$),of the sequentially uniformly continuous mapping f and,finally,of the uniformly continuous mapping d .Therefore, h is sequentially uniformly continuous and,hence (uniformly) continuous by our assumption $\kappa<u$.If r is a point of ${}^{\kappa}2$ with only finitely many non-zero coordinates,then $hr=0$ (indeed,if $pr_\alpha r=1$ and $\alpha\notin J$,then $fy=f\tilde{y}$,where $\tilde{y}$ has all but the αth coordinate as y has and the other equals to $pr_\alpha x$;by induction one gets $fy=f(r\cdot x)$ and,clearly,$y=r_J\cdot x$).Since the point of ${}^{\kappa}2$ with all coordinates 1 is an accumulation point of just examined points r,the value of h at $\{1\}$ must be zero,too.But,$h\{1\}=d(fx,fy)$ and,thus, $fx=fy$,which was to be proved.

COROLLARY. *It is consistent with ZFC that the class of sequential uniform spaces is productive.*

Proof. There are models of set-theory with ZFC having no (or not enough) weakly inaccessible cardinals,so that u does not exist in them.

Now,we are able to provide the promissed example that the class of sequential uniform spaces is not closed-hereditary.Of course,such an example may be given without using Proposition 5,but then it is more difficult to show the required properties.

EXAMPLE 3. The power ${}^{\omega_1}2$ of the two-point uniformly discrete space is sequential,by Proposition 5,and contains as a closed subspace the topological ordered space ω_1+1 of countable ordinals and of the first uncountable one.The compact uniform space ω_1+1 is certainly not sequential (its sequential modification has the point ω_1 as an isolated point). Since ${}^{\omega_1}2$ is not a sequential topological space but it is a sequential uniform space which is compact (hence fine),the assertion in Proposition 2 cannot be converted.The same example is used to show that the class of functionally sequential spaces is not closed-hereditary: by [M],${}^{\omega_1}2$ is functionally sequential and its closed subspace ω_1+1 is not.

We have completed a discussion about our table.Naturally,there are much more interesting properties of sequential uniform spaces to be discussed and while we shall look at some of them there are certainly many more which wait for somebody to investigate them.We shall suggest some possible investigations as Problems (but I am not quite sure that all of them are nontrivial).We shall leave to the readers to find out relations between the sequentiality and various modifications (e.g.find a nonsequential uniform space such that its precompact or fine or another modification is sequential,and conversely:the given space is sequential but its modification is not).Notice that the Example 1 provides an example of the situation that inverse limits do not preserve sequentiality of uniform spaces.Indeed,$X-N=\bigcap_n\{X-\{0,1,\dots,n\}\}$, $X-N$ is not sequential but all the spaces $X-\{0,1,\dots,n\}$ are sequential uniform spaces.

In topological spaces,sequential ones are just quotients of metric spaces. This fact is caused by preserving metrizability by sums, which does not hold in uniform spaces. To prove that quotients of metrizable uniform spaces do not cover all sequential spaces one can use the result proved in [HP], namely that the uniform weight of any quotient of metrizable space does not exceed the cardinality 2^{ω} of continuum (more precisely, the cofinality of the ordered set ${}^{\omega}\omega$ of mappings on ω into ω); clearly, the uniform weight of a sum of metrizable uniform spaces may be as big as we wish. (Recall that the uniform weight of a uniform space (X,u) is the least cardinality of a base of u.)

It is easy to find examples that the completion of a nonsequential uniform space is sequential (e.g. the space X from Example 1 is the completion of $N \times N$), and conversely that the completion of a sequential uniform space is not sequential (e.g. the completion of the ordered space ω_1 is ω_1+1). Following functional analysis, a uniform space is said to be *sequentially complete* if every its Cauchy sequence converges. The class of sequentially complete separated spaces is productive and (sequentially) closed-hereditary in $\mathcal{U}nif$, so that every uniform space X has a sequential completion (it is the sequential closure of $\pi(X)$ in the completion $\pi : X \to \pi X$ of X, i.e. the sequential closure of X in πX if X is separated). The above using of Example 1 shows that the sequential completion of X may be a sequential uniform space without X being sequential. If X is a sequential uniform space then its sequential completion is a sequential uniform space as well (indeed, if f is sequentially uniformly continuous on the sequential completion of X, then its restriction g to X is uniformly continuous and hence the uniform extension of g coincides with f on the sequential completion of X).

PROBLEM 1. Investigate properties of sequential completions. For instance, consider the class of spaces such that their usual completion is already the sequential completion; investigate the relation of the sequential completion to extensions of sequential spaces in $\mathcal{T}op$, $\mathcal{T}op_{c\ell}$, etc.; when the sequential completion commutes with basic constructions or modifications?

Going from uniform spaces to topological ones, we obtain a generalization of sequentiality in completely regular spaces, i.e. the spaces the topology of which is induced by a sequential uniformity. This new class is "almost" productive.

PROBLEM 2. Investigate the class of just defined generalized sequential spaces. Is it possible to define them internally without using uniform spaces? What are the properties of this new class? How to extend this class in a reasonable way to topological spaces which are not completely regular?

One way to answer the last question is to take the sequential semiuniform spaces, the closure structure induced by them and then the topological modifications. Another way would be to take the so-called quasiuniform spaces (they induce exactly all topological spaces), to define sequentiality there and then take the induced topology (in this case the sequence-spaces must be changed since uniform spaces are both coreflective and epireflective in quasiuniform spaces, so that one must take a "nonsymmetric" sequence-space). Of course, one may go via quite different general structures inducing topological spaces. In any case it is useful to look at the sequentiality defined in those general structures (for the case of nearness spaces see the contribution of H.Herrlich in these Proceedings).

PROBLEM 3. Investigate sequential spaces in various general structures like quasiuniform spaces, nonsymmetric proximity spaces, nearness spaces, etc., defined by means of convenient sequence-spaces. What are the relations between such sequential spaces? What are their properties?

It is known that uniform groups (i.e. groups with a compatible uniformity, or equivalently, topological groups with the identical left and right uniformities) have some extra properties that are not possessed by all topological groups.

PROBLEM 4. Is it possible to develop a reasonable theory of sequential uniform groups? What is its relation to the sequential topological groups?

At the end of my lecture, V.Koutník asked a very interesting question which we put here as

PROBLEM 5. Is there a counterpart of Fréchet spaces in sequential uniform spaces? One way could be to define them as hereditarily sequential uniform spaces. Is such a definition convenient as in topological spaces? Is there a nice internal characterization of this class? What about the uniform spaces, where one does not need to take normal sequences for describing the uniformity from the adjacent sequences? (Clearly, such spaces are hereditarily sequential as quotients in *SUnif* but the description is certainly not nice.)

LITERATURE

[AČu] Antonovskii, M.J., D.V.Čudnovskii: *Some questions in general topology and Tichonov semifields*. Uspechi Mat.Nauk 31 (1976), 71-128.

[B] Biesterfeldt, H.J.Jr.: *Uniformization of convergence spaces I*. Math. Ann. 177 (1968), 31-42.

[CF] Cook, C.H., and H.R.Fischer: *Uniform convergence structures*. Math. Ann. 173 (1967), 290-306.

[Č] Čech, E.: *Topological spaces* (revised by Z.Frolík and M.Katětov) Academia, Praha 1966.

[Ču] Čudnovskii, D.V.: *Sequentially continuous mappings and real-valued measurable cardinals*. Infinite and finite sets, Budapest 1973.

[E] Engelking, R.: *General topology*. PWN, Warszawa 1977.

[EŠ] Efremovič, V.A., and A.S.Švarc: *A new definition of uniform spaces*. Dokl.Akad.Nauk SSSR 89 (1953), 393-396.

[F] Franklin,S.P.: *Spaces in which sequences suffice.* Fund.Math.57 (1965),107-115.

[Fr_1] Frič,R.: *On the sequential uniform structures of A.Goetz.* Topology and its appl.Budva 1972,Beograd 1973,90-93.

[Fr_2] Frič,R.: *On the completion of sequential structures.* Topology and its appl.,Budva 1972,Beograd 1973,94-96.

[FK] Frič,R.,and V.Koutník: *Sequential structures.* Abh.Akad.Wiss.DDR Abt.Math.-Naturwiss.-Technik (1979) 4,37-56.

[Gä] Gähler,W.: *Über Limes- und Pseudouniformisierbarkeit.* Demonstratio Math.6 (1973),613-632.

[Ga] Gazik,R.J.: *On Biesterfeldt's completion axiom spaces.* Proc.Amer. Math.Soc.52 (1975),401-405.

[GK] Gazik,R.J.,and D.C.Kent: *Coarse uniform convergence spaces.* Pacific J.Math.61 (1975),143-150.

[Go_1] Goetz,A.: *A notion of uniformity for L-spaces of Fréchet.* Proc.Top. Symp.Prague 1961,Academia,Praha 1962,177-178.

[Go_2] Goetz,A.: *Proximity-like structures in convergence spaces.* Top.Conf. Arizona St.Univ.1967,Tempe,Arizona 1968,104-137.

[H_1] Hušek,M.: *Generalized proximity and uniform spaces I.* Comment.Math. Univ.Carolinae 5 (1964),247-266.

[H_2] Hušek,M.: *Generalized proximity and uniform spaces II.* Comment. Math.Univ.Carolinae 6 (1965),119-139.

[H_3] Hušek,M.: *Products of uniform spaces.* Czech.Math.J.29 (1979),130-141.

[HP] Hušek,M.,and J.Pelant: *Uniform weight of uniform quotients.* Comment. Math.Univ.Carolinae 24 (1983),335-340.

[HR] Hušek,M.,and M.D.Rice: *Productivity of coreflective subcategories of uniform spaces.* Gen.Top.Appl.9 (1978),295-306.

[I] Isbell,J.R.: *Uniform spaces.* Amer.Math.Soc.,Providence 1964.

[Kl] Keller,H.H.: *Die Limes-Uniformisierbarkeit der Limesräume.* Math.Ann. 176 (1968),334-341.

[Kn] Kent,D.C.: *On convergence groups and convergence uniformities.* Fund. Math.60 (1967),213-222.

[Ko_1] Koutník,V.: *On sequentially regular convergence spaces.* Czech.Math. J.17 (1967),232-247.

[Ko_2] Koutník,V.: *Sequential envelopes and completeness.* Proc.Symp.Ext. Theory Top.Struct.Berlin 1967,Berlin 1969,141-143.

[M] Mazur,S.: *On continuous mappings in product spaces.* Fund.Math.39 (1952),229-238.

[N] Noble,N,: *The continuity of functions on cartesian products.* Trans. Amer.Math.Soc.149 (1970),187-198.

[P] Poljakov,V.Z.: *On uniform convergence spaces.*(Russian.) Colloq.Math. 13 (1965),167-179.

[R] Reed,E.E.: *Proximity convergence structures.* Pacific J.Math.46 (1973),471-485.

[S] Sjöberg,B.: *Zur Theorie der uniformen Limesräume.* Soc.Sci.Fenn. Comment.Phys.-Math.36 (1969),25-41.

ORDERED CAUCHY SPACES

D. C. Kent[1] and R. Vainio[2]

Introduction.

Cauchy spaces provide a completion theory much more general than that obtained from metric or uniform spaces, and much simpler than that arising from uniform convergence spaces. R. Ball [1], [2] has shown the applicability of Cauchy spaces in the study of order-related completions, but until now no effort has been made to define "compatibility" between an order structure and a Cauchy structure on the same set. In the case of a lattice, one can, of course, demand that the lattice operations be Cauchy continuous. However this criterion is not appropriate for partially ordered sets, and may be too restrictive even for some lattice applications.

L. Nachbin [9] has defined an "ordered uniform space" by requiring that the uniform and order structures must arise from a common source, that being a quasi-uniform structure on the same underlying set. We adapt Nachbin's method to define an ordered uniform convergence space, and from this the definition of "ordered Cauchy structure" follows naturally. The resulting notion of "ordered Cauchy space" has an internal characterization reminiscent of Keller's original Cauchy space axioms [7].

In developing a completion theory for ordered Cauchy spaces, we follow E. Reed [10]. In this way, we obtain for each ordered Cauchy space a family of ordered Cauchy completions.

Many additional questions remain: for example, under what conditions do ordered Cauchy completions preserve uniformizability, regularity, or total boundedness?

For brevity, all proofs are omitted.

1. Preliminaries.

Let X be a set, and F(X) the set of all (proper) filters on X. If $\mathfrak{F}, \mathfrak{G} \in F(X)$, and $F \cap G \neq \phi$ for all $F \in \mathfrak{F}$, $G \in \mathfrak{G}$, then $\mathfrak{F} \vee \mathfrak{G}$ denotes the filter on X generated by $\{F \cap G: F \in \mathfrak{F}, G \in \mathfrak{G}\}$. On the other hand, if $F \cap G = \phi$ for some $F \in \mathfrak{F}$ and some $G \in \mathfrak{G}$, we say that $\mathfrak{F} \vee \mathfrak{G}$ fails to exist. If $A \subseteq X$, $A \neq \phi$, we denote by $\mathring{A}$ the filter of all oversets of A; in particular, the fixed ultrafilters are denoted by $\dot{x}$, for $x \in X$.

Turning to the product set $X \times X$, the diagonal $\{(x,x): x \in X\}$ is designated by Δ. Compositions are defined as follows: for subsets A, B of $X \times X$, $A \circ B = \{(x,y)$: there is $z \in X$ such that $(x,z) \in A$ and $(z,y) \in B\}$. Compositions and inverses of filters are defined and designated in the obvious way.

---

[1]Dept. of Pure and Appl. Math., Washington State University, Pullman, WA 99164, U.S.A.
[2]Dept. of Math., Åbo Akademi, SF-20500 Åbo 50, Finland

Definition 1.1. A set σ of filters on $X \times X$ is called a quasi-uniform convergence structure on X if:

(u_1) $\dot{\Delta} \in \sigma$

(u_2) $\mathfrak{F} \cap \mathfrak{G} \in \sigma$ whenever $\mathfrak{F}, \mathfrak{G} \in \sigma$

(u_3) $\mathfrak{F} \circ \mathfrak{G} \in \sigma$ whenever $\mathfrak{F}, \mathfrak{G} \in \sigma$ and $\mathfrak{F} \circ \mathfrak{G}$ exists .

(u_4) $\mathfrak{F} \in \sigma$ and $\mathfrak{F} \subseteq \mathfrak{G}$ implies $\mathfrak{G} \in \sigma$.

If σ satisfies the additional condition

(u_5) $\mathfrak{F} \in \sigma$ implies $\mathfrak{F}^{-1} \in \sigma$

then σ is called a uniform convergence structure on X.

Quasi-uniform convergence structures are natural generalizations of quasi-uniformities; the latter have been studied extensively (for example, see [4], whereas the former have not, to our knowledge, been previously investigated. Uniform convergence structures were introduced by Cook and Fischer [3]); another term often used for the same concept is "pseudouniforme Struktur" (see [6]).

A subset σ' of a quasi-uniform convergence structure σ is called a base for σ if, for each $\mathfrak{G} \in \sigma$, there is $\mathfrak{B} \in \sigma'$ such that $\mathfrak{B} \subseteq \mathfrak{G}$. One convenient base for any quasi-uniform convergence structure σ is $\sigma_\Delta = \{\mathfrak{G} \in \sigma : \mathfrak{G} \subseteq \dot{\Delta}\}$.

The set $Q_u(X)$ of all quasi-uniform convergence structures on X forms a complete lattice under the natural ordering (the dual of set inclusion). If $\rho, \sigma \in Q_u(X)$, then $\rho \wedge \sigma$ has for a base all finite compositions $\mathfrak{G}_1 \circ \ldots \circ \mathfrak{G}_n$, where, for each j, $\mathfrak{G}_j \in \rho_\Delta$ or $\mathfrak{G}_j \in \sigma_\Delta$. If $\sigma \in Q_u(X)$, then $\sigma \vee \sigma^{-1}$ is the uniform convergence structure on X induced by σ. The lattice operations "$\vee$" and "$\wedge$", applied to quasi-uniform convergence structures (including uniform convergence structures) will always be taken relative to the lattice $Q_u(X)$.

Definition 1.2. A set $\mathbf{C}$ of filters on X is a Cauchy structure on X if:

(c_1) $\dot{x} \in \mathbf{C}$, for all $x \in X$

(c_2) $\mathfrak{F} \in \mathbf{C}$ and $\mathfrak{F} \subseteq \mathfrak{G}$ implies $\mathfrak{G} \in \mathbf{C}$

(c_3) If $\mathfrak{F}, \mathfrak{G} \in \mathbf{C}$ and $\mathfrak{F} \vee \mathfrak{G}$ exists, then $\mathfrak{F} \cap \mathfrak{G} \in \mathbf{C}$.

With every uniform convergence structure μ on X, there is associated the induced Cauchy structure $\mathbf{C}_\mu = \{ \mathfrak{F} \in F(X) : \mathfrak{F} \times \mathfrak{F} \in \mu\}$. On the other hand, one can associate with each Cauchy structure $\mathbf{C}$ a finest compatible uniform convergence structure $\mu^{\mathbf{C}}$, with base consisting of all finite intersections of filters of the form $(\mathfrak{F} \times \mathfrak{F}) \cap \dot{\Delta}$, where $\mathfrak{F} \in \mathbf{C}$ (one can show that this set of finite intersections is closed under finite compositions). Also associated with a Cauchy structure $\mathbf{C}$ on X is the convergence structure $q_{\mathbf{C}}$, defined by: $\mathfrak{F} \to x$ in $(X, q_{\mathbf{C}})$ iff $\mathfrak{F} \cap \dot{x} \in \mathbf{C}$.

If $f : (X_1, \mathbf{C}_1) \to (X_2, \mathbf{C}_2)$ is a function from one Cauchy space to another, then f is Cauchy-continuous if $f(\mathfrak{F}) \in \mathbf{C}_2$ whenever $\mathfrak{F} \in \mathbf{C}_1$.

We next consider the notation and terminology associated with an order relation $\leq$ on X. We shall always assume that $(X, \leq)$ is a poset, and we identify the order relation with its graph $\mathfrak{G} = \{(x,y): x \le y\} \subseteq X \times X$. Indeed, it will usually be convenient to designate the poset by $(X,\mathfrak{G})$ rather than the more conventional notation $(X,\le)$.

Let $(X,\mathfrak{G})$ be a poset; let $x \in X$, $A \subseteq X$, and $\mathfrak{F} \in F(X)$. Let $i(x) = \{y: x \le y\}$, let $i(A) = \cup\{i(x): x \in A\}$, and let $i(\mathfrak{F})$ be the filter generated by $\{i(F): F \in \mathfrak{F}\}$. The corresponding "decreasing" concepts are defined dually and denoted by $d(x)$, $d(A)$, and $d(\mathfrak{F})$, respectively. Furthermore, let $\hat{A} = d(A) \cap i(A)$, and let $\hat{\mathfrak{F}} = d(\mathfrak{F}) \vee i(\mathfrak{F})$. A set A (respectively, filter $\mathfrak{F}$) is convex if $A = \hat{A}$ (respectively, $\mathfrak{F} = \hat{\mathfrak{F}}$).

If $(X_1, \mathfrak{G}_1)$ and $(X_2, \mathfrak{G}_2)$ are posets, a function $f:(X_1, \mathfrak{G}_1) \to (X_2, \mathfrak{G}_2)$ is increasing if $f(\mathfrak{G}_1) \subseteq \mathfrak{G}_2$. (Here and elsewhere we shall use the same symbol f for a function from X_1 into X_2 and also for the induced function mapping $X_1 \times X_1$ into $X_2 \times X_2$; the latter function is, of course, given by $f(x,y) = (f(x), f(y))$.)

One last concept which we shall mention in this preliminary section is the convergence ordered space; by this we mean a triple $(X, \mathfrak{G}, q)$, where $\mathfrak{G}$ is a partial order, and q a convergence structure (or "Limitierung" in the sense of Fischer) subject to the "convexity" condition: $\hat{\mathfrak{F}} \to x$ whenever $\mathfrak{F} \to x$. A convergence ordered space is T_2-ordered if $\mathfrak{G}$ is a closed subset of $X \times X$. This latter property is stronger than the usual T_2 (or Hausdorff) property and plays an important role in the study of ordered topological and convergence spaces (see, for instance, [4], [8], and [9]).

2. Ordered Cauchy Spaces.

Throughout this section, $(X, \mathfrak{G})$ will be an arbitrary poset.

Definition 2.1. Let (X,μ) be a uniform convergence space. Then $(X, \mathfrak{G}, \mu)$ is an ordered uniform convergence space if there is a quasi-uniform convergence structure σ on X such that $\dot{\mathfrak{G}} \in \sigma$, $\mu = \sigma \vee \sigma^{-1}$, and $\mathfrak{G} = \cup\{\cap \mathfrak{a}: \mathfrak{a} \in \sigma\}$.

Definition 2.2. Let $(X, \mathbf{C})$ be a Cauchy space. Then $(X, \mathfrak{G}, \mathbf{C})$ is an ordered Cauchy space (abbreviated o.c.s.) if there is a uniform convergence structure μ on X such that $(X, \mathfrak{G}, \mu)$ is an ordered uniform convergence space and $\mathbf{C} = \mathbf{C}_\mu$.

We see that $(X, \mathfrak{G}, \mathbf{C})$ is an o.c.s. iff $\mathbf{C}$ and $\mathfrak{G}$ are both determined by some quasi-uniform convergence structure σ on X (in the sense described in the preceding definitions). We shall obtain a more precise characterization of this concept by making use of a particular quasi-uniform convergence structure $\sigma_{\mathfrak{G},\mathbf{C}}$ constructed directly from $\mathfrak{G}$ and $\mathbf{C}$. Let $[\mathfrak{G}] = \{\mathfrak{L} \in F(X \times X): \dot{\mathfrak{G}} \le \mathfrak{L}\}$ be the quasi-uniform convergence structure on X generated by $\mathfrak{G}$. If $\mathbf{C}$ is a Cauchy structure on $(X, \mathfrak{G})$, we define $\sigma_{\mathfrak{G},\mathbf{C}} = [\mathfrak{G}] \wedge \mu^{\mathbf{C}}$; we also define the associated uniform convergence structure $\mu_{\mathfrak{G},\mathbf{C}} = (\sigma_{\mathfrak{G},\mathbf{C}}) \vee (\sigma_{\mathfrak{G},\mathbf{C}})^{-1}$.

If $\mathbf{C}$ is a Cauchy structure on a poset $(X, \mathfrak{G})$, we define an order relation on $\mathbf{C}$ as follows: $\mathfrak{F} \lesssim \mathfrak{L}$ if $\mathfrak{F} \times \mathfrak{L} \in \sigma_{\mathfrak{G},\mathbf{C}}$. It is a simple matter to verify that $\lesssim$ is reflexive and transitive, but not generally anti-symmetric.

Proposition 2.3. Let $\mathbf{C}$ be a Cauchy structure on a poset $(X, \mathfrak{G})$. If $\mathfrak{F}, \mathfrak{L} \in \mathbf{C}$, then $\mathfrak{F} \lesssim \mathfrak{L}$ iff there is a finite set of filters $\mathfrak{H}_1, \ldots, \mathfrak{H}_n$ in $\mathbf{C}$ such that $i(\mathfrak{F}) \vee d(\mathfrak{H}_1)$, $i(\mathfrak{H}_1) \vee d(\mathfrak{H}_2), \ldots, i(\mathfrak{H}_n) \vee d(\mathfrak{L})$ all exist.

Proposition 2.4. Let $\mathbf{C}$ be a Cauchy structure on $(X, \mathfrak{G})$. Then $\sigma_{\mathfrak{G},\mathbf{C}}$ has a base of filters of the form $\bigcap_{j=1}^{n} (d(\mathfrak{F}_j) \times i(\mathfrak{G}_j)) \cap \dot{\mathfrak{G}}$, where $\mathfrak{F}_1,\ldots,\mathfrak{F}_n, \mathfrak{G}_1,\ldots,\mathfrak{G}_n$ are all in $\mathbf{C}$, and $\mathfrak{F}_j \lesssim \mathfrak{G}_j$, $i = 1,\ldots,n$.

Proposition 2.5. If $(X, \mathfrak{G}, \mathbf{C})$ is an o.c.s., then $\mathfrak{F} \in \mathbf{C}$ iff $d(\mathfrak{F}) \times i(\mathfrak{F}) \in \sigma_{\mathfrak{G},\mathbf{C}}$.

Proposition 2.6. Let $\mathbf{C}$ be a Cauchy structure on $(X, \mathfrak{G})$. Then $(X, \mathfrak{G}, \mathbf{C})$ is an o.c.s. iff the following conditions are satisfied:

(1) $\mu_{\mathfrak{G},\mathbf{C}}$ is compatible with $\mathbf{C}$.

(2) $\mathfrak{G} = \bigcup \{\bigcap \mathfrak{A}: \mathfrak{A} \in \sigma_{\mathfrak{G},\mathbf{C}}\}$.

Each of the conditions of Proposition 2.6 has a relatively simple characterization.

Proposition 2.7. Let $\mathbf{C}$ be a Cauchy structure on a poset $(X, \mathfrak{G})$. Then $\mu_{\mathfrak{G},\mathbf{C}}$ is compatible with $\mathbf{C}$ iff the following conditions are satisfied:

(oc_1) $\mathfrak{F} \in \mathbf{C}$ implies $\hat{\mathfrak{F}} \in \mathbf{C}$

(oc_2) If $\mathfrak{F}, \mathfrak{G} \in \mathbf{C}$, $\mathfrak{F} \lesssim \mathfrak{G}$, and $\mathfrak{G} \lesssim \mathfrak{F}$, then $\mathfrak{F} \cap \mathfrak{G} \in \mathbf{C}$.

Proposition 2.8. Let $\mathbf{C}$ be a Cauchy structure on $(X, \mathfrak{G})$. Then $\mathfrak{G} = \bigcup \{\bigcap \mathfrak{A}: \mathfrak{A} \in \sigma_{\mathfrak{G},\mathbf{C}}\}$ iff $\dot{x} \lesssim \dot{y}$ implies $x \le y$.

We say that a Cauchy structure $\mathbf{C}$ on a poset $(X, \mathfrak{G})$ satisfies condition (oc_3) if $\dot{x} \lesssim \dot{y}$ implies $x \le y$. The three preceding propositions can be summarized as follows.

Corollary 2.9. Let $\mathbf{C}$ be a Cauchy structure on $(X, \mathfrak{G})$. Then $(X, \mathfrak{G}, \mathbf{C})$ is an o.c.s. iff the conditions (oc_1), (oc_2), and (oc_3) are all satisfied.

Proposition 2.10. Let $\mathbf{C}$ be a T_2 Cauchy structure on $(X, \mathfrak{G})$. If $\mathfrak{G}$ is a total order, then (oc_2) $\Rightarrow$ (oc_3). If $(X, \mathbf{C})$ is complete, then (oc_3) $\Rightarrow$ (oc_2).

Proposition 2.11. Every o.c.s. is T_2-ordered.

3. Ordered Cauchy Completions.

In this section, $(X, \mathfrak{G}, \mathbf{C})$ will designate an ordered Cauchy space. Two filters $\mathfrak{F}, \mathfrak{G}$ in $\mathbf{C}$ are equivalent if $\mathfrak{F} \cap \mathfrak{G} \in \mathbf{C}$. Let $X^* = \{[\mathfrak{F}] : \mathfrak{F} \in \mathbf{C}\}$ be the set of Cauchy equivalence classes; let $j : X \to X^*$ be the natural injection defined by $j(x) = [\dot{x}]$. If $[\mathfrak{F}], [\mathfrak{G}] \in X^*$, we define $[\mathfrak{F}] \le^* [\mathfrak{G}]$ if there is $\mathfrak{F}_1 \in [\mathfrak{F}]$ and $\mathfrak{G}_1 \in [\mathfrak{G}]$ such that $\mathfrak{F}_1 \lesssim \mathfrak{G}_1$; let $\mathfrak{G}^* = \{([\mathfrak{F}], [\mathfrak{G}]) \subseteq X^* \times X^* : [\mathfrak{F}] \le^* [\mathfrak{G}]\}$.

Proposition 3.1. $\mathfrak{G}^*$ is a partial order on X^*. Furthermore, $[\mathfrak{F}] \le^* [\mathfrak{G}]$ iff $\mathfrak{F}_1 \lesssim \mathfrak{G}_1$ for all $\mathfrak{F}_1 \in [\mathfrak{F}]$ and $\mathfrak{G}_1 \in [\mathfrak{G}]$.

An ordered Cauchy completion $((Y, \mathfrak{G}', \mathfrak{D}), \varphi)$ of an o.c.s. $(X, \mathfrak{G}, \mathbf{C})$ is a pair consisting of a complete o.c.s. $(Y, \mathfrak{G}', \mathfrak{D})$ and an ordered Cauchy embedding $\varphi : (X, \mathfrak{G}, \mathbf{C}) \to (Y, \mathfrak{G}', \mathfrak{D})$ such that $\varphi(X)$ is dense in Y. The completion is order strict if $\mathfrak{G}'$ is the smallest closed partial order on Y which contains $j(\mathfrak{G})$.

Two ordered Cauchy completions $((Y_1, \mathfrak{G}_1, \mathfrak{D}_1), \varphi_1)$ and $((Y_2, \mathfrak{G}_2, \mathfrak{D}_2), \varphi_2)$ of an o.c.s. $(X, \mathfrak{G}, \mathbf{C})$ are equivalent if there is an ordered Cauchy isomorphism $\psi: Y_1 \to Y_2$ which makes the following diagram commute:

$$\begin{array}{ccc} & & (Y_1, \mathcal{O}_1, \mathcal{D}_1) \\ & \overset{\varphi_1}{\nearrow} & \\ (X, \mathcal{O}, \mathcal{C}) & & \downarrow \psi \\ & \underset{\varphi_2}{\searrow} & \\ & & (Y_2, \mathcal{O}_2, \mathcal{D}_2) \end{array}$$

Finally, a T_2 completion $((Y, \mathcal{C}'), \varphi)$ of a T_2 Cauchy space $(X, \mathcal{C})$ is said to be in <u>standard form</u> if $Y = X^*$ and $\varphi = j$.

<u>Proposition 3.2</u>. Every order-strict, ordered Cauchy completion of an o.c.s. $(X, \mathcal{O}, \mathcal{C})$ is equivalent to a completion of the form $((X^*, \mathcal{O}^*, \mathcal{C}'), j)$.

In this paper, we limit our investigation to order-strict, ordered Cauchy completions, which we shall henceforth refer to simply as <u>order-strict completions</u>. In [10], Reed constructed a family of T_2-completions for any T_2-Cauchy space $(X, \mathcal{C})$. We shall adopt her procedure to obtain a family of completions for an arbitrary o.c.s.

Starting with an arbitrary o.c.s. $(X, \mathcal{O}, \mathcal{C})$, we define a <u>selection function</u> to be a function $\lambda: X^* \to F(X)$ with the following properties: (1) for each $p \in X^*$, there is $\mathcal{G} \in p$ such that $\lambda(p) \leq \mathcal{G}$; (2) $\lambda([\dot{x}]) = \dot{x}$, for all $x \in X$. Let Λ be the set of all selection functions, $\lambda \in \Lambda$, and let $A \subseteq X$; we define $A^\lambda = \{p \in X^*: A \in \lambda(p)\}$. If $\mathcal{F} \in F(X)$, let $\mathcal{F}^\lambda$ be the filter generated by $\{A^\lambda : A \in \mathcal{F}\}$. Note that $j(A) \subseteq A^\lambda$ and $\mathcal{F}^\lambda \subseteq j(\mathcal{F})$ is valid for all $A \subseteq X$, $\mathcal{F} \in F(X)$, and $\lambda \in \Lambda$.

With each nonempty subset Γ of Λ, Reed associates a T_2-completion $((X^*, \mathcal{C}_\Gamma), j)$ of $(X, \mathcal{C})$, where $\mathcal{C}_\Gamma = \{ \mathcal{P} \in F(X^*)$: there is $p \in X^*$ such that, for all $\lambda \in \Gamma$, there is $\mathcal{F}_\lambda \in p$ such that $(\mathcal{F}_\lambda)^\lambda \cap \dot{p} \leq \mathcal{P}\}$. It will follow that $((X^*, \mathcal{O}^*, \dot{\mathcal{C}}_\Gamma), j)$ is an order-strict completion of $(X, \mathcal{O}, \mathcal{C})$ whenever $(X^*, \mathcal{O}^*, \mathcal{C}_\Gamma)$ is an o.c.s., but there is no reason to suppose that this will be true in general. Indeed, the definition of "selection function" depends only on the Cauchy space $(X, \mathcal{C})$; we must restrict our attention to selection functions which are, in some sense, compatible also with the partial order $\mathcal{O}$ on X.

Let Λ_0 be the set of all selection functions for $(X, \mathcal{O}, \mathcal{C})$ which satisfy the following order-compatibility conditions:

(os_1) $p \leq^* s$ in $(X^*, \mathcal{O}^*) \Rightarrow i(\lambda(p)) \subseteq \lambda(s)$ and $d(\lambda(s)) \subseteq \lambda(p)$.

(os_2) If $p \in X^*$ and $\mathcal{F} \in p$, then there is $\mathcal{G} \in p$, $\mathcal{G} \subseteq \mathcal{F}$, such that $i(\mathcal{G}) \subseteq \lambda(p)$ or $d(\mathcal{G}) \subseteq \lambda(p) \Rightarrow i(\mathcal{G}) \vee d(\mathcal{G}) \subseteq \lambda(p)$.

<u>Lemma 3.3</u>. If $\lambda \in \Lambda_0$, and $A \subseteq X$, then $i^*(A^\lambda) \subseteq (i(A))^\lambda$ and $d^*(A^\lambda) \subseteq (d(A))^\lambda$.

<u>Lemma 3.4</u>. Let $\mathcal{F} \in p \in X^*$ and $\mathcal{G} \in s \in X^*$; assume $\lambda \in \Lambda_0$. Then the existence of each of the following filters on X^* implies $\mathcal{F} \lesssim \mathcal{G}$: (1) $i^*(\mathcal{F}^\lambda) \vee d^*(\mathcal{G}^\lambda)$; (2) $i^*(\dot{p}) \vee d^*(\mathcal{G}^\lambda)$; (3) $i^*(\mathcal{F}^\lambda) \vee d^*(\dot{s})$, (4) $i^*(\dot{p}) \vee d^*(\dot{s})$.

<u>Theorem 3.5</u>. If $(X, \mathcal{O}, \mathcal{C})$ is an o.c.s. and $\Gamma \subseteq \Lambda_0$, then $((X^*, \mathcal{O}^*, \mathcal{C}_\Gamma), j)$ is an order-strict completion of $(X, \mathcal{O}, \mathcal{C})$.

References

[1] Ball, R. N.: "Convergence and Cauchy Structures on Lattice Ordered Groups", Trans. Amer. Math. Soc. 259 (1980), 357-392.

[2] Ball, R. N.: "Cut Completions of Lattice Ordered Groups by Cauchy Constructions", Ordered Groups, M. Dekker, Lecture Notes in Pure and Appl. Math., Vol. 62 (1980), 81-92.

[3] Cook, C. H. and H. R. Fischer: "Uniform Convergence Structures", Math. Annalen 173 (1967), 290-306.

[4] Fletcher, P. and W. Lindgren: Quasi-uniform Spaces, Lecture Notes in Pure and Appl. Math., Vol. 77 (1982).

[5] Fric, R. and D. C. Kent; "Completion Functors for Cauchy Spaces", Internat. J. Math. & Math. Sci., 4 (1979), 589-604.

[6] Gähler, W.: Grundstrukturen der Analysis, Band I, Akademie-Verlag, Berlin (1977).

[7] Keller, H. H.: "Die Limes -Uniformisierbarkeit der Limesräume," Math. Annalen, 176 (1968), 334-341.

[8] Kent, D. C. and G. D. Richardson: "A Compactification for Convergence Ordered Spaces", Canadian Math. Bull. (to appear).

[9] Nachbin, L.: Topology and Order, Van Nostrand Math. Studies, No. 4, Princeton, N. J. (1965).

[10] Reed, E. E.: "Completions of Uniform Convergence Spaces", Math. Annalen, 194 (1971), 83-108.

ON SUBSPACES OF COMPLETE CONVERGENCE SPACES

Gert Kneis [1)]

The subsequent note deals with some external characterization of subspaces of special complete convergence spaces, which is helpful for constructing completions in the categorical setting.

For the terminology on convergence spaces, the reader is referred to W. GÄHLER [4] and to the author's papers [7] and [8] and for the categorical terminology also to W. GÄHLER [4] and to H. HERRLICH [6].

1. Starting with the standard construction of a categorical completion of pseudo-uniform spaces in the sense of COOK and FISCHER by O. WYLER [14] in 1970, several completion and compactification functors were constructed as composition of convenient more easy functors, e. g. epireflectors or bireflectors. The constructions of the completion functors for the category LUN of uniform limit spaces (see W. GÄHLER [4]) and for the category PVEC of pseudo-topological vector spaces (see S. GÄHLER, W. GÄHLER, and G. KNEIS [5]) and of the regular ČECH-STONE compactification for the category LIM of limit spaces (see O. WYLER [13] and W. GÄHLER [4]) are further such examples.

The question as to the existence of reflectors may be reduced to the more easy question as to the existence of epireflectors by means of the well-known Theorem of KENNISON and BARON (see H. HERRLICH [6]):

Let $\mathcal{C}$ be a complete category. The equalizers of $\mathcal{C}$ are denoted as <u>subobjects</u> of $\mathcal{C}$ and the natural morphism $g : K \longrightarrow X$ of some equalizer K of two morphisms $g_1 : X \longrightarrow Y$ and $g_2 : X \longrightarrow Y$ is denoted as an <u>embedding</u>. Any morphism f of $\mathcal{C}$ is assumed to be the composition gh of some epimorphism h and some embedding g. Finally, $\mathcal{C}$ is assumed to be locally small, i. e., for an arbitrary object X of $\mathcal{C}$, the class of all non-isomorphic monomorphisms $j : Z \longrightarrow X$ has to be a set. Then, an arbitrary reflector $r : \mathcal{C} \longrightarrow \mathcal{A}$ into some full subcategory $\mathcal{A}$ of $\mathcal{C}$ can be represented as the composition of two epireflectors $\mathcal{C} \longrightarrow \mathcal{B}$ and $\mathcal{B} \longrightarrow \mathcal{A}$. The smallest of all such full subcategories $\mathcal{B}$ of $\mathcal{C}$ containing $\mathcal{A}$ as full subcategory consists of the $\mathcal{C}$-subobjects of objects of $\mathcal{A}$.

This smallest category is designated as $\mathcal{B}_{\mathcal{C}\mathcal{A}}$ and we call the representation of the reflector r as composition of the epireflectors $\mathcal{C} \longrightarrow \mathcal{B}_{\mathcal{C}\mathcal{A}}$

1) Akademie der Wissenschaften der DDR, Institut für Mathematik, 1086 Berlin, DDR, Mohrenstraße 39

and $\mathcal{B}_{\mathcal{C}\mathcal{A}} \longrightarrow \mathcal{A}$ the standard factorization of r.

As an immediate consequence of this theorem, the question as to the existence of some reflector of the category $\mathcal{C}$=PTOP of pseudo-topological spaces into the subcategory $\mathcal{A}$=$PTOP_{sep,c}$ of separated compact spaces turns out to be equivalent to the question as to the existence of some epireflector of the category $PTOP_{sep}$ of separated pseudo-topological spaces into $PTOP_{sep,c}$ because every separated pseudo-topological space is a pseudo-topological subspace of some separated compact space (see G. D. RICHARDSON [11], e. g.). The fact that $PTOP_{sep,c}$ is not epireflective in $PTOP_{sep}$ was proved by H.-P. BUTZMANN and the author [1].

Let us remark here that the representation of the completion functors mentioned above for the categories LUN, PUN, and PVEC by means of some saturation functors have not the standard factorization. But from the construction of these functors in [4], [14], and [5], respectively, the standard factorization immediately can be derived. In the following we list these standard factorizations and the standard factorization for the completion functor for the category PAB of pseudo-topological abelian groups (see R. FRIČ and D. C. KENT [2]), - $\mathcal{A}$ means the full subcategory of separated complete objects of the respective category $\mathcal{C}$:

$\mathcal{C}$	$\mathcal{B}_{\mathcal{C}\mathcal{A}}$
LUN	LUN_{sep}
PUN	PUN_{sep}
PVEC	$PVEC_{sep,Cq}$ 1)
PAB	PAB_{sep}

2. For the construction of reflectors $\mathcal{C} \longrightarrow \mathcal{A}$, some convenient description of the objects of the category $\mathcal{B}_{\mathcal{C}\mathcal{A}}$ is needed. Such a description is suggested by the standard factorization of the ČECH-STONE functor of the category $\mathcal{C}$=TOP of topological spaces into the category $\mathcal{A}$=$TOP_{sep,c}$ of separated compact spaces. In this case, $\mathcal{B}_{\mathcal{C}\mathcal{A}}$ consists of the separated completely regular spaces, and they are characterized as those separated spaces X which carry the initial topology of all objects of $\mathcal{A}$ with respect to the class of all morphisms of X to objects of $\mathcal{A}$.

In the following, this characterization of the objects of the category $\mathcal{B}_{\mathcal{C}\mathcal{A}}$ in the case of the ČECH-STONE functor is generalized to the case of convergence spaces. The respective generalization is helpful for constructing completion functors for categories of convergence spaces.

---

1) The property Cq means that every Cauchy filter is quasi-bounded in the sense of A. FRÖLICHER and W. BUCHER [3].

Let $\mathcal{C}$ be some concrete category over the category $\mathcal{D}$ given by some faithful functor $V : \mathcal{C} \longrightarrow \mathcal{D}$ where $\mathcal{D}$ is assumed to be one of the categories LUN or LIM. Let $\mathcal{A}$ be some full subcategory of $\mathcal{C}$. For any object X of $\mathcal{C}$, by $K_{\mathcal{A}}(X)$ we denote the union of all sets of $\mathcal{C}$-morphisms of X to Y where Y runs over the objects of $\mathcal{A}$. A $\mathcal{C}$-object X is said to be $\mathcal{A}$-acceptable if the space VX carries the initial uniform limit structure (in the case $\mathcal{D}$=LUN) or the initial limit structure (in the case $\mathcal{D}$=LIM), respectively, of all spaces VY with respect to all mappings $Vf : VX \longrightarrow VY$ where $f : X \longrightarrow Y$ runs over $K_{\mathcal{A}}(X)$. Let $\mathcal{C}^{\mathcal{A}}$ denote the full subcategory of $\mathcal{C}$ of all $\mathcal{A}$-acceptable objects X for which VX is separated.

Remark. The categories PGR of pseudo-topological groups and PVEC of pseudo-topological vector spaces are concrete categories over LUN as well as over LIM. Then the two possible notions of acceptability coincide.

The subsequent theorem is concerned with concrete situations. For the constructions used here, instead of LUN the full subcategory UCON of uniform convergence spaces in the sense of O. WYLER is needed. The assumptions are such that in particular the categories of topological spaces, uniform spaces, topological groups, and topological vector spaces are included.

Theorem 1. Let $\mathcal{C}$ be some full isomorphism-closed bireflective subcategory of one of the categories TOP, UCON, PGR, or PVEC. Let $\mathcal{A}$ be some full subcategory of $\mathcal{A}$ the objects of which are separated. Then the following assertions hold:

(i) $\mathcal{C}^{\mathcal{A}}$ contains $\mathcal{B}_{\mathcal{C}\mathcal{A}}$ as full subcategory.

(ii) $\mathcal{C}^{\mathcal{A}}$ is epireflective in $\mathcal{C}$.

Remark. The notion of acceptability may be more generally introduced by using the concept of topological categories (see G. RICHTER [12], e. g.). In consequence thereof, acceptability can be moreover defined for sequential convergence spaces, in particular for the category $\mathcal{L}^*$GR of $\mathcal{L}^*$-groups (see J. NOVÁK [10]). Then Theorem 1 is also valid for $\mathcal{C}=\mathcal{L}^*$GR and $\mathcal{A}$ being some full subcategory of $\mathcal{C}$ the objects of which are equipped with single-valued convergence.

The characterization we have in mind of the objects of $\mathcal{B}_{\mathcal{C}\mathcal{A}}$ asserts the identity of the categories $\mathcal{B}_{\mathcal{C}\mathcal{A}}$ and $\mathcal{C}^{\mathcal{A}}$, which can be proved in special cases. In [7], [8], and [9], in several such special cases additionally some epireflector of $\mathcal{C}^{\mathcal{A}}$ into $\mathcal{A}$ is constructed. By combining this epireflector with the epireflector given in Theorem 1, $\mathcal{A}$ turns out to be reflective in $\mathcal{C}$. By this construction, the objects of $\mathcal{C}^{\mathcal{A}}$ are embedded as $\mathcal{C}$-subobjects into objects of $\mathcal{A}$.

Theorem 2. $\mathcal{C}^{\mathcal{A}} = \mathcal{B}_{\mathcal{C}\mathcal{A}}$ holds for the subsequent categories $\mathcal{C}$ and full subcategories $\mathcal{A}$[1]:

$\mathcal{C}$	$\mathcal{A}$ (characterized by properties of objects)
UCON	separated + complete + uniformly regular
PVEC	separated + complete + regular
PVEC	separated + complete + locally convex (+ regular)
PVEC	separated + complete + equable [2] (+ regular)
PGR	separated + complete
PGR	separated + complete + regular
$\mathcal{L}^*$GR	single-valued convergence + complete

References

[1] BUTZMANN, H.-P., and G. KNEIS: Čech-Stone compactification of pseudo-topological spaces. To be published.

[2] FRIČ, R., and D. C. KENT: Completion of pseudo-topological groups. Math. Nachr. 99 (1980), 99-103.

[3] FRÖLICHER, A., and W. BUCHER: Calculus in vector spaces without norm. Lecture Notes in Math. 30, Berlin, Heidelberg, New York 1966.

[4] GÄHLER, W.: Grundstrukturen der Analysis. Berlin, Basel, Bd. I 1977, Bd. II 1978.

[5] -, S. GÄHLER, and G. KNEIS: Completion of pseudo-topological vector spaces. Math. Nachr. 75 (1976), 185-206.

[6] HERRLICH, H.: Topologische Reflexionen und Coreflexionen. Lecture Notes in Math. 78, Berlin, Heidelberg, New York 1968.

[7] KNEIS, G.: A general completion theory. Preprint P-Math-07/83 (1983), Inst. f. Math. Akad. Wiss. DDR, 49 p.

[8] -: Completion for some concrete categories over the category of uniform limit spaces. Proc. Internat. School on Convergence Structures and Appl. II, Schwerin (1983), Abh. Akad. Wiss. DDR, 119-130, Berlin 1984.

[9] -: Eine allgemeine Theorie der Vervollständigung und Čech-Stone-Kompaktifizierung. Dissertation B, 135p., Univ. Greifswald 1984.

[10] NOVÁK, J.: On convergence groups. Czech. Math. J. 20 (1970), 357-374.

[11] RICHARDSON, G. D.: A Stone-Čech compactification for limit spaces. Proc. Amer. Math. Soc. 25 (1970), 403-404.

[12] RICHTER, G.: Kategorielle Algebra. Berlin 1979.

[13] WYLER, O.: The Stone-Čech compactification for limit spaces. Notices Amer. Math. Soc. 15 (1968), 169.

[14] -: Ein Komplettierungsfunktor für uniforme Limesräume. Math. Nachr. 46 (1970), 1-12.

---

1) The assertion $\mathcal{C}^{\mathcal{A}} = \mathcal{B}_{\mathcal{C}\mathcal{A}}$ also holds for the categories $\mathcal{C}$ and $\mathcal{A}$ listed at the end of the first part.

2) See [3].

CLOSURE AND TOPOLOGICAL SEQUENTIAL CONVERGENCE

Václav Koutník [1)]

0. Introduction

In this paper the problem which sequential convergences are convergences in a topological or, more generally, in a closure space is considered. After giving a characterization of topological and closure convergences the class of convergence spaces whose convergences are closure convergences is investigated. It is shown that some basic theorems valid for single-valued convergences can be extended to this class and, moreover, that this is the largest class of many-valued convergences for which they remain true. Finally, a modification $\mathcal{L}^+$ of a given convergence $\mathcal{L}$ is introduced and a generalization of these theorems to the class of all convergence spaces is obtained.

First we recall some basic definitions ([5], [18]). A <u>convergence space</u> $(X, \mathcal{L}, \lambda)$ is a closure space (X, λ) where the closure operator λ for X is induced by a convergence $\mathcal{L} \subset X^N \times X$ which satisfies the axioms $(\mathcal{L}_1)$ (constant sequences converge) and $(\mathcal{L}_2)$ (subsequences of convergent sequences converge). If $(\langle x_n \rangle, x) \in \mathcal{L}$ then we say that the sequence $\langle x_n \rangle$ $\mathcal{L}$-converges to x and write $x_n \xrightarrow{\mathcal{L}} x$. Hence for each $A \subset X$ we have $\lambda A = \{x : x_n \xrightarrow{\mathcal{L}} x, \cup\{x_n\} \subset A\}$. A convergence is <u>single-valued</u> if it satisfies the axiom $(\mathcal{L}_0)$ (limits are unique); it is <u>maximal</u> if it satisfies the Urysohn axiom $(\mathcal{L}_3)$. A mapping $f : (X, \mathcal{L}, \lambda) \to (Y, \mathcal{M}, \mu)$ is <u>continuous</u> if $f[\lambda A] \subset \mu f[A]$ for each $A \subset X$; i.e. if $f : (X, \lambda) \to (Y, \mu)$ is continuous in the usual sense. To every convergence $\mathcal{L}$ there corresponds a unique maximal convergence denoted by $\mathcal{L}^*$ and inducing the same closure. It is defined by: $x_n \xrightarrow{\mathcal{L}^*} x$ if each subsequence $\langle x_{n_i} \rangle$ of $\langle x_n \rangle$ contains a subsequence $\langle x_{n_{i_j}} \rangle$ such that $x_{n_{i_j}} \xrightarrow{\mathcal{L}} x$.

A convergence $\mathcal{L}$ for X naturally induces a topology for X and, conversely, a topological or, more generally, a closure space induces a convergence. We have the following two definitions.

<u>Definition 1</u> ([18]). A <u>topological modification</u> of a convergence space $(X, \mathcal{L}, \lambda)$ is the topological space (X, u) such that u is the finest topological closure for X coarser than λ. We denote it by (X, λ^{ω_1}).

Note that (X, λ^{ω_1}) is a <u>sequential space</u> in the sense of Franklin ([4]) and $(X, \mathcal{L}, \lambda)$ is a <u>Fréchet space</u> ([4]) iff $\lambda = \lambda^{\omega_1}$, i.e.

---

1) Institute of Mathematics, Czechoslovak Academy of Sciences, Žitná 25, 115 67 Praha 1, Czechoslovakia

iff λ is a topology for X.

Definition 2 ([14]). A convergence modification of a closure space (X,u) is the convergence space $(X, \mathcal{L}(u), \lambda(u))$ where the convergence $\mathcal{L}(u)$ is defined as follows:

$$x_n \xrightarrow{\mathcal{L}(u)} x \text{ if for each u-neighborhood } U(x) \text{ of } x \text{ there is } n_0 \text{ such that } x_n \in U(x) \text{ for } n \geq n_0$$

and $\lambda(u)$ is the corresponding convergence closure.

In terms of the closure operator we have $x_n \xrightarrow{\mathcal{L}(u)} x$ iff $x \in u(\cup\{x_{n_i}\})$ for each subsequence $\langle n_i \rangle$ of $\langle n \rangle$ ([12], cf. [7]).

Let $(X, \mathcal{L}, \lambda)$ be a convergence space. We denote $(X, \overline{\mathcal{L}}, \overline{\lambda})$ the convergence modification of (X, λ) and $(X, \overline{\overline{\mathcal{L}}}, \overline{\overline{\lambda}})$ the convergence modification of (X, λ^{ω_1}), i.e. $\overline{\mathcal{L}} = \mathcal{L}(\lambda)$ and $\overline{\overline{\mathcal{L}}} = \mathcal{L}(\lambda^{\omega_1})$.

Proposition 1 (cf. [14]). Let $(X, \mathcal{L}, \lambda)$ be a convergence space. Then

(i) $\mathcal{L} \subset \mathcal{L}^* \subset \overline{\mathcal{L}} \subset \overline{\overline{\mathcal{L}}}$.

(ii) $\lambda = \lambda^* = \overline{\lambda} < \overline{\overline{\lambda}}$.

There are examples to show that no other equalities hold in general. Hence a convergence space induces the following spaces.

$$\begin{array}{ccccc} (X, \mathcal{L}, \lambda) & \longrightarrow & (X, \lambda) & \longrightarrow & (X, \lambda^{\omega_1}) \\ \downarrow & & \downarrow & & \downarrow \\ (X, \mathcal{L}^*, \lambda) & & (X, \overline{\mathcal{L}}, \lambda) & & (X, \overline{\overline{\mathcal{L}}}, \overline{\overline{\lambda}}) \end{array}$$

Definition 3. A convergence $\mathcal{L}$ for X is a closure (topological) convergence if there exists a closure (topological space) (X,u) such that $\mathcal{L} = \mathcal{L}(u)$.

The notion of a topological convergence was introduced in [2]. Clearly every closure and hence also every topological convergence is maximal.

1. Topological convergence

The problem to characterize, in terms of convergence, topological convergences was put forward by M. Dolcher in 1960. In his paper [3] he showed that

1. Every single-valued maximal convergence is topological.
2. A convergence $\mathcal{L}$ is topological iff $\mathcal{L} = \overline{\overline{\mathcal{L}}}$.
3. There exists a non-topological convergence.

The first assertion was proved also by J. Kisyński ([11]) in the same year and for the special case of linear semi-ordered spaces in [10]. Let us mention here the example showing the third result.

<u>Example 1</u> ([3]). Let $X = \bigcup\{x_n\} \cup \bigcup\{y_n\} \cup \{y\}$ and let a convergence $\mathcal{L}$ for X be defined as follows.

$z \xrightarrow{\mathcal{L}} z$ for each $z \in X$.

$x_{n_i} \xrightarrow{\mathcal{L}} y_m$ for each $m \in N$ and each $\langle n_i \rangle$.

$y_{n_i} \xrightarrow{\mathcal{L}} y$ for each $\langle n_i \rangle$.

Clearly, $x_n \not\xrightarrow{\mathcal{L}^*} x$ while $x_n \xrightarrow{\bar{\bar{\mathcal{L}}}} y$. Hence $\mathcal{L}^*$ is not topological. However, we have $\mathcal{L}^* = \bar{\mathcal{L}}$ and therefore $\mathcal{L}^*$ is a closure convergence.

Sufficient conditions for a convergence to be topological were first given by P. Antosik ([1]). His results were improved by A. Kamiński ([8], cf. [9]) who showed that if, for some α , $U_\alpha \wedge C_\alpha$ holds, then $\mathcal{L}$ is topological, where

U_α : If $\langle x_n \rangle$ does not converge to x , then there is a subsequence $\langle n_i \rangle$ of $\langle n \rangle$ such that $x \in X - \lambda^\alpha(\cup\{x_{n_i}\})$.

C_α : For each $\langle y_n \rangle \in X^N$ there is a subsequence $\langle n_i \rangle$ of $\langle n \rangle$ such that $\lambda^\alpha(\cup\{y_{n_i}\})$ is λ-closed.

However, while every topological convergence satisfies U_α for all $\alpha \le \omega_1$ and C_{ω_1} always holds, A. Kamiński constructed for each β , $2 \le \beta \le \omega_1$, a topological convergence which does not satisfy C_α for $\alpha < \beta$. He gave the following characterization: $\mathcal{L}$ is topological iff U_{ω_1} holds, i.e. if $\langle x_n \rangle$ does not converge to x then there is a subsequence $\langle n_i \rangle$ of $\langle n \rangle$ such that $x \in X - \lambda^{\omega_1}(\cup\{x_{n_i}\})$.

The following condition was introduced in [14].

($\mathcal{B}$) Let $\langle x_n \rangle \in X^N$ and let $x \in X$. If for each subsequence $\langle n_p \rangle$ of $\langle n \rangle$ there are points $x^{\langle n_p \rangle}_{m_1 m_2 \ldots m_j} \in X$, $m_i \in N$, $i=1,2,\ldots,j$, $j \in N$, such that

(i) $x^{\langle n_p \rangle}_{m_1} \xrightarrow{\mathcal{L}} x$ for $m_1 \to \infty$,

(ii) $x^{\langle n_p \rangle}_{m_1 m_2 \ldots m_{j+1}} \longrightarrow x^{\langle n_p \rangle}_{m_1 m_2 \ldots m_j}$ for $m_{j+1} \to \infty$, $j \in N$,

(iii) for each $\langle r_s \rangle \in N^N$ there is $k = k(\langle r_s \rangle)$ such that
$$\left|\{m_k : \left|\{m_{k+1} : x^{\langle n_p \rangle}_{r_1 r_2 \ldots r_{k-1} m_k m_{k+1}} \in \cup\{x_{n_p}\}\right| = \omega\}\right| = \omega$$
then
$$x_n \xrightarrow{\mathcal{L}^*} x .$$

The next theorem characterizes, in terms of convergence, topological convergences.

Theorem 1 (cf. [14]). The following statements are equivalent.

(i) $\mathcal{L}$ is topological.

(ii) $\mathcal{L} = \bar{\bar{\mathcal{L}}}$.

(iii) $\mathcal{L}$ satisfies $(\mathcal{L}_3) \wedge (\mathcal{B})$.

The equivalence of (i) and (ii) is due to M. Dolcher ([3]).

2. Closure convergence

The first question to consider is whether all maximal convergences are closure convergences. Since all topological convergences are closure convergences as well it follows from the result of M. Dolcher that all single-valued maximal convergences are closure convergences. As the non--topological convergence $\mathcal{L}^*$ in Example 1 is a closure convergence we need another example.

Example 2. Let $X = \bigcup\{x_n\} \cup \{x\}$ and let a convergence $\mathcal{L}$ be defined as follows:

$z \xrightarrow{\mathcal{L}} z$ for each $z \in X$,

the constant sequence $\langle x_m, x_m, \ldots \rangle$ converges to x for each $m \in N$.

We have $x_n \overset{\mathcal{L}^*}{\not\to} x$ while $x_n \xrightarrow{\bar{\mathcal{L}}} x$. Hence $\mathcal{L}^* \subsetneq \bar{\mathcal{L}} = \bar{\bar{\mathcal{L}}}$ and by Theorem 2 given later $\mathcal{L}^*$ is not a closure convergence.

The following condition was introduced in [13] (cf. [6]).

($\mathcal{A}$) If the constant sequence $\langle x_m \rangle$ $\mathcal{L}$-converges to x for each $m \in N$, then $\langle x_n \rangle$ $\mathcal{L}$-converges to x.

The closure convergences are characterized by the next theorem.

Theorem 2 (cf. [6]). The following statements are equivalent.

(i) $\mathcal{L}$ is a closure convergence.

(ii) $\mathcal{L} = \bar{\mathcal{L}}$.

(iii) $\mathcal{L}$ satisfies $(\mathcal{L}_3) \wedge (\mathcal{A})$.

The relations between conditions ($\mathcal{A}$) and ($\mathcal{B}$) and the axiom ($\mathcal{L}_0$) are as follows.

Proposition 2. $(\mathcal{L}_0) \Rightarrow (\mathcal{A}) \Rightarrow (\mathcal{B})$.

Hence for single-valued convergences ($\mathcal{A}$) and ($\mathcal{B}$) are always satisfied and we obtain from Theorem 1 the result of Dolcher ([3]) and from Theorem 2 the result of Novák ([18], Theorem 2) as special cases.

We are now going to show some interesting properties of the class of convergence spaces satisfying the condition ($\mathcal{A}$).

For single-valued convergences we have the following well-known theorems.

Let $(X,\mathcal{L},\lambda)$, $(X,\mathcal{M},\mu)$ and $(Y,\mathcal{N},\nu)$ be single-valued convergence spaces. Then

A ([7], [18]). $\mathcal{L} = \overline{\mathcal{L}}$ iff $\mathcal{L}$ satisfies $(\mathcal{L}_3)$.

B ([18]). $\mathcal{L}^* \subset \mathcal{M}^*$ iff $\lambda < \mu$.

C ([17]). A mapping $f : (X,\mathcal{L},\lambda) \rightarrow (Y,\mathcal{N},\nu)$ is continuous iff $x_n \xrightarrow{\mathcal{L}} x$ implies that there is a subsequence $\langle n_i \rangle$ of $\langle n \rangle$ such that $f(x_{n_i}) \xrightarrow{\mathcal{N}} f(x)$.

Example 2 can be used to show that all these statements are in general false for many-valued convergence spaces. However, they remain true for convergences satisfying the condition $(\mathcal{A})$. We have already seen (Theorem 2) that condition $(\mathcal{A})$ is not only sufficient but also necessary for statement A to hold. The following theorems deal with statements B and C.

<u>Theorem 3.</u> Let $(X,\mathcal{M},\mu)$ be a convergence space. Then the following statements are equivalent.

(i) $(X,\mathcal{M},\mu)$ satisfies condition $(\mathcal{A})$.

(ii) For each convergence space $(X,\mathcal{L},\lambda)$ we have $\mathcal{L}^* \subset \mathcal{M}^*$ iff $\lambda < \mu$.

<u>Theorem 4.</u> Let $(Y,\mathcal{N},\nu)$ be a convergence space. Then the following statements are equivalent.

(i) $(Y,\mathcal{N},\nu)$ satisfies condition $(\mathcal{A})$.

(ii) For every convergence space $(X,\mathcal{L},\lambda)$ and for every mapping $f : (X,\mathcal{L},\lambda) \rightarrow (Y,\mathcal{N},\nu)$ the mapping f is continuous iff $x_n \longrightarrow x$ implies there is a subsequence $\langle n_i \rangle$ of $\langle n \rangle$ such that $f(x_{n_i}) \xrightarrow{\mathcal{N}} f(x)$.

Therefore the class of convergence spaces satisfying condition $(\mathcal{A})$ is the largest class to which statements A, B and C can be extended. This class contains all natural many-valued convergence spaces, in particular all pseudometric spaces.

To obtain a generalization of statements A, B and C to the class of all convergence spaces we need the notion of an $\mathcal{L}^+$ modification.

<u>Definition 4.</u> Let $(X,\mathcal{L},\lambda)$ be a convergence space. Denote $\mathcal{L}^+$ the following convergence for X : $x_n \xrightarrow{\mathcal{L}^+} x$ if $x_n \xrightarrow{\mathcal{L}} x$ or the constant sequence $\langle x_m, x_m, x_m, \dots \rangle$ $\mathcal{L}$-converges to x for each $m \in N$.

<u>Proposition 3.</u> Let $(X,\mathcal{L},\lambda)$ be a convergence space. Then

(i) $(\mathcal{L}^+)^* = \overline{\mathcal{L}}$ and therefore $\lambda^+ = \lambda$.

(ii) $\mathcal{L}^+ = \mathcal{L}$ iff $(X,\mathcal{L},\lambda)$ satisfies condition $(\mathcal{A})$.

<u>Theorem 5.</u> Let $(X,\mathcal{L},\lambda)$, $(X,\mathcal{M},\mu)$ and $(Y,\mathcal{N},\nu)$ be converg-

ence spaces. Then

(i) $\mathcal{L}^+ = \overline{\mathcal{L}}$ iff $\mathcal{L}$ satisfies $(\mathcal{L}_3)$.

(ii) $(\mathcal{L}^+)^* \subset (\mathcal{M}^+)^*$ iff $\lambda < \mu$.

(iii) A mapping $f : (X, \mathcal{L}, \lambda) \rightarrow (Y, \mathcal{N}, \nu)$ is continuous iff $x_n \xrightarrow{\mathcal{L}} x$ implies there exists a subsequence $\langle n_i \rangle$ of $\langle n \rangle$ such that $f(x_{n_i}) \xrightarrow{\mathcal{N}^+} f(x)$.

The assertion (iii) of Theorem 5 is relevant when we consider many-valued convergences compatible with a given algebraic structure, in particular many-valued convergence groups (cf. [13]).

The proofs and details are contained in forthcoming papers ([15], [16]).

References

[1] Antosik, P.: On a topology of convergence. Colloq.Math. 21(1970), 205-209.

[2] Bednarek, A.R., and J. Mikusiński: Convergence and topology. Bull.Acad.Polon.Sci.Sér.Sci.Math.Astronom.Phys. 17(1969), 437-42.

[3] Dolcher, M.: Topologie e strutture di convergenza. Ann. Scuola Norm.Super. Pisa Ser III 14(1960), 63-92.

[4] Franklin, S.P.: Spaces in which sequences suffice. Fund.Math. 57 (1965), 107-115.

[5] Frič, R., and V. Koutník: Sequential structures. Convergence Structures and Applications to Analysis, Abh.Akad.Wiss.DDR, Abt. Math.-Naturwiss.-Technik, 1979, Nr. 4 N. Akademie-Verlag, Berlin 1980, 37-56.

[6] Frič, R., and V. Koutník: Sequential convergence since Kanpur conference.General Topology and its Relations to Modern Analysis and Algebra, V (Proc. Fifth Prague Topological Sympos., Prague 1981). Heldermann Verlag, Berlin 1982, 193-205.

[7] Hausdorff, F.: Gestufte Räume. Fund.Math. 25(1935), 486-502.

[8] Kamiński, A.: On characterization of topological convergence. Proc.Conf. on Convergence (Szczyrk, 1979). Polska Akad. Nauk, oddzial w Katowicach, Katowice 1980, 50-70.

[9] Kamiński, A.: Remarks on multivalued convergences. General Topology and its Relations to Modern Analysis and Algebra, V (Proc. Fifth Prague Topological Sympos., Prague 1981). Heldermann Verlag, Berlin 1982, 418-422.

[10] Kantorovich, L.V., B.Z. Vulih, and A.G. Pinsker: Functional analysis in semi-ordered spaces (Russian). Moskva 1950.

[11] Kisyński, J.: Convergence du type L. Colloq.Math. 7(1960), 205-11.

[12] Koutník, V.: On convergence in closure spaces. Proc.Internat. Sympos. on Topology and its Appl. (Herceg Novi, 1968). Beograd 1969, 226-230.

[13] Koutník, V.: Many-valued convergence groups. Proc.Conf. on Convergence (Szczyrk, 1979). Polska Akad. Nauk, oddzial w Katowicach, Katowice 1980, 71-75.

[14] Koutník, V.: On topological convergence. Proc.Conf. on Convergence and Generalized Functions (Katowice 1983).(To appear.)

[15] Koutník, V.: On many-valued convergence. (To appear.)

[16] Koutník, V.: Topological convergence. (To appear.)

[17] Kuratowski, K.: Topologie I. Warszawa 1933.

[18] Novák, J.: On convergence spaces and their sequential envelopes. Czechoslovak Math.J. 15(90) (1965), 74-100.

MULTISEQUENCES AND THEIR STRUCTURE IN SEQUENTIAL SPACES

Petr Kratochvil [1)]

1. Introduction

E. Borel has used the sequence convergence property of set algebras and proposed a natural idea of measure extension ([1]): If $A = \lim A_n$, if a measure μA_n has been determined for each set A_n, $n=0,1,2,\ldots$, and if μA has not been determined yet, then put

(1) $\mu A = \lim \mu A_n$.

The idea of E. Borel has been realized by J. Novák ([4]) who raised a problem to eliminate the outer measure from his solution. The author of this paper introduced a notion of a multisequence ([3]) and used multisequences in a solution of the problem.

In this paper we present fundamental properties of a multisequence. The sequential spaces provide a suitable structure in which the fundamental results can be described. Hence we restrict our study to the framework of sequential spaces.

A multisequence is a generalization of a double sequence to higher multiplicity. Let a sequential convergence closure operator λ be given. The operator λ need not be idempotent in general, i.e. it can occur $\lambda A \neq \lambda^2 A$ for a set A. Given an element $m = m(\Theta)$ of $\lambda^2 A$, there exist points $m(i) \in \lambda A$ such that $m = \lim_{i\to\infty} m(i)$. For each natural number i, there exist $m(i,j) \in A$ such that $m(i) = \lim_{j\to\infty} m(i,j)$, j being a natural number. We can say that the double sequence $m(i,j)$, $i,j=0,1,2,\ldots$, converges (in a s u i t a b l e sense) to the point m. Similarly we construct a triple sequence $m(\alpha) \in A$, $\alpha = \langle i,j,k\rangle \in \omega^3$, in the case of a point from $\lambda^3 A$, ω denotes the first infinite ordinal number and also the set of all natural numbers. There are many examples of λ such that $\lambda^\xi \neq \lambda^{\xi+1}$ for an arbitrary countable ordinal number ξ, e.g. the set of all Baire measurable real functions equipped with the pointwise sequential convergence. It seems that an index $\alpha \in \omega^\xi$ is necessary for a general ξ in the previous construction. However we propose a notion of a multisequence using indices $\alpha = \langle i_0, i_1, \ldots\rangle \in \sum = \omega^\omega$ and a corresponding notion of a multisequence convergence (see Definition 3). For the double sequence and a number i, $M_i = \langle m(i,j)\rangle_{j=1}^{\infty}$ represents a simple sequence which can be called a restriction of the double sequence to the index i. On the

---

1) Institute of Mathematics, Czechoslovak Academy of Sciences, Žitná 25, 115 67 Praha 1, Czechoslovakia

other hand, for simple sequences M_i, $i=0,1,\dots$, we can define a double sequence such that every M_i is a restriction of the double sequence to the index i. The resulting double sequence can be called the composition of the family $\{M_i ; i=0,1,\dots\}$. Corresponding generalizations are given in Definition 6 and Corollary 2.

In the sequel we use the notion of an order of a point with respect to a set introduced by J. Novák ([5]). Analogously we introduce an order of a multisequence (see Definition 8). The main result can be formulated as follows. If x is of an order ξ with respect to A, then there exists a multisequence in A of an order ξ and converging to x but no multisequence in A of an order less than ξ converges to x (see Corollary 3).

An elementary filter is an image of a Fréchet filter under a sequence. Therefore we can regard elementary filters and sequences as equivalent notions, and the theory of sequential convergence spaces as a special case of the more general theory of convergence spaces. However sequences are more easier to handle than filters and that is the reason why sequential convergence is intensively investigated at the present time. We generalize the notion of a sequence to that of a multisequence (Definition 1). Filter of sections of a multisequence (see Definition 2) can be considered as a generalization of an elementary filter. Hence the situation is quite similar to the described relation between elementary filters and sequences. Multisequences and their fundamental properties are studied in this paper but powers of the Fréchet filter and their relation to the filter of sections are not introduced and the corresponding statements are not proved.

2. Multisequences in sequential spaces

Let (X,u) be a sequential space in the sence of S.P. Franklin ([2]), i.e. a topological space such that a subset A of X is closed if and only if it is sequentially closed. Denote λ the convergence closure operator which is determined by the topological closure operator u. Recall that λA is the set of all limits of sequences in A. The power of λ is defined recurrently, $\lambda^{-1}A = \emptyset$, $\lambda^{0}A = A$ and

(2) $$\lambda^{\xi}A = \bigcup_{\eta<\xi} \lambda(\lambda^{\eta}A)$$

for an ordinal number $\xi > 0$.

In every sequential space X,

(3) $$\lambda^{\Omega}A = uA$$

for each subset A of X, where Ω denotes the first uncountable ordinal number. If $x \in uA$, then there is

(4) the least ξ such that $x \in \lambda^{\xi}A$.

The ordinal number ξ is less than Ω and it is not limit if $\xi > 0$. The ordinal number ξ is called the order of x with respect to A in the space X (see [5]). Clearly $x \in \lambda^{\xi}A - \lambda^{\xi-1}A$. Denote

$S = \bigcup \{\omega^n;\ n \in \omega\}$ the set of all finite sequences of natural numbers and $\sum = \omega^\omega$ the Baire space, i.e. the set of all infinite sequences of nonnegative integers with the natural product topology, taking ω discrete. The basic neighborhoods are of the form $N(k_0, \dots, k_n) = \{\alpha \in \sum;\ \alpha(0) = k_0, \dots, \alpha(n) = k_n\}$, one for each tuple $k_0, \dots, k_n$. They are closed and open subsets of the Baire space $\sum$.

Recall that ω^0 contains only one element. We shall denote it by Θ, $\omega^0 = \{\Theta\}$. Therefore $\Theta \in \omega^0 \subset S$. If $m : S \rightarrow X$ is a mapping of S into a set X, then we can write $m(\ )$ instead of $m(\Theta)$, $m(\ ) = m(\Theta)$. Similarly $f(\ ) = f(\Theta)$ and $\xi(\ ) = \xi(\Theta)$ for mappings $f : S \rightarrow \omega$ and $\xi : S \rightarrow \{\eta;\ \eta < \Omega\}$, etc. Let k be a natural number. We define a mapping $\tau_k : \sum \rightarrow \sum$. Given an element $\alpha = \langle i_0, i_1, \dots \rangle \in \sum$, then $\tau_k(\alpha) = \langle j_0, j_1, j_2, \dots \rangle$, where $j_0 = k$ and $j_{p+1} = i_p$ for $p \in \omega$. The mapping τ_k is a homeomorphism of $\sum$ onto the basic neighborhood $N(k)$. For a tuple $k_0, k_1, \dots, k_{n-1}$ of natural numbers, $n \in \omega$, we define the composition $\tau_{k_0, k_1, \dots, k_{n-1}} = \tau_{k_0} \tau_{k_1} \dots \tau_{k_{n-1}}$, i.e. $\tau_{k_0, k_1, \dots, k_{n-1}}(\alpha) = \langle k_0, k_1, \dots, k_{n-1}, i_0, i_1, \dots \rangle \in \sum$. The mapping $\tau_{k_0, k_1, \dots, k_n}$ is a homeomorphism of $\sum$ onto the basic neighborhood $N(k_0, k_1, \dots, k_n)$. It is called a shift (cf. [3]).

Remark 1. Sometimes we use $k_0, k_1, \dots, k_n$ instead of the precise notation of a tuple of numbers $\langle k_0, k_1, \dots, k_n \rangle$. Accordingly we write $M_{k_0, k_1, \dots, k_n}$ instead of M_s where $s = \langle k_0, k_1, \dots, k_n \rangle$, etc. We denote $st = \langle k_0, k_1, \dots, k_n, i_0, i_1, \dots, i_p \rangle$ if $t = \langle i_0, i_1, \dots, i_p \rangle$ and s is defined as above.

The notion of a multisequence has been introduced by the author. We recall

Definition 1 (see [3]). Let A be a set and let $M : \sum \rightarrow A$. The mapping M is said to be a multisequence in the set A if M is a continuous mapping of $\sum$ into the set A equipped with the discrete topology. (A multisequence in a space (X,u) or in a group $(X,+)$ means a multisequence in X, etc.)

The following auxiliary notions have been defined (cf. [3]). Let $f : S \rightarrow \omega$ be a mapping. Denote

(5) $$\sum_f = \{\alpha;\ \alpha \in \sum \text{ and } f(\alpha(0), \dots, \alpha(n-1)) \leqq \alpha(n) \text{ for each natural number } n\}$$

the set of all sequences from $\sum$ such that the components $\alpha(n)$ are sufficiently large.

The definition of $\sum_f$ can be reformulated as follows. An element $\alpha = \langle i_0, i_1, \dots \rangle \in \sum$ belongs to $\sum_f$ if and only if

the first component of α is greater or equal than the number $f(\ )$, $f(\ ) \leqq i_0$,

the second component of α fulfills $f(i_0) \leqq i_1$,

the third component of α fulfills $f(i_0, i_1) \leqq i_2$,

etc.

Now we prove that the family

(6) $\mathcal{B} = \{\sum_f ; f \in \omega^S\}$ is a filterbase in $\sum'$.

Let $f \in \omega^S$, $g \in \omega^S$ and let $h(s) = \max(f(s), g(s))$ for each $s \in S$. If $\alpha \in \sum_h$, then by (5), $h(\alpha(0), \ldots, \alpha(n-1)) \leqq \alpha(n)$ for each natural number n . Since $f(s) \leqq h(s)$ for $s \in S$, we have $f(\alpha(0), \ldots, \alpha(n-1)) \leqq \alpha(n)$ for each natural number n , i.e. $\alpha \in \sum'_f$. Similarly we can prove $\alpha \in \sum'_g$. Since α is an arbitrary element of $\sum'_h$, we get $\sum_h \subset \sum_f \cap \sum_g$. Hence (6) is true since $\emptyset \notin \mathcal{B} \neq \emptyset$ is easy to prove.

Definition 2. Let M be a multisequence in a set A . The image of the filterbase $\mathcal{B}$ under M is, of course, a filterbase of a filter $\mathcal{F}$ in A . The filter $\mathcal{F}$ is called the filter of sections of the multisequence M .

The author had defined a limit of a multisequence in the set of all real numbers (see Definition 4.1 in [3]). It is easy to see that this definition is a special case of the following

Definition 3. Let (X,u) be a topological space, let M be a multisequence in X and let $x \in X$ be a point. The multisequence M is said to converge to x and the point x to be a limit of M , $x = \lim M$, if the filter of sections of M converges to x .

Theorem 1. Let (X,u) be a sequential space, let A be a subset of X and let $x \in X$. The point x lies in the closure of A , $x \in uA$, if and only if there exists a multisequence in A converging to the point x .

Proof. Sufficiency. Let M be a multisequence in A converging to x . According to Definition 2 and Definition 3, the filter of sections of M is the filter in A converging to x i.e. $x \in uA$.

Let $x \in uA$ and let λ denote the corresponding sequential closure operator in (X,u) . The proof of the necessity is divided into seven parts.

a) A construction. We construct points $m(s) \in X$, $s \in S$. The order of $m(s)$ with respect to A is denoted by $\xi(m(s))$ or simply by $\xi(s)$. Let α be the order of the point x with respect to A . Put $m(\Theta) = x$, we have $\xi(\Theta) = \alpha$.

If $\xi(\Theta) = 0$, i.e. $x \in A$, then put $m(j) = x$ for each natural number j . We have $\xi(j) = \xi(\Theta) = 0$ for each $j \in \omega$; $m(\Theta) = \lim m(j)$.

If $\xi(\Theta) > 0$, then the definition of λ implies that there exist $m(j) \in X$, $j \in \omega$, such that $m(\Theta) = \lim m(j)$ and $\xi(\Theta) > \xi(j)$ for each $j \in \omega$.

b) Let j be given.

If $\xi(j) = 0$, i.e. $m(j) \in A$, then put $m(j,k) = m(j)$ for each natural number j . We have $\xi(j,k) = \xi(j) = 0$ for each $k \in \omega$ and $m(j) = \lim_{k \to \infty} m(j,k)$.

If $\xi(j) > 0$, then the definition of λ implies that there exist $m(j,k) \in X$, $k \in \omega$, such that $m(j) = \lim_{k \to \infty} m(j,k)$ and $\xi(j) > \xi(j,k)$

for each $k \in \omega$.

In the sequel we write i_0 instead of j and i_1 instead of k . Hence we have constructed points $m(i_0,i_1)$, $m(i_0)$, $m(\Theta)$ in X such that $\xi(\Theta) \geq \xi(i_0) \geq \xi(i_0,i_1)$, $i_0 \in \omega$, $i_1 \in \omega$.

c) Let a number $j > 1$ be given. For each k , $0 \leq k \leq j$, and for each tuple $s = \langle i_0,i_1,\dots,i_{k-1} \rangle$ let points $m(s) \in X$ and corresponding ordinal numbers $\xi(s)$ have been defined such that

(7) $\quad m(i_0,i_1,\dots,i_{k-1}) = \lim\limits_{i_k \to \infty} m(i_0,i_1,\dots,i_k)$ if $0 \leq k < j$,

(8) $\quad \xi(i_0,i_1,\dots,i_{k-1}) \geq \xi(i_0,i_1,\dots,i_k)$, $0 \leq k < j$,
whereas the inequality is strong iff $\xi(i_0,i_1,\dots,i_{k-1}) > 0$.

Let a tuple $s = \langle i_0,i_1,\dots,i_{j-1} \rangle$ be given.

If $\xi(s) = 0$, i.e. $m(s) \in A$, then put $m(i_0,i_1,\dots,i_j) = m(s)$ for each $i_j \in \omega$. We have $\xi(i_0,i_1,\dots,i_j) = \xi(s) = 0$ for each i_j and $m(s) = \lim\limits_{i_j \to \infty} m(i_0,i_1,\dots,i_j)$.

If $\xi(s) > 0$, then according to (4), $m(s) \in \lambda^{\eta} A - \lambda^{\eta-1} A$, where $\eta = \xi(s)$. Hence there are points $y_k \in \lambda^{\eta-1} A$ such that $m(s) = \lim y_k$. Put $m(i_0,i_1,\dots,i_j) = y_{i_j}$ for each $i_j \in \omega$. We have $\xi(s) > \xi(i_0,i_1,\dots,i_j)$ for any $i_j \in \omega$ and $m(s) = \lim\limits_{i_j \to \infty} m(i_0,i_1,\dots,i_j)$.

Now we have points $m(s)$, $s \in \bigcup\limits_{k=0}^{j+1} \omega^k$, such that (7) and (8) are fulfilled with $0 \leq k < j+1$.

d) Induction Principle implies that points $m(s)$, and their orders $\xi(s)$ are determined for each $s \in S$ such that (7) and (8) are fulfilled for each $j \in \omega$.

Given $\alpha = \langle i_0,i_1,\dots \rangle \in \sum$, write a nonincreasing sequence of ordinal numbers $\xi(0) \geq \xi(i_0) \geq \xi(i_0,i_1) \geq \dots$.

The fundamental property of ordinal numbers implies that there is the least number q such that

(9) $\quad \xi(i_0,i_1,\dots,i_{q-1}) = \xi(i_0,i_1,\dots,i_q)$.

Put $M(\alpha) = m(i_0,i_1,\dots,i_q)$. Now $M(\alpha)$ is defined for each $\alpha \in \sum$.

e) The mapping M is the mapping of $\sum$ into A . Proof of (e): The equality in (9) implies $\xi(i_0,i_1,\dots,i_q) = 0$ by (8) , which means $m(i_0,i_1,\dots,i_q) \in \lambda^0 A = A$.

f) M is a multisequence in A . Proof of (f): The equality in (9) implies $0 = \xi(i_0,i_1,\dots,i_q) = \xi(i_0,i_1,\dots,i_q,i_{q+1}) = \dots$ which means, by the construction in (a), (b) and (c), that the sequences $\langle m(i_0,i_1,\dots,i_q,k_{q+1}) \rangle_{k_{q+1}=0}^{\infty}$, $\langle m(i_0,i_1,\dots,i_q,k_{q+1},k_{q+2}) \rangle_{k_{q+2}=0}^{\infty}$, are all constant and identically equal to $m(i_0,i_1,\dots,i_q)$. According to (9)

(10) $\quad M(\alpha) = m(i_0,i_1,\dots,i_q)$

for each $\alpha = \langle i_0,i_1,\dots \rangle \in N(i_0,i_1,\dots,i_q)$.

The mapping M is continuous in α since it is constant on a neighborhood $N(i_0,i_1,\dots,i_n)$.

g) The multisequence M converges to the point x in the sequential space (X,u) . Proof of (g): Let U be an open neighborhood of x in the space (X,u) . We show that U contains an element of the filter of sections of M . Since $x = m(\Theta) = \lim m(i_0)$, there exists an index i'_0 such that $m(i_0)\in U$ if $i_0 \geqq i'_0$. Put $f(\Theta) = i'_0$.

Let $f(s)$ be defined for each $s\in \bigcup_{k=0}^{q-1} \omega^k$, where q be a positive integer number, such that

(11) $\quad m(i_0,i_1,\dots,i_k)\in U$ if $f(i_0,i_1,\dots,i_{k-1}) \leqq i_k$

for each k , $0 \leqq k<q$, and each tuple $s = \langle i_0,i_1,\dots,i_k \rangle$.

Given a tuple $s = \langle i_0,i_1,\dots,i_q \rangle$ such that (11) is true for each k , $0\leqq k<q$, the relation (7) implies the existence of an index i'_q such that $m(s)\in U$ if $i'_q \leqq i_q$. If we put $f(i_0,i_1,\dots,i_{q-1}) = i'_q$, then (11) is also true for $k = q$.

Induction principle implies that f is defined for each finite tuple from S and (11) is true for all k .

Let $\alpha = \langle k_0,k_1,\dots \rangle \in \sum'_f$ be given, i.e. the condition in (11) is fulfilled (see (5)). The value $M(\alpha) = m(i_0,i_1,\dots,i_q)$ is defined by (10) and $M(\alpha)\in U$ since (11). Therefore the open neighborhood U contains the element $M\left[\sum'_f\right]$ of the filter of sections of M ; M converges to x by Definition 3.

Proof of Theorem 1 is finished.

3. The structure of the multisequence

Recall the idea of the proof of Theorem 1. We constructed a set of points $m(i_0,i_1,\dots,i_n)$ which generated a desired multisequence M . The mapping $m : S \to X$ will be called the generator of M in (X,u) . We give an exact definition of a generator and prove that generators are uniquely determined in Hausdorff spaces and that there is a multisequence without any generator.

Definition 4. Let (X,u) be a topological space and let M be a multisequence in a subset $A\subset X$. A mapping $m : S \to X$ is said to be a generator of M in the space (X,u) and M is said to possess a generator m in (X,u) if the following two conditions are fulfilled.

(i) $m(i_0,i_1,\dots,i_{n-1}) = \lim_{i_n\to\infty} m(i_0,i_1,\dots,i_n)$ for each tuple $\langle i_0,i_1,\dots,i_{n-1}\rangle \in S$,

(ii) if $\alpha = \langle i_0,i_1,\dots \rangle \in \sum'$, then $m(i_0,i_1,\dots,i_n) = M(\alpha)$ for all but finitely many n .

Proposition 1. Let (X,u) be a sequential space, let A be a subset and x a point of X . The point x lies in the closure uA of A if and only if there exists a multisequence in A converging to x and possessing a generator $m : S \to X$. (We shall see in Lemma 1 below that $x = m(\Theta)$).

Proof. The same as that of Theorem 1. The set $\{m(s);\ s\in S\}$ constructed in the proof of Theorem 1 is the generator $m : S \to X$ of the multisequence M .

Remark 2. Let (X,u) be a sequential space, let A be a subset and x a point in X . If there exists a multisequence in A converging to x, then $x\in uA$ and so there exists a multisequence in A which possesses a generator in (X,u) and which converges to the point x .

Proposition 2. Let (X,u) be a Hausdorff space. A multisequence M in A , $A\subset X$, possesses at most one generator m in (X,u) .

Proof. Suppose that, on the contrary, m and r are two generators of M in the space (X,u) such that for $s = \langle i_o, i_1, \dots, i_{n-1}\rangle \in S$ the relation $m(s) \neq r(s)$ is true.

(a) The condition (i) of Definition 4 implies that the sequences $\mu = \langle m(i_o,i_1,\dots,i_n)\rangle_{i_n=0}^{\infty}$ and $\varrho = \langle r(i_o,i_1,\dots,i_n)\rangle_{i_n=0}^{\infty}$ cannot be equal since limits are uniquely determined in Hausdorff spaces. Therefore there exists a natural number i_n such that $r(s,i_n) \neq$ $\neq m(s,i_n)$.

(b) Let a tuple $t = \langle i_n, i_{n+1}, \dots, i_{k-1}\rangle$ have been defined such that

$$(12) \qquad m(s,i_n,i_{n+1},\dots,i_{p-1}) \neq r(s,i_n,i_{n+1},\dots,i_{p-1})$$

for each $p = n, n+1, \dots, k$. Similarly as in (a), we get sequences

$$\langle m(s,i_n,i_{n+1},\dots,i_k)\rangle_{i_k=0}^{\infty}, \quad \langle r(s,i_n,i_{n+1},\dots,i_k)\rangle_{i_k=0}^{\infty}$$

with distinct limits and the existence of i_k such that (12) is true for $p = k+1$.

(c) Induction principle implies that there exists $\alpha =$ $= \langle i_o, i_1, \dots\rangle \in \sum$ such that (12) is true for all $p \geq n$. We meet a contradiction to (ii) of Definition 4 since the sequences $\langle m(i_o,i_1,\dots,i_k)\rangle_{k=0}^{\infty}$ and $\langle r(i_o,i_1,\dots,i_k)\rangle_{k=0}^{\infty}$ are almost constant and equal to $M(\alpha)$ by (ii) of Definition 4.

Remark 3. Let x be a point in a sequential space (X,u) . Evidently a constant mapping $M : \sum \to X$, $M(\alpha) = x$, $\alpha\in\sum$, is a multisequence in X which converges to x and which possesses a generator $m : S \to X$ in (X,u) , $m(s) = x$, $s\in S$. All the conditions of Definition 1, Definition 3 and Definition 4 are easy to verify.

Definition 5. A constant mapping of $\sum$ into a set A is called a constant multisequence in A .

Lemma 1. Let M be a multisequence in a sequential space (X,u) and let M possess a generator m in (X,u) . Then M converges to the point $m(\Theta)$.

Proof. Denote $x = m(\Theta)$ and let U be an open neighborhood of x . According to (i) of Definition 4, there exists a number $f(\Theta)\in\omega$ such that $f(\Theta) \leq i_o$ implies $m(i_o)\in U$. If $i_o \geq f(\Theta)$ is given, then by the same consideration we conclude that there is $f(i_o)\in\omega$ such that $m(i_o,i_1)\in U$ whenever $f(i_o) \leq i_1$.

It is easy to see that in such a way as in (g) of Proof of Theorem 1 we can continue and define $f : S \rightarrow \omega$ such that (11) is true for all k and all tuples $\langle i_0, i_1, \ldots, i_k \rangle$ of natural numbers. Define $\sum_f$ by (5) and let $\alpha = \langle i_0, i_1, \ldots \rangle \in \sum_f$. According to (ii) of Definition 4, there exists n such that $M(\alpha) = m(i_0, i_1, \ldots, i_n)$. The relation (11) implies $m(i_0, i_1, \ldots, i_n) \in U$ since $\alpha \in \sum_f$. Therefore the element $M\left[\sum_f\right]$ of the filterbase of the filter of sections of M is contained in the open set U and so M converges to x. Q.E.D.

Let n be a natural number and $s = \langle k_0, k_1, \ldots, k_{n-1} \rangle$ a tuple of natural numbers. Consider a multisequence M in a set A and denote $P_s = M|_{N(s)}$ the restriction of the mapping M to the basic neighborhood $N(s) \subset \sum$, $P_s(\alpha) = M(\alpha)$ for any $\alpha \in N(s)$. For a shift τ_s, a composition $M_s = P_s \tau_s$ of two continuous mappings is a continuous mapping of $\sum$ into the discrete space A, hence a multisequence in A.

Definition 6. Let n be a natural number and let M be a multisequence in a set A. For a tuple $s = \langle k_0, k_1, \ldots, k_{n-1} \rangle$ of natural numbers, denote $M_s = M|_{N(s)} \tau_s$ the composition of two mappings $M|_{N(s)}$ and τ_s, i.e. $M_{k_0, k_1, \ldots, k_{n-1}}(i_0, i_1, \ldots) = M(k_0, k_1, \ldots, k_{n-1}, i_0, i_1, \ldots)$ for each $\langle i_0, i_1, \ldots \rangle \in \sum$. The multisequence M_s is called the restriction of the multisequence M to $N(s)$ or shortly to s.

Lemma 2. Let $r = \langle i_0, i_1, \ldots, i_{p-1} \rangle \in S$ and $s = \langle k_0, k_1, \ldots, k_{q-1} \rangle \in S$ be tuples of natural numbers and let M be a multisequence in a set A. Denote the tuple $t = \langle i_0, i_1, \ldots, i_{p-1}, k_0, k_1, \ldots, k_{q-1} \rangle$ and M_r the restriction of M to r. Then the restriction of M_r to s is equal to a restriction M_t, $(M_r)_s = M_t$.

Proof. For an arbitrary $\alpha \in \sum$ we get by Definition 6: $M_t(\alpha) = M(t\alpha) = M(rs\alpha) = M_r(s\alpha) = (M_r)_s(\alpha)$ which finishes the proof.

Lemma 3. Let M be a multisequence in a set A and let $\alpha = \langle k_0, k_1, \ldots \rangle \in \sum$ be a sequence of natural numbers. Then there exists a natural number n such that the restriction of M to the tuple $s = \langle k_0, k_1, \ldots, k_{n-1} \rangle$ is a constant multisequence in A.

Proof. According to Definition 1, M is a continuous mapping of $\sum$ into A equipped with the discrete topology. This implies the existence of a basic neighborhood $N(s)$, $s = \langle k_0, k_1, \ldots, k_{n-1} \rangle$, of the point α in the Baire space $\sum$ such that M is constant on $N(s)$, $M|_{N(s)}$ is constant, i.e. the restriction of M to s is a constant multisequence in the set A, $M_s = M|_{N(s)} \tau_s$ is constant.

Proposition 3. Let a natural number n be given and let for each tuple $s = \langle k_0, k_1, \ldots, k_{n-1} \rangle \in \omega^n$, $Q^{(s)}$ be an arbitrary multisequence in a set A. Define a mapping $M : \sum \rightarrow A$. For $\alpha = \langle k_0, k_1, \ldots \rangle \in \sum$, put

$$(13) \qquad M(\alpha) = Q^{(s)}(\beta) = Q^{(k_0, k_1, \ldots, k_{n-1})}(k_n, k_{n+1}, \ldots),$$

where $s = \langle k_0, k_1, \ldots, k_{n-1} \rangle$ and $\beta = \tau_s^{-1}(\alpha)$ be the inverse image of α. Then M is a multisequence in the set A and $Q^{(s)}$ is equal to the restriction of M to s for each $s \in \omega^n$, $Q^{(s)} = M_s$ for

$s \in \omega^n$.

Proof. Since the tuple s is uniquely determined by α, M is a well-defined mapping. The restriction of M to the basic neighborhood $N(k_o,k_1,\dots,k_n)$ is the composition of two continuous mappings τ_s^{-1} and $Q^{(s)}$ according to (13). Hence M is continuous on $N(k_o,k_1,\dots,k_n)$. Since $N(k_o,k_1,\dots,k_{n-1})$ was a basic neighborhood and since a disjoint union $\sum = \cup\{N(s);\ s\in\omega^n\}$, the mapping M is continuous on $\sum$. The relation $Q^{(s)} = M_s$ is evident.

Definition 7. The multisequence M defined in Proposition 3 is called the composition of the family $\{Q^{(s)};\ s\in\omega^n\}$.

Corollary 1. There exists a composition of an arbitrary family $\{Q^{(s)};\ s\in\omega^n\}$ of multisequences in a set A.

Lemma 4. Let $n\in\omega$, let $\{M^{(s)};\ s\in\omega^n\}$ be a family of multisequences in a sequential space (X,u) and let M be a composition of $\{M^{(s)};\ s\in\omega^n\}$. If the multisequence M possesses a generator m in (X,u), then each $M^{(s)}$, $s = \langle k_o,k_1,\dots,k_{n-1}\rangle\in\omega^n$, possesses also a generator m_s in (X,u) and

(14) $\quad m_{k_o,k_1,\dots,k_{n-1}}(i_o,i_1,\dots,i_{p-1}) = m(k_o,k_1,\dots,k_{n-1},i_o,i_1,\dots,i_{p-1})$

for any $\langle i_o,i_1,\dots,i_{p-1}\rangle\in S$.

Proof. Given $s = \langle k_o,k_1,\dots,k_{n-1}\rangle\in\omega^n$ and $\langle i_o,i_1,\dots,i_{p-1}\rangle\in S$, we verify the conditions of Definition 4.

(i) $$m_s(i_o,i_1,\dots,i_{p-1}) = m(s,i_o,i_1,\dots,i_{p-1}) = \lim_{i_p\to\infty} m(s,i_o,i_1,\dots,i_p) = \lim_{i_p\to\infty} m_s(i_o,i_1,\dots,i_p)\ .$$

(ii) The multisequence $M^{(s)}$ is the restriction of M to s by Proposition 3. According to Definition 6, $M^{(s)}(i_o,i_1,\dots) = M(s,i_o,i_1,\dots) = m(s,i_o,i_1,\dots,i_p) = m_s(i_o,i_1,\dots,i_p)$ for all but finitely many p. The proof is finished.

Notice that it is not necessary for a multisequence to possess a generator even if it converges and even if the restrictions possess generators. This will be shown in the following

Example 1. We define a multisequence in the set X of all real numbers equipped with the usual sequential topology.

For each pair of natural numbers u, v define a real number

(15) $\quad x(u,v) = (-1)^v.2^{-u}+ 3^{-v}$.

We observe that $\lim_{w\to\infty} x(u,2w) = + 2^{-u}$, $\quad \lim_{w\to\infty} x(u,2w+1) = - 2^{-u}$.

On the other hand, the limit

(16) $\quad \lim_{v\to\infty} x(u,v)$ does not exist

since there are two subsequences with unequal limits.

Let a real number $\varepsilon > 0$ be given. Denote w the integral part of $2 - \frac{\log\varepsilon}{\log 2}$. Clearly,

(17) $\quad$ if $u\geq w$ and $v\geq w$, then $|x(u,v)| < \varepsilon$.

For each pair $s = \langle u,v \rangle$ denote a constant multisequence $Q^{(s)}$ in X,

$Q^{(s)}(\beta) = x(s)$ for any $\beta \in \sum$.

Apply Corollary 1 and Proposition 3 and denote the composition of $\{Q^{(s)};\ s \in \omega^2\}$ by M.

Lemma 5. The multisequence M defined in Example 1 converges to zero.

Proof. Let $\varepsilon > 0$ be an arbitrary number and let w fulfil (17). Put $f(s) = w$ for $s \in \omega^0 \cup \omega^1$ and $f(s) = 0$ for $s \in S - (\omega^0 \cup \omega^1)$. Now let $\alpha = \langle k_0, k_1, \dots \rangle \in \sum'_f$, where $\sum_f$ is defined by (5). Apply (13) for $n = 2$. Then $|M(\alpha)| = |Q^{(s)}(\beta)| = |Q^{(k_0,k_1)}(k_2,k_3,\dots)| = |x(k_0,k_1)| < \varepsilon$.

Therefore the element $M[\sum'_f]$ of the filter of sections of M is a subset of the open neighborhood $(-\varepsilon, +\varepsilon)$ of zero, i.e. M converges to zero.

Lemma 6. The multisequence M defined in Example 1 possesses no generator in the sequential space R of all real numbers.

Proof. Suppose that, on the contrary, m is a generator of M in R. For $n = 1$, according to (i) of Definition 4, Remark 3 and (14), $m(k_0) = \lim_{k_1 \to \infty} m(k_0,k_1) = \lim_{k_1 \to \infty} m_{k_0,k_1}(\Theta) = \lim_{k_1 \to \infty} x(k_0,k_1)$ which is a contradiction to (16).

Definition 8.

(i) Every constant multisequence is said to have the order zero.

(ii) Let ξ be a positive isolated ordinal number. Let a class of multisequences which have the order less than ξ have been defined. Let a multisequence M in a sequential space be given such that an order η_k of each restriction M_k of M to $\langle k \rangle$ has been defined. Denote a positive isolated ordinal number

(18) $\qquad \eta = 1 + \limsup \eta_k$.

Then the multisequence M is said to have the order η.

Lemma 7. Every multisequence has some order.

Proof. Suppose that, on the contrary, there exists a multisequence M that has no order. The multisequence M is not constant by (i) of Definition 8 and according to (ii) of the same Definition 8, there exists a restriction M_{k_0} that has no order, i.e. by (ii) of Lemma 2, a restriction M_{k_0,k_1} of M_{k_0} to $\langle k_1 \rangle$ that has no order, etc. We get a sequence $k_0, k_1, \dots$ of natural numbers such that $M_{k_0,k_1,\dots,k_n}$ has no order for each n. We meet a contradiction to Lemma 3 and (i) of Definition 8.

Remark 4. Similarly as in Proof of Lemma 7 we can show that there is no multisequence of an uncountable order. Clearly the positive order cannot be a limit ordinal number.

Proposition 4. Let (X,u) be a sequential space, let λ denote the corresponding sequential convergence closure operator, let $x \in X$, $A \subset X$ and ξ be a countable isolated ordinal number or $\xi = 0$. The order of x with respect to A is less than or equal to ξ, i.e. $x \in \lambda^{\xi} A$, if and only if there exists a multisequence M in the set A which possesses a generator m in (X,u), M converges to x and M has the order ξ.

<u>Proof.</u> We use Induction principle.

(a) $\xi = 0$. In this case $x \in \lambda^{0} A = A$ is clearly equivalent to the existence of a constant multisequence M in (X,u) such that $M(\alpha) = x$ for all $\alpha \in \sum$.

(b) Let an isolated ordinal number $\eta > 0$ be given and let Proposition 4 be true for each $\xi < \eta$.

<u>Sufficiency.</u> Let M be a multisequence in A converging to x and having the order η. Let m be a generator of M in (X,u). According to Lemma 1, Lemma 4 and (i) of Definition 4, the restrictions M_k converge to $m(k)$ and $x = \lim m(k)$. Denote the order of M_k by η_k. Then (18) implies that $\eta_k < \eta$ for almost all k and $m(k) \in \lambda^{\eta_k} A$ since Proposition 4 is supposed to be true for ordinal numbers less than η. Therefore $x \in \lambda^{\eta} A$ since $x = \lim m(k)$.

<u>Necessity.</u> Let $x \in \lambda^{\eta} A$, i.e. there exist points x_k such that $x = \lim x_k$, $x_k \in \lambda^{\eta_k} A$ and $\eta = 1 + \lim\sup \eta_k$. There exist multisequences $Q^{(k)}$ in A which have orders η_k, possess generators m_k and converge to x_k. Denote M the composition of $\{Q^{(k)};\ k \in \omega\}$, M possesses a generator m, $m(\Theta) = x$, $m(k,s) = m_k(s)$ for $s \in S$. According to Lemma 1 and Lemma 4, M converges to x and has the order $1 + \lim\sup \eta_k = \eta$.

<u>Corollary 2.</u> Let (X,u) be a sequential space, $x \in X$, $A \subset X$. For an ordinal number η, denote C_η the set of all multisequences in A which have a generator in (X,u) converging to x and have the order η. Then x is of the order ξ with respect to A if and only if $C_\xi \neq \emptyset$ and $C_\eta = \emptyset$ for any $\eta < \xi$.

4. Multisequences in sequential convergence groups

In this section we apply multisequences to sequential convergence groups. We suppose that a compatible sequential convergence in a group is given and prove that our notion of convergence (see Definition 3) gives also a compatible convergence with the group operation. It is possible to introduce convergence structure using multisequences. This interesting idea is due to J. Novák who uses double sequences, i.e. multisequences of a multiplicity equal two (see [6]).

<u>Proposition 5.</u> Let $(G,u,\circ)$ be a sequential group, i.e. a sequential space (G,u) equipped with a continuous group operation. Let P be a multisequence in G with a generator p and let Q be a multisequence in G with a generator q. Denote M, $M(\alpha) = P(\alpha) \circ [Q(\alpha)]^{-1}$ for each $\alpha \in \sum$, a mapping of $\sum$ into G. Denote m, $m(s) = p(s) \circ [q(s)]^{-1}$ a mapping of S into G. Then

(i) M is a multisequence in G with a generator m, and

(ii) if P converges to a point x and Q converges to a point y, then M converges to $x \circ y^{-1}$.

<u>Proof.</u> (i) Given $\alpha = \langle k_0, k_1, \dots \rangle \in \sum$, there exist r, s and basic neighborhoods $U = N(k_0, k_1, \dots, k_r)$ and $V = N(k_0, k_1, \dots, k_s)$ such that P is constant on U and Q is constant on V, these constants will be denoted by a and b, i.e. $P(\alpha) = a$ for all $\alpha \in U$ and $Q(\alpha) = b$ for all $\alpha \in U$, $a \in G$, $b \in G$. Put $j = \max(r, s)$, $W = U \cap V = N(k_0, k_1, \dots, k_j)$. Then $\alpha \in W$ implies $M(\alpha) = P(\alpha) \circ [Q(\alpha)]^{-1} = c = a \circ b^{-1}$. Hence M is constant on the basic neighborhood W which means that M is continuous at the element α. Since α was an arbitrary element of $\sum$, M is a multisequence in G.

Now we verify the first condition (i) of Definition 4. Since p and q fulfil this condition, we get

$$m(i_0, i_1, \dots, i_{n-1}) = p(i_0, i_1, \dots, i_{n-1}) \circ [q(i_0, i_1, \dots, i_{n-1})]^{-1} =$$

$$= [\lim_{i_n \to \infty} p(i_0, i_1, \dots, i_n)] \circ [\lim_{i_n \to \infty} q(i_0, i_1, \dots, i_n)]^{-1} =$$

$$= \lim_{i_n \to \infty} [p(i_0, i_1, \dots, i_n) \circ (q(i_0, i_1, \dots, i_n))^{-1}] = \lim_{i_n \to \infty} m(i_0, i_1, \dots, i_n)$$

for a tuple $\langle i_0, i_1, \dots, i_{n-1} \rangle \in S$.

The condition (ii) of Definition 4 is a consequence of the fact that it is true for p and q: There are numbers r and s such that, for $\alpha = \langle i_0, i_1, \dots \rangle \in \sum$, $p(i_0, i_1, \dots, i_n) = P(\alpha)$ for $n \geq r$ and $q(i_0, i_1, \dots, i_n) = Q(\alpha)$ for $n \geq s$. Then $n \geq \max(r, s)$ implies

$$m(i_0, i_1, \dots, i_n) = p(i_0, i_1, \dots, i_n) \circ [q(i_0, i_1, \dots, i_n)]^{-1} =$$
$$= P(\alpha) \circ [Q(\alpha)]^{-1} = M(\alpha).$$

The multisequence M converges to $m(\Theta) = p(\Theta) \circ [q(\Theta)]^{-1} = x \circ y^{-1}$ by Lemma 2. The proof of Proposition 5 is finished.

References

[1] Borel, E.: Leçons sur la théorie des fonctions. Gauthier-Villars, Paris 1914.

[2] Franklin, S.P.: Spaces in which sequences suffice. Fund.Math. <u>57</u> (1965), 107-115.

[3] Kratochvíl, P.: Multisequences and measures. General Topology and its Relations to Modern Analysis and Algebra IV, Proc. Fourth Prague Topol.Symp., 1976, Part B, 237-244.

[4] Novák, J.: Über die eindeutigen stetigen Erweiterungen stetiger Funktionen. Czechoslovak Math.J. <u>8</u>(83) (1958), 344-355.

[5] Novák, J.: On convergence spaces and their sequential envelopes. Czechoslovak Math.J. <u>15</u>(90) (1963), 74-100.

[6] Novák, J.: On convergence of double sequences. These Proceedings.

CONVERGENGE IN A LOCALIZABLE FUZZY TOPOLOGY ON SPACES OF PROBABILITY MEASURES

R. Lowen [1)]

As in [7], [8] we let X be a separable metrizable topological space, $\mathcal{B}(X)$ the Borel σ-algebra, $\mathcal{T}(X)$ the topology, $\Gamma(X)$ the measurable fuzzy sets and $\mathcal{M}(X)$ the probability measures on X.
It is our purpose in this paper to introduce a fuzzy topology on $\mathcal{M}(X)$, which will be finer than the structures of [7], [8] and the main advantage of which will be the fact that it is localizable in the sense of being generated by a system of fuzzy neighborhoods. We shall prove some basic results of this new structure, demonstrate its relation with the structures of [7], [8] and prove a fuzzy version of the so-called "Portmanteau theorem" [1].

We recall that I denotes the unit interval and that T_m and S_m denote the t-norm and its dual defined by $T_m(a,b) := \max\{a+b-1,0\}$, $S_m(a,b) := \min\{a+b,1\}$ for all $a,b \in I$. We also recall that if $\mathcal{T}$ is a topology then $\omega(\mathcal{T})$ denotes the generated fuzzy topology i.e. the set of all l.s.c. fuzzy sets [3].
Finally we also recall that if Δ is a fuzzy topology the α-level topology is $\iota_\alpha(\Delta) := \{\mu^{-1}]\alpha,1] \mid \mu \in \Delta\}$ for any $\alpha \in [0,1[$ and that the topological modification is defined as $\iota(\Delta) := \sup_{\alpha\in[0,1[} \iota_\alpha(\Delta)$ [3], [4].

For any $P \in \mathcal{M}(X)$, $B \in \mathcal{B}(X)$ and $\gamma \in \Gamma(X)$ we define $\nu_P^B \in I^{\mathcal{M}(X)}$ and $\nu_P^\gamma \in I^{\mathcal{M}(X)}$ by

$$\nu_P^B(Q) := (P(X\setminus B) + Q(B)) \wedge 1$$

$$\nu_P^\gamma(Q) := \left(\int(1-\gamma)dP + \int \gamma dQ\right) \wedge 1,$$

for finite collections $\mathcal{A} \subset \mathcal{B}(X)$ and $\Lambda \subset \Gamma(X)$ we put

$$\nu_P^{\mathcal{A}} := \inf_{A\in\mathcal{A}} \nu_P^A \quad \text{and} \quad \nu_P^\Lambda := \inf_{\lambda\in\Lambda} \nu_P^\lambda.$$

1) Dienst Wiskundige Analyse, University of Antwerp
RUCA, Groenenborgerlaan 171, 2020 Antwerp, Belgium

For the definition of T_m- and other fuzzy neighborhood systems we refer to [2], [6].

We recall that if N is a basis for a fuzzy neighborhood prefilter then the fuzzy neighborhood prefilter itself is given by

$$N^{\sim} := \{\mu \mid \forall \varepsilon > 0 \quad \exists \nu \in N : \nu - \varepsilon \leq \mu\},$$

[2], [6].

PROPOSITION 1. If we put, for all $P \in M(X)$:

$$N_1(P) := \{\nu_P^A \mid A \text{ finite, } A \subset T(X)\}$$

$$N_2(P) := \{\nu_P^\Lambda \mid \Lambda \text{ finite, } \Lambda \subset \omega(T(x))\}$$

then $(N_1(P))_{P \in M(X)}$ and $(N_2(P))_{P \in M(X)}$ are bases for a T_m-fuzzy neighborhoodsystem on $M(X)$ and moreover, for each $P \in M(X)$ we have $N_1(P)^{\sim} = N_2(P)^{\sim}$ i.e. the induced fuzzy topologies coincide.

PROOF. We shall omit the tiresome details of the verification that $N_1(P)$ and $N_2(P)$ are indeed bases for T_m-fuzzy neighborhoodsystems, and concentrate on proving that $N_1(P)^{\sim} = N_2(P)^{\sim}$.

By $\nu_P^G = \nu_P^{1_G}$ for any $G \in T(X)$ and $P \in M(X)$ the inclusion $N_1(P)^{\sim} \subset N_2(P)^{\sim}$ is clear. To show the other inclusion let $\lambda \in \omega(T(X))$ and $\varepsilon > 0$ and choose $k \in \mathbb{N}$ such that $\frac{1}{k} < \varepsilon$. An easy verification shows that if we put $G_i := \lambda^{-1}]\frac{i}{k},1]$, $i = 0,\dots,k-1$ then for any $R \in M(X)$:

$$\frac{1}{k}\sum_{i=1}^{k-1} R(G_i) \leq \int \lambda dR \leq \frac{1}{k}\sum_{i=0}^{k-1} R(G_i)$$

and consequently for any $P,Q \in M(X)$:

$$\int \lambda dP - \int \lambda dQ \leq \frac{1}{k}\Big(\sum_{i=1}^{k-1} P(G_i)-Q(G_i)\Big) + \frac{1}{k} P(G_0).$$

If we now put $G := \{G_1,\dots,G_{k-1}\}$ then we obtain

$$\begin{aligned}\nu_P^G(Q) &= \inf_{i=1}^{k-1}(P(X\setminus G_i) + Q(G_i)) \wedge 1\\ &= (1 - \sup_{i=1}^{k-1}(P(G_i)-Q(G_i))) \wedge 1\\ &\leq (1 - \frac{1}{k}\sum_{i=1}^{k-1}(P(G_i)-Q(G_i))) \wedge 1\end{aligned}$$

$$\leq (1 - \int \lambda dP + \int \lambda dQ + \varepsilon) \wedge 1$$
$$\leq \nu_P^{\lambda}(Q) + \varepsilon,$$

which proves the remaining inclusion. □

We shall denote the fuzzy topology generated by

$$(N_1(P)^{\sim})_{P \in M(X)} = (N_2(P)^{\sim})_{P \in M(X)}$$

by $\Delta^1(X)$. We recall [2], [6] that this fuzzy topology is then determined by the following closure- and interior operators.
For any $\mu \in I^{M(X)}$, and $P \in M(X)$:

$$\bar{\mu}(P) = \inf_{\substack{G \subset T(X) \\ G \text{ finite}}} \sup_{R \in M(X)} T_m(\nu_P^G(R), \mu(R)) \tag{1}$$

$$= \inf_{\substack{\Lambda \subset \omega(T(X)) \\ \Lambda \text{ finite}}} \sup_{R \in M(X)} T_m(\nu_P^{\Lambda}(R), \mu(R)) \tag{2}$$

and

$$\overset{\circ}{\mu}(P) = \sup_{\substack{G \subset T(X) \\ G \text{ finite}}} \inf_{R \in M(X)} S_m(1 - \nu_P^G(R), \mu(R)) \tag{3}$$

$$= \sup_{\substack{\Lambda \subset \omega(T(X)) \\ \Lambda \text{ finite}}} \inf_{R \in M(X)} S_m(1 - \nu_P^{\Lambda}(R), \mu(R)) \tag{4}$$

We also recall that $\Delta(X)$ denotes the fuzzy topology on $M(X)$ generated by the subbasis (for open fuzzy sets) $\{\delta_G | G \in T(X)\}$ where $\delta_G \in I^{M(X)}$ is defined by $\delta_G(P) := P(G)$. Further $\Delta^*(X)$ denotes the fuzzy topology on $M(X)$ generated by the subbasis (for open fuzzy sets) $\{\delta_\lambda | \lambda \in \omega(T(X))\}$ where $\delta_\lambda \in I^{M(X)}$ is defined by $\delta_\lambda(P) := \int \lambda dP$, [7], [8].

<u>PROPOSITION</u> 2. $\Delta(X) \subset \Delta^*(X) \subset \Delta^1(X)$ and moreover all inclusions are strict.

<u>PROOF</u>. The first inclusion and its strictness were shown in [7]. Let $\lambda \in \omega(T(X))$ then by (4) we obtain

$$\overset{\circ}{\delta}_\lambda(P) = \sup_{\substack{\Lambda \subset \omega(T(X)) \\ \Lambda \text{ finite}}} \inf_{R \in M(X)} S_m(1 - \nu_P^{\Lambda}(R), \delta_\lambda(R))$$

$$\geq \inf_{R \in M(X)} S_m(1 - \nu_P^{\lambda}(R), \delta_{\lambda}(R))$$

$$= \inf_{R \in M(X)} (1 - (1 - \int \lambda dP + \int \lambda dR) \wedge 1 + \int \lambda dR) \wedge 1$$

$$\geq \int \lambda dP = \delta_{\lambda}(P)$$

which proves the second inclusion. Strictness here follows from the remark that $\Delta^1(X)$ (being generated by a T_m-fuzzy neighborhoodsystem) is translation invariant [2] and thus all level topologies $\iota_{\alpha}(\Delta^1(X))$ coincide whereas $\iota_0(\Delta^*(X)) \neq \iota_{\alpha}(\Delta^*(X))$ for any $\alpha > 0$ [7]. □

THEOREM 3. The topological modification, $\iota(\Delta^1(X))$, coincides with the weak topology on $M(X)$.

PROOF. That $\iota(\Delta^1(X))$ is finer than the weak topology follows from Proposition 2 and the fact that $\iota(\Delta(X))$ coincides with the weak topology [7]. Conversely suppose that $\mu \in \Delta^1(X)$, $\alpha \in [0,1[$ and $P \in \mu^{-1}]\alpha,1]$ then by (3) there exist $\mathcal{G} \subset T(X)$, $\mathcal{G}$ finite, and $\varepsilon > 0$ such that for all $R \in M(X)$:

$$S_m(1 - \nu_P^{\mathcal{G}}(R), \mu(R)) > \alpha + \varepsilon \tag{5}$$

Put $V := \{R \in M(X) | R(G) > P(G) - \varepsilon \quad \forall G \in \mathcal{G}\}$, then V is a weak neighborhood of P [1] and by (5) it follows that for any $R \in V$:

$$\begin{aligned} \mu(R) &> \alpha + \varepsilon - 1 + \nu_P^{\mathcal{G}}(R) \\ &= \alpha + \varepsilon - 1 + \inf_{G \in \mathcal{G}}(1 - P(G) + R(G)) \wedge 1 \\ &\geq \alpha + \varepsilon - 1 + \inf_{G \in \mathcal{G}}(1 - P(G) + P(G) - \varepsilon) \wedge 1 \\ &= \alpha \end{aligned}$$

Consequently $V \subset \mu^{-1}]\alpha,1]$ which proves that $\mu^{-1}]\alpha,1]$ is a weak neighborhood of each of its points i.e. is weakly open. □

Remark that as a consequence all α-level topologies, $\iota_{\alpha}(\Delta^1(X))$, $\alpha \in [0,1[$, coincide with the weak topology.

We recall that if X,Y are separable metrizable spaces and $f : X \to Y$ is measurable then it has a natural extension $\hat{f} : M(X) \to M(Y)$ defined by $\hat{f}(P)(B) := P(f^{-1}(B))$ for any $P \in M(X)$, $B \in \mathcal{B}(Y)$ [9].

THEOREM 4. The assignment $\mathrm{Ext}^{I}(X) := (M(X),\Delta^{I}(X))$, $\mathrm{Ext}^{1}(f) := \hat{f}$ determines a covariant functor from the category of separable metrizable spaces and continuous maps to the category of fuzzy topological spaces and continuous maps.

PROOF. The only verification necessary is that $\hat{f}$ is continuous if f is. It will suffice to show that $\hat{f}$ is continuous in each point [2], [6].
Let $H \in T(Y)$, $P \in M(X)$ and $Q \in M(X)$ then

$$\begin{aligned}
&\hat{f}^{-1}\left(\nu^{H}_{\hat{f}(P)}\right)(Q) \\
&= \nu^{H}_{\hat{f}(P)}(\hat{f}(Q)) \\
&= (\hat{f}(P)(Y\setminus H) + \hat{f}(Q)(H)) \wedge 1 \\
&= (P(X\setminus f^{-1}(H)) + Q(f^{-1}(H))) \wedge 1 \\
&= \nu_{P}^{f^{-1}(H)}(Q)
\end{aligned}$$

and consequently for any $\mathcal{H} \subset T(X)$, $\mathcal{H}$ finite, we have

$$\hat{f}^{-1}\left(\nu^{\mathcal{H}}_{\hat{f}(P)}\right) = \nu_{P}^{f^{-1}(\mathcal{H})}. \qquad \square$$

For any $P \in M(X)$, $\mathcal{G} \subset T(X)$, $\mathcal{G}$ finite and $\varepsilon > 0$ we put $V(P,\mathcal{G},\varepsilon) := \{Q \in M(X) \mid \forall G \in \mathcal{G} : Q(G) > P(G) - \varepsilon\}$. It is well known [1] that the set of all $V(P,\mathcal{G},\varepsilon)$ where $\mathcal{G} \subset T(X)$, $\mathcal{G}$ finite and $\varepsilon > 0$, is a basis for the neighborhoods at P in the weak topology. For fixed $\beta \in [0,1[$ we now put $\mathcal{V}_{\beta}(P)$ the filter generated by the set of all $V(P,\mathcal{G},\varepsilon)$ where again $\mathcal{G} \subset T(X)$, $\mathcal{G}$ finite but now $\varepsilon > \beta$.

THEOREM 5. If $\mathcal{F}$ is a filter on $M(X)$ and $P \in M(X)$ then $\lim \omega(\mathcal{F})(P) \geq \alpha$ if and only if $\mathcal{V}_{1-\alpha}(P) \subset \mathcal{F}$, ($\omega(\mathcal{F}) := [\{1_F \mid F \in \mathcal{F}\}]$ [5]).

LEMMA. If $A \subset M(X)$ and $P \in M(X)$ then $\overline{1_A}(P) \geq \alpha$ if and only if for all $\mathcal{H} \in \mathcal{V}_{1-\alpha}(P) : \mathcal{H} \cap A \neq \phi$.

PROOF of Lemma. From (1) it follows that

$$\overline{1_A}(P) = \inf_{\substack{\mathcal{G}\subset T(X)\\ \mathcal{G} \text{ finite}}} \sup_{R\in A} \nu_P^{\mathcal{G}}(R)$$

and consequently $\overline{1_A}(P) \geq \alpha$ iff :

$$\forall G \subset T(X),\ G \text{ finite},\ \forall \delta > 0\ \ \exists R \in A : \nu_P^G(R) > \alpha - \delta. \tag{6}$$

Now remark that $\nu_P^G(R) > \alpha - \delta \Leftrightarrow \forall \dot{G} \in G : R(G) > P(G) - (1-\alpha+\delta)$

$\Leftrightarrow\ R \in V(P,G,1-\alpha+\delta)$,

and consequently (6) holds iff

$$\forall G \subset T(X),\ G \text{ finite};\ \forall \varepsilon > 1 - \alpha : V(P,G,\varepsilon) \cap A \neq \phi$$

which proves the lemma.

PROOF of Theorem. From the definition of convergence in fuzzy topological spaces [5] we easily obtain

$$\lim \omega(F)(P) = \inf_{\substack{U \text{ ultra} \\ U \supset F}} \inf_{A \in U} \overline{1_A}(P)$$

Thus by the lemma we have that $\lim w(F)(P) \geq \alpha$ iff for any U ultra, $U \supset F$, $V_{1-\alpha}(P)$ and U have a common finer filter which holds iff $V_{1-\alpha}(P) \subset F$. □

In the sequel we shall consider sequences on $M(X)$ rather than filters. If $(P_n)_n$ is a sequence on $M(X)$ and $[(P_n)_n]$ denotes the filter generated by this sequence then if $\lim \omega[(P_n)_n](P) \geq \alpha$ we shall denote this shortly by $P_n \xrightarrow{\alpha} P$.
Futher we denote $C(X,I)$ (resp. $U(X,I)$) the continuous (resp. uniformly continuous) fuzzy sets on X.

THEOREM 6 (Portmanteau theorem). Given a sequence $(P_n)_n$ on $M(X)$ the following are equivalent :

(i) $P_n \xrightarrow{\alpha} P$

(ii) $\underline{\lim}\ P_n(G) \geq P(G) - 1 + \alpha$ $\quad \forall$ G open

(iii) $\overline{\lim}\ P_n(F) \leq P(F) + 1 - \alpha$ $\quad \forall$ F closed

(iv) $\overline{\lim}|P_n(A)-P(A)| \leq 1 - \alpha$ $\quad \forall$ A P-continuity set

(v) $\overline{\lim}|\int f dP_n - \int f dP| \leq 1 - \alpha$ $\quad \forall f \in C(X,I)$

(vi) $\overline{\lim}|\int f dP_n - \int f dP| \leq 1 - \alpha$ $\quad \forall f \in U(X,I)$

PROOF. We prove the following implications :

$$\begin{array}{ccccccc} (i) & \Leftrightarrow & (ii) & \Leftrightarrow & (iii) & \Leftrightarrow & (iv) \\ & & & & \nearrow \;\; \searrow & & \\ & & & (vi) & \Leftarrow & (v) & \end{array}$$

(i) $\Leftrightarrow$ (ii). By Theorem 5 we have $P_n \xrightarrow{\alpha} P$ iff :

$$\forall G \subset T(X),\ G \text{ finite},\ \forall \varepsilon > 1-\alpha \quad \exists n_0 \quad \forall n \geq n_0 : P_n \in V(P,G,\varepsilon)$$

which by the finiteness of G holds iff :

$$\forall G \in T(X),\ \forall \varepsilon > 1-\alpha \quad \exists n_0 \quad \forall n \geq n_0 : P_n(G) > P(G) - \varepsilon$$

which in turn holds iff :

$$\inf_n \sup_{k \geq n} (P(G) - P_k(G)) \leq 1 - \alpha$$

i.e. iff $\underline{\lim}\, P_n(G) \geq P(G) - 1 + \alpha$.

(ii) $\Leftrightarrow$ (iii). By complementation.

(iii) $\Leftrightarrow$ (iv). Since we already know that (ii) and (iii) are equivalent we can write for any P-continuity set A :

$$\begin{aligned} P(\bar{A}) &\geq \overline{\lim}\, P_n(\bar{A}) - 1 + \alpha \\ &\geq \overline{\lim}\, P_n(A) - 1 + \alpha \\ P(\mathring{A}) &\leq \underline{\lim}\, P_n(\mathring{A}) + 1 - \alpha \\ &\leq \underline{\lim}\, P_n(A) + 1 - \alpha \end{aligned} \qquad \text{and}$$

Since $P(A) = P(\bar{A}) = P(\mathring{A})$ we obtain

$$\begin{aligned} &\overline{\lim}(P(A) - P_n(A)) \\ &\quad = P(A) - \underline{\lim}\, P_n(A) \\ &\quad \leq 1 - \alpha \end{aligned} \qquad \text{and}$$

$$\begin{aligned} &\overline{\lim}(P_n(A) - P(A)) \\ &\quad = \overline{\lim}\, P_n(A) - P(A) \\ &\quad \leq 1 - \alpha \end{aligned}$$

which proves (iv).

(iv) $\Rightarrow$ (iii). It is easily seen that there exists a sequence $\delta_k \downarrow 0$ such that for all k

$$F_k := \{x \mid d(x,F) < \delta_k\}$$

is a P-continuity set. By (iv) we then obtain for all k

$$\overline{\lim}|P_n(F_k)-P(F_k)| \leq 1-\alpha \quad \Rightarrow$$

$$\overline{\lim}\, P_n(F_k) \leq P(F_k) + 1 - \alpha \quad \Rightarrow$$

$$\overline{\lim}\, P_n(F) \leq P(F_k) + 1 - \alpha$$

and consequently

$$\overline{\lim}\, P_n(F) \leq \inf_k P(F_k) + 1 - \alpha$$
$$= P(F) + 1 - \alpha$$

(iii) $\Rightarrow$ (v). We may suppose that $0 < f(x) < 1$ for all $x \in X$. For all $k \in \mathbb{N}_0$ put

$$F_i^k := \{x|\tfrac{i}{k} \leq f(x)\} \qquad i = 0,\dots,k$$

then for any $Q \in M(X)$ we have (as before)

$$\frac{1}{k}\sum_{i=1}^{k} Q(F_i^k) \leq \int f dQ \leq \frac{1}{k} + \frac{1}{k}\sum_{i=1}^{k} Q(F_i^k) \tag{7}$$

By (iii) we have

$$\overline{\lim}\, P_n(F_i^k) \leq P(F_i^k) + 1 - \alpha \qquad \forall k \in \mathbb{N}_0,\ \forall i = 0,\dots,k$$

and thus applying the righthand-side of (7) to P_n and the lefthand-side to P we obtain

$$\overline{\lim} \int f dP_n \leq \overline{\lim}\left(\frac{1}{k} + \frac{1}{k}\sum_{i=1}^{k} P_n(F_i^k)\right)$$
$$\leq \frac{1}{k} + \frac{1}{k}\sum_{i=1}^{k} \overline{\lim}\, P_n(F_i^k)$$
$$\leq \frac{1}{k} + \frac{1}{k}\sum_{i=1}^{k} (P(F_i^k) + 1 - \alpha)$$
$$= \frac{1}{k} + \frac{1}{k}\sum_{i=1}^{k} P(F_i^k) + 1 - \alpha$$
$$\leq \frac{1}{k} + \int f dP + 1 - \alpha$$

Letting $k \to \infty$ we obtain

$$\overline{\lim} \int f dP_n \leq \int f dP + 1 - \alpha$$

Applying an analogous reasoning to -f we obtain

$$\underline{\lim} \int f dP_n \geq \int f dP - 1 + \alpha$$

which together with the previous inequality proves (v).

(v) ⇒ (vi). Trivial.

(vi) ⇒ (iii). If F is closed then, by regularity, for any $\varepsilon > 0$ there exists $\delta > 0$ such that if we put

$$G_\delta := \{x \mid d(x,F) < \delta\}$$

then $P(G_\delta) < P(F) + \varepsilon$.

Now put

$$f(x) := (1 - \frac{1}{\delta} d(x,F)) \vee 0$$

(i.e. $f = \psi_F^\delta$ in [8]) then clearly $f \in \mathcal{U}(X,I)$ and thus from (vi) we have

$$\overline{\lim} \left| \int f dP_n - \int f dP \right| \leqslant 1 - \alpha$$

which implies

$$\begin{aligned} \overline{\lim}\, P_n(F) &\leqslant \overline{\lim} \int f dP_n \\ &\leqslant \int f dP + 1 - \alpha \\ &\leqslant P(G_\delta) + 1 - \alpha \\ &\leqslant P(F) + 1 - \alpha + \varepsilon \end{aligned}$$

which by the arbitrariness of ε shows (iii). □

Concluding remarks

(1) An important advantage $\Delta^1(X)$ has over $\Delta(X)$ and $\Delta^*(X)$ is its tractability. This is already shown by Theorem 6 were we are actually able to calculate degrees of convergence as is maybe more clearly seen if we write the results of this theorem in a different but easily derivable form. Let again $(P_n)_n$ be a sequence on $\mathcal{M}(X)$ and $P \in \mathcal{M}(X)$ then we obtain

$$\begin{aligned} \lim \omega[(P_n)_n](P) &= \inf_{G \text{ open}} (\underline{\lim}\, P_n(G) - P(G) + 1) \\ &= \inf_{F \text{ closed}} (P(F) - \overline{\lim}\, P_n(F) + 1) \\ &= 1 - \sup_{A \text{ P-continuity set}} \overline{\lim} |P_n(A) - P(A)| \\ &= 1 - \sup_{f \in C(X,I)} \overline{\lim} \left| \int f dP_n - \int f dP \right| \\ &= 1 - \sup_{f \in \mathcal{U}(X,I)} \overline{\lim} \left| \int f dP_n - \int f dP \right| . \end{aligned}$$

(2) The calculative advantage of $\Delta^1(X)$ does not occur at the detriment of many other good properties of $\Delta(X)$ and $\Delta^*(X)$. It would take us to far to show this here. A small example however is that Theorem 3.7 [7] is preserved in Δ^1 with proof which moreover is much simpler than the one for $\Delta(X)$ and $\Delta^*(X)$.

REFERENCES

[1] P. Billingsley : Convergence of probability measures. John Wiley and sons (1968).

[2] U. Höhle : Probabilistische Topologien. Manuscripta Math. 26, 223-245 (1978).

[3] R. Lowen : Fuzzy topological spaces and fuzzy compactness. J. Math. Anal. Appl. 56, 621-633 (1976).

[4] R. Lowen : A comparison of different compactness in fuzzy topological spaces. J. Math. Anal. Appl. 64, 446-454 (1978).

[5] R. Lowen : Convergence in fuzzy topological spaces. General Topology Appl. 10, 147-160 (1979).

[6] R. Lowen : Fuzzy neighborhood spaces. Fuzzy sets and systems 7, 165-189 (1982).

[7] R. Lowen : On the existence of natural fuzzy topologies on spaces of probability measures. Math. Nachr. 115, 33-57 (1984).

[8] R. Lowen : Fuzzy convergence versus weak convergence in spaces of probability measures. Abhandlungen Akad. Wissenschaften (to appear).

[9] J. Neveu : Bases mathématiques du calcul des probabilités Masson (1984).

ON THE FUNCTION CLASS OF CAUCHY CONTINUOUS MAPS

Eva Lowen-Colebunders[1)]

§ 1. Introduction and preliminaries

A Hausdorff Cauchy structure $\mathcal{C}$ on a set X is a collection of filters satisfying the following conditions :

(i) $\dot{x} \in \mathcal{C}$ for every $x \in X$.

(ii) If $\mathcal{F} \in \mathcal{C}$ and $\mathcal{G} \supset \mathcal{F}$ then $\mathcal{G} \in \mathcal{C}$.

(iii) If $\mathcal{F} \in \mathcal{C}$ and $\mathcal{G} \in \mathcal{C}$ and $\mathcal{F} \vee \mathcal{G}$ exists, then $\mathcal{F} \cap \mathcal{G} \in \mathcal{C}$.

(iv) If $\dot{x} \cap \dot{y} \in \mathcal{C}$ then $x = y$.

The filters in $\mathcal{C}$ are called Cauchy filters.

$\Gamma(\mathcal{C})$ is the collection of real valued Cauchy continuous maps, i.e. functions mapping Cauchy filters in $\mathcal{C}$ to Cauchy filters on $\mathbb{R}$ (for the usual metric). The collections $\Gamma(\mathcal{C})$ occur in many well known situations, in fact in all situations where extensions are involved.

Example (1.1)

When X has a Hausdorff uniformity $\mathcal{U}$ and $\mathcal{C}$ is the collection of $\mathcal{U}$-Cauchy filters then $\Gamma(\mathcal{C})$ is exactly the class of functions on X with a continuous extension to the uniform completion. $\Gamma(\mathcal{C})$ has nicer properties than the collection $U(\mathcal{U})$ of all uniformly continuous functions, in particular, $\Gamma(\mathcal{C})$ is an algebra while $U(\mathcal{U})$ does not have this property in general.

Example (1.2)

When (X,q) is a Hausdorff convergence space and it has a Hausdorff extension (Y,p) containing (X,q) as a strictly dense subset [3] (in particular when (Y,p) is a Hausdorff topological space in which (X,q) is a dense subspace) then (Y,p) induces a Cauchy structure $\mathcal{C}$ on X, namely the collection of all filters on X generating a convergent filter on (Y,p). $\Gamma(\mathcal{C})$ again is the collection of all functions on X having a continuous extension to (Y,p).

Example (1.3)

If (X,q) is a Hausdorff convergence space let (Y,p) = (X,q) in the previous example. Then $\mathcal{C}$ is the collection of all convergent filters and $\Gamma(\mathcal{C})$ is equal to the collection $\mathcal{C}(q)$ of all continuous functions.

So every collection of continuous maps on a Hausdorff space is a collection $\Gamma(\mathcal{C})$ for a suitable Cauchy structure $\mathcal{C}$. In this paper we will characterize the collections of continuous maps among the classes $\Gamma(\mathcal{C})$. For the theory of Cauchy spaces we refer to [3], [5], [6] or [10], for function class theory to [1] and [2] and for supplementary information on Cauchy continuous maps and the classes $\Gamma(\mathcal{C})$ to [11], [7] and [8].

We recall some notations : If $(x_n)_{n\in\mathbb{N}}$ is a sequence, let $\mathcal{F}<x_n>$ be the

---

1) Department of Mathematics, Vrije Universiteit Brussel, Pleinlaan 2, 1050 Brussel, Belgium.

Fréchet filter generated. If Φ is a class of funtions on X then $t(\Phi)$ is the weak topology of Φ, $c(\Phi)$ is the weak Cauchy structure, $u(\Phi)$ is the weak uniformity. The filters in $c(\Phi)$ are called *Φ-filters* [1].
A sequence $(x_n)_n$ such that $F\langle x_n\rangle \in c(\Phi)$ is called a *Φ-sequence*. Φ^* is the collection of bounded functions in Φ. A function class Φ is *filter closed* [1] if $\Gamma c(\Phi) = \Phi$, i.e. $f \in \Phi$ whenever $f(F)$ converges for every Φ-filter F.
Throughout the paper we make the assumption that the function class $\Gamma(C)$ separates the points of X.

§ 2. Characterization of the collections of continuous maps among the classes $\Gamma(C)$

In this section we answer the following question : When is a collection $\Gamma(C)$ of Cauchy continuous maps on X a collection $\mathcal{C}(q)$ of continuous maps for a suitable convergence structure q on X ? An equivalent formulation is the following : When is $\Gamma(C)$ equal to $\mathcal{C}(t(\Gamma(C)))$?
Every collection $\Gamma(C)$ is a lattice and an algebra over $\mathbb{R}$ for the pointwise operations, $\Gamma(C)$ contains the constants and is uniformly complete. Moreover $\Gamma(C)$ is filter closed. (In fact filter closedness characterizes the collections $\Gamma(C)$ among all function classes [7]). Considering $\Gamma(C)$ as a subalgebra of $\mathcal{C}(t(\Gamma(C)))$ and applying the Mrówka approximation theorem [9] it follows that $\Gamma(C)$ is a collection of continuous functions if and only if $\Gamma(C)$ is inversion closed and separates the zerosets of $t(\Gamma(C))$.
We introduce another "closure condition" on $\Gamma(C)$, which is a strengthening of filter closedness, and we will give another characterization of the collections $\mathcal{C}(q)$.

Definition (2.1)
A function class $\Phi \subset \mathbb{R}^X$ is *sequentially closed* if $f \in \Phi$ whenever f is $t(\Phi)$-continuous and $(f(x_n))_{n\in\mathbb{N}}$ converges for every Φ-sequence $(x_n)_{n\in\mathbb{N}}$ on X.
From the definition it is clear that every collection $\mathcal{C}(q)$ is sequentially closed. Other examples of sequentially closed function classes arise from the following result.

Theorem (2.2)
If every non convergent Cauchy filter in $c(\Gamma(C))$ contains a Cauchy filter with a countable base then $\Gamma(C)$ is sequentially closed.
Proof : Suppose f is $t(\Gamma(C))$-continuous and $(f(x_n))_{n\in\mathbb{N}}$ converges for every $\Gamma(C)$-sequence $(x_n)_{n\in\mathbb{N}}$. Let $F \in c\Gamma(C)$. Either F $t(\Gamma(C))$-converges, and then $f(F)$ converges, or F contains a filter $G \in c\Gamma(C)$ having a countable base $\{G_n \mid n \in \mathbb{N}\}$. For every sequence $(x_n)_{n\in\mathbb{N}}$ with $F\langle x_n\rangle \supset G$ the sequence $(f(x_n))_{n\in\mathbb{N}}$ converges and clearly all such sequences have to converge to the same limit.
Since $G = \cap\{F\langle x_n\rangle \mid F\langle x_n\rangle \supset G\}$ it follows that $f(G)$ converges. Then

$f(F)$ converges aswell. We can conclude that $f \in \Gamma(c(\Gamma(C)))$ and by the filterclosedness of $\Gamma(C)$ that $f \in \Gamma(C)$.

Corollary 2.3

Assuming nonmeasurable cardinals, every uniform space for which every non convergent Cauchy filterclass has a minimum with a countable base, has a sequentially closed collection of Cauchy continuous maps ; in particular every metric space has a sequentially closed collection of Cauchy continuous maps.

There is no relation between inversion closedness and sequentially closedness.

Examples 2.4

Let M be the filter of countable complements on $\mathbb{P}$ and let U be the uniformity $U = [\Delta] \cap M \times M$. The collection of U-Cauchy filters is $C = \{\dot{x} \mid x \in X\} \cup \{F \mid F \supset M\}$.

The class $\Gamma(C) = \{f : \mathbb{R} \to \mathbb{R} \mid f(M) \text{ converges}\}$ is inversion closed but not sequentially closed.

By corollary 2.3 the collection of Cauchy continuous maps on $]0,1]$ is sequentially closed. However it is not inversion closed.

Theorem 2.5

$\Gamma(C)$ is a collection of continuous maps if and only if $\Gamma(C)$ is sequentially closed and separates the zerosets of $t\Gamma(C)$.

Proof : Suppose $\Gamma(C)$ is sequentially closed and separates the zerosets of $t(\Gamma(C))$. If $\Gamma(C) \neq \mathcal{C}(t(\Gamma(C)))$, choose a function $f \in \mathcal{C}(t(\Gamma(C)))$, $f \notin \Gamma(C)$ and a $\Gamma(C)$-sequence $(x_n)_{n\in\mathbb{N}}$ such that $(f(x_n))_{n\in\mathbb{N}}$ does not converge. If $(f(x_n))_{n\in\mathbb{N}}$ is bounded, choose $M > 0$ such that $|f(x_n)| \leq M$ for $n \in \mathbb{N}$. Let $h = f \wedge M \vee (-M)$. Then we clearly have $h \in \mathcal{C}^*(t(\Gamma(C)))$ but $h \notin \Gamma(C)$. In the other case, when $(f(x_n))_{n\in\mathbb{N}}$ is unbounded, for instance unbounded from above, choose a subsequence $(x_{k_n})_{n\in\mathbb{N}}$ such that $(f(x_{k_n}))_{n\in\mathbb{N}}$ is strictly increasing to $+\infty$. Let $g : \mathbb{R} \to \mathbb{R}$ be defined by

$$g(x) \begin{cases} 1 & \text{for } x \leq f(x_{k_0}) \\ \dfrac{2(x-f(x_{k_{2i}}))}{f(x_{k_{2i}}) - f(x_{k_{2i+1}})} + 1 & \text{for } f(x_{k_{2i}}) \leq x \leq f(x_{k_{2i+1}}) \\ \dfrac{-2(x-f(x_{k_{2i+1}}))}{f(x_{k_{2i+1}}) - f(x_{k_{2i+2}})} - 1 & \text{for } f(x_{k_{2i+1}}) \leq x \leq f(x_{k_{2i+2}}) \end{cases}$$

Then $g \circ f \in \mathcal{C}^*(t\Gamma(C))$ but $g \circ f \notin \Gamma(C)$. Applying the Hewitt approximation theorem [4] to the subalgebra $\Gamma^*(C)$ of $\mathcal{C}^*(t\Gamma(C))$, a contradiction follows.

Corollary 2.6

Assuming nonmeasurable cardinals, if on a metric space the Cauchy continuous maps separate the closed sets then every continuous function is Cauchy continuous.

References

[1] Császár, A. : Function classes, compactifications, real-compactifications. Ann. Univ. Sci. Budapest. Eötvös Sect. Math. 17 (1974), 139-156.

[2] Császár, A. : Some problems concerning C(X), General Topology and its relations to modern Analysis and Algebra IV, Lecture Notes in Mathematics, 609, 41-55, Springer-Verlag, Berlin 1976.

[3] Gazik, R. and Kent, D. : Regular completions of Cauchy spaces via function algebras. Bull. Austral. Math. Soc. 11 (1974), 77-88.

[4] Hewitt, E. : Certain generalizations of the Weierstrass Approximation theorem. Duke Math. J. 14 (1947), 419-427.

[5] Keller, H. : Die Limes-Uniformisierbarkeit der Limesraüme. Math. Ann. 1976 (1968), 334-341.

[6] Lowen-Colebunders, E. : Uniformizable Cauchy spaces. Internat. J. Math. and Math. Sci. 5 (1982), 513-527.

[7] Lowen-Colebunders, E. : On composition closed function classes. Acta Math. Acad. Sci. Hung., in press.

[8] Lowen-Colebunders, E. : Algebras of Cauchy continuous maps. J. Math. Anal. Appl., in press.

[9] Mrówka, S. : Some approximation theorems for rings of unbounded functions. Notices of Amer. Math. Soc. 11 (1964), 666.

[10] Reed, E. : Completion of uniform convergence spaces, Math. Ann. 194 (1971), 83-108.

[11] Snipes, R. : Cauchy regular functions, J. Math. Anal. Appl. 79 (1981), 18-25.

ON THE LOGARITHM OF THE DIFFERENTIAL OPERATOR

Jan Mikusiński [1])

We show that the limit $\lim\limits_{\alpha\to 0}\frac{s^{\alpha}-1}{\alpha}$ exists not only, if s is a positive number, but also if s is the differential operator and the limit is meant in the operational sense. It is known that the differential operator s is the convolutional inverse $s=\frac{1}{l}$ of the function l whose value is 1 for non negative arguments and 0 elsewhere. We shall show that then the limit is

$$\lim_{\alpha\to 0}\frac{s^{\alpha}-1}{\alpha} = -C - s\{\ln t\} \qquad (C \text{ Euler's constant}).$$

By an operator we mean a convolutional quotient $\frac{p}{q}$ of two functions p,q which are continuous in $[0,\infty)$ and vanish in $(-\infty,0)$: the function in the denominator is supposed not to vanish identically. We say that a sequence of operators converges to a limit, if it can be represented as a sequence with a common denominator and the sequence of numerators tends almost uniformly to the numerator of the limit while the denumerator remains the same. In symbols

$$\frac{a_n}{b_n}\to\frac{a}{b} \text{ iff } \frac{a_n}{b_n}=\frac{p_n}{q},\ p_n\rightrightarrows p,\ \frac{a}{b}=\frac{p}{q} \quad (a,b,p,q,a_n,b_n,p_n\in C[0,\infty)).$$

How to understand the limit of operational function $f(\lambda)=\{f(\lambda,t)\}$ is obvious. In the proof we use the well known equations

$$l^{\alpha}=\left\{\frac{t^{\alpha-1}}{\Gamma(\alpha)}\right\}(\alpha>0) \text{ and } l^{\alpha}l^{\beta}=l^{\alpha+\beta}.$$

We have

$$\frac{s^{\alpha}-1}{\alpha}=\frac{1-l^{\alpha+1}}{\alpha l^{\alpha+1}}=\frac{l^{2-\alpha}\cdot\frac{1-l^{\alpha+1}}{\alpha}}{l^{3}}.$$

Since the denominator in the last fraction is constant, it suffices to determine the limit of the numerator. We have

$$l^{2-\alpha}\cdot\frac{1-l^{\alpha+1}}{\alpha}=\left\{\frac{t^{1-\alpha}}{(2-\alpha)}*\frac{1}{\alpha}\left(1-\frac{t^{\alpha}}{\Gamma(1+\alpha)}\right)\right\}.$$

If each factor of a convolution converges almost uniformly, then also the convolution converges uniformly to the convolution of the limits. Evidently, the first factor $\frac{t^{1-\alpha}}{\Gamma(2-\alpha)}$ converges uniformly to $\frac{t}{\Gamma(2)}$. To find the limit of the second factor we write

$$\frac{1}{\alpha}\left(1-\frac{t^{\alpha}}{\Gamma(1+\alpha)}\right)=\frac{1}{\alpha}(1-t^{\alpha})+\frac{1}{\alpha}\left(t^{\alpha}-\frac{t^{\alpha}}{\Gamma(1+\alpha)}\right)=-\frac{t^{\alpha}-1}{\alpha}+\frac{t^{\alpha}}{\Gamma(1+\alpha)}\frac{\Gamma(1+\alpha)-\Gamma(1)}{\alpha}.$$

---

[1]) Mathematical Institut of the Polish Academy of Sciences, University of California, Santa Barbara, Special Year of Analysis 1983/1984.

Hence

$$\lim_{\alpha\to 0} \frac{1}{\alpha}\left(1 - \frac{t^{\alpha}}{\Gamma(1+\alpha)}\right) = -\ln t + 1 \cdot \Gamma'(1) = -\ln t - C$$

and

$$\lim_{\alpha\to 0} \frac{s^{\alpha}-1}{\alpha} = \frac{l^2\{-\ln t - C\}}{l^3} = -s\{\ln t + C\} = -C - s\{\ln t\}.$$

Having this limit we can show that $-C-s\{\ln t\}=\ln s$. In fact, using the functional equation $l^{\lambda_1}.l^{\lambda_2}=l^{\lambda_1+\lambda_2}$ we may write

$$\frac{s^{\lambda+\alpha}-s^{\lambda}}{\alpha} = \frac{s^{\lambda}(s^{\alpha}-1)}{\alpha} = s^{\lambda}\,\frac{s^{\alpha}-1}{\alpha}$$

and thus

$$(s^{\lambda})' = \lim_{\alpha\to 0} \frac{s^{\lambda+\alpha}-s^{\lambda}}{\alpha} = s^{\lambda}(-C-s\{\ln t\})$$

which implies

$$-C-s\{\ln t\} = \ln s ,$$

according to the general definition of a logarithm of an operator.

CONVERGENCE OF DOUBLE SEQUENCES

J. Novák [1)]

In this article we give a definition and basic properties of a convergence of double sequences of points. The theory of such convergences seems to be useful for solving problems concerning, for instance, products of Fréchet spaces or completions of convergence groups etc.

Let P be a set. Elements $(x_n) \in P^N$ are called sequences of points. It follows that a sequence (x_n) is a map $\phi : N \to P$ such that $\phi(n) = x_n$. In this sense we speak of constant sequences, one-to-one sequences etc. *Increasing* sequences of natural numbers will be denoted (n_k), (m_i), etc. A sequence (y_i) is a subsequence of a sequence (x_n), in symbols $(y_i) \prec (x_n)$, if there is (n_i) (increasing) such that $y_i = x_{n_i}$, $i \in N$. A subsequence $(x_{n_i}) \prec (x_n)$ is called of the first kind provided that the set $N - \cup n_i$ [2)] is finite. We define one-to-one sequences (x_n) and (y_n) to be almost disjoint (a.d.) if the set $\cup x_n \cap \cup y_n$ is finite. A system of one-to-one sequences every two distinct sequences of which are a.d. is called an *almost disjoint system* of sequences.

Let $(x_{mn})_{n=1}^{\infty}$ be a sequence of points $x_{mn} \in P$. Then $((x_{mn})_{n=1}^{\infty})_{m=1}^{\infty}$ is a sequence of sequences $(x_{mn})_{n=1}^{\infty}$, $m \in N$. It is called a twofold sequence (cf. [6]) and denoted $((x_{mn})_n)_m$ or simply (x_{mn}). Notice that infinite matrices $[x_{mn}]$, $(m,n) \in N \times N$, are called double sequences. Since $((x_{mn})_n)_m \in (P^N)^N$ and $[x_{mn}] \in P^{N\times N}$ there is, from the logical point of view, a difference between these two notions. But to each twofold sequence (x_{mn}) there corresponds, in a unique way, a matrix $[y_{mn}]$ such that $x_{mn} = y_{mn}$, $(m,n) \in N \times N$. For this reason we use the usual term of double sequence instead of twofold sequence.

To distinguish sequences (x_m) and (x_{mn}) we sometimes call (x_m) simple sequences. Let (x_{mn}) be a double sequence. We define a cross-subsequence in (x_{mn}) to be a simple sequence (z_{m_i}) such that $z_{m_i} = x_{m_i f(i)}$, $i \in N$, where $f : N \to N$. A cross-subsequence $(x_{mf(m)})_m$ is called a cross-sequence in (x_{mn}). A simple sequence $(x_{mn})_n$ is called a row-sequence in (x_{mn}), more precisely, the m-th row-sequence. Each subsequence of it is called a row-subsequence in (x_{mn}).

<u>Definition 1</u>. Let (x_{mn}) and (y_{ij}) be double sequences. We say that (y_{ij}) is a *subsequence* of (x_{mn}), in symbols $(y_{ij}) \prec (x_{mn})$, if

---

1) Institute of Mathematics of the Czechoslovak Academy of Sciences, Žitná 25, 115 67 Praha 1, Czechoslovakia

2) For the sake of simplicity we write $\cup x_n$ instead of $\cup\{x_n\}$.

there are sequences [1] of natural numbers $(m_i)_i$ and $(n_j^i)_j$, $i \in N$, such that $y_{ij} = x_{m_i n_j^i}$, $(i,j) \in N \times N$. If $(z_{rs}) \prec (y_{ij})$ and $(y_{ij}) \prec (x_{mn})$ then, clearly, $(z_{rs}) \prec (x_{mn})$.

We say that a double subsequence $(y_{ij}) \prec (x_{mn})$ is of the *first kind* if $(y_{ij}) = (x_{m_i n_j^i})$ where the sets of natural numbers $N - \bigcup m_i$ and $N - \bigcup n_j^i$, $i \in N$, are finite.

Remark. If (y_{ij}) is a double subsequence of (x_{mn}) and $(z_k)_k$ a cross-subsequence (a row-subsequence) in (y_{ij}) then it is a cross-subsequence (a row-subsequence) in (x_{mn}) as well.

Definition 2. Let P be a set and let $\mathcal{L}''$ be a collection of pairs $[(x_{mn}),x]$ where (x_{mn}) is a double sequence of points of P and x is a point of P. We say that $\mathcal{L}''$ is a *double convergence* or a convergence of double sequences for P and write $\operatorname{Lim} x_{mn} = x$ for $[(x_{mn}),x] \in \mathcal{L}''$ if the axioms $(\mathcal{L}_0)$, $(\mathcal{L}_1)$, $(\mathcal{L}_2)$ are true:

$(\mathcal{L}_0)$ If $\operatorname{Lim} x_{mn} = x$ and $\operatorname{Lim} x_{mn} = y$ then $x = y$.

$(\mathcal{L}_1)$ If $x_{mn} = x$, $m \in N$, $n \in N$, then $\operatorname{Lim} x_{mn} = x$.

$(\mathcal{L}_2)$ If $\operatorname{Lim} x_{mn} = x$ and $(y_{ij}) \prec (x_{mn})$ then $\operatorname{Lim} y_{ij} = x$.

If $\operatorname{Lim} x_{mn} = x$ we say that (x_{mn}) converges to x or that x is a limit of the double sequence (x_{mn}).

If, moreover, the axiom $(\mathcal{L}_3)$ is true:

$(\mathcal{L}_3)$ If x is a point and (x_{mn}) a double sequence of points of P such that if $(y_{ij}) \prec (x_{mn})$ then there is a double subsequence of (y_{ij}) which converges to x then $\operatorname{Lim} x_{mn} = x$

we say that $\mathcal{L}''$ is a star double convergence.

1. Double convergence in convergence spaces

Now, we shall define a double convergence for closure (pretopological) spaces. The class of closure spaces includes all topological and convergence spaces. By a closure space we understand a set P and a closure operator w which satisfies the following properties

$$w\emptyset = \emptyset, \quad w\{x\} = \{x\}, \quad A \subset wA, \quad w(A \cup B) = wA \cup wB,$$

i.e. axiom T_1 is assumed. Recall the definition of an $\mathcal{L}$-space and of a convergence space ([4], [1]). An $\mathcal{L}$-space (Fréchet $\mathcal{L}$-class) is a set P and a convergence $\mathcal{L}$ satisfying Fréchet axioms $(\mathcal{L}_0)$, $(\mathcal{L}_1)$, $(\mathcal{L}_2)$ of

[1] Notice that, according to our notation, the sequences $(m_i)_i$ and $(n_j^i)_j$ are increasing.

convergence. A convergence space $(P, \mathcal{L}, \lambda)$ is a closure space (P, λ) where the closure operator λ for P is induced by an $\mathcal{L}$-space $(P, \mathcal{L})$ in the following way: the closure λA of a set $A \subset P$ is a set of limits of sequences (x_n) such that $\bigcup x_n \subset A$. In convergence spaces a neighborhood of a point x is defined to be a set W such that

$$\text{if } \lim x_n = x \text{ then the set } \bigcup x_n - W \text{ is finite.}$$

Hausdorff (T_2) and regular (T_3) convergence spaces are defined analogously to Hausdorff and regular topological spaces.

Definition 3. Let (P,w) be a closure space. We define a collection $\mathcal{M}''$ to be a set of elements $((x_{mn}),x)$, $x \in P$, $x_{mn} \in P$, such that

$$\text{if } (y_{ij}) < (x_{mn}) \text{ then } x \in w \bigcup\bigcup y_{ij} .$$

Theorem 1. Let (P,w) be a closure space. Let the collection $\mathcal{M}''$ satisfy $(\mathcal{L}_0)$. Then $\mathcal{M}''$ is a star double convergence for P.

Proof. Since $\{x\} = w\{x\}$, $x \in P$, it follows that $((x),x) \in \mathcal{M}''$. Hence $(\mathcal{L}_1)$ is true. Let $((x_{mn}),x)$ be an element of $\mathcal{M}''$ and (y_{ij}) a subsequence of (x_{mn}). If $(z_{rs}) < (y_{ij})$ then $(z_{rs}) < (x_{mn})$ and by Definition 3 $x \in w \bigcup\bigcup z_{rs}$. Consequently, $((y_{ij}),x) \in \mathcal{M}''$ and so $(\mathcal{L}_2)$ holds true. Now, suppose that x is a point and (x_{mn}) a double sequence of points of P. Let (y_{ij}) be any subsequence of (x_{mn}) and (z_{rs}) a subsequence of (y_{ij}) such that $((z_{rs}),x) \in \mathcal{M}''$. Then $x \in w\bigcup\bigcup z_{rs}$, by Definition 3. Since $\bigcup\bigcup z_{rs} \subset \bigcup\bigcup y_{ij}$ we have $x \in w\bigcup\bigcup y_{ij}$. Consequently $((x_{mn}),x) \in \mathcal{M}''$. It follows that $(\mathcal{L}_3)$ is true.

Definition 4. Let $(X, \mathcal{L})$ be an $\mathcal{L}$-space. We define a collection $\mathcal{L}''$ to be a set of elements $((x_{mn}),x)$, $x \in X$, $x_{mn} \in X$, such that the condition (i) is true.

(i) If $(y_{ij}) < (x_{mn})$ then either
   (i1) there is a cross-subsequence in (y_{ij}) which $\mathcal{L}$-converges to x
   (i2) or there is a natural number i_0 and in each row-sequence $(y_{ij})_j$, $i \geq i_0$, a subsequence which $\mathcal{L}$-converges to x.

Theorem 2. Let $(X, \mathcal{L})$ be an $\mathcal{L}$-space. Let the collection $\mathcal{L}''$ for X satisfy $(\mathcal{L}_0)$. Then $\mathcal{L}''$ is a star double convergence for $(X, \mathcal{L})$.

Proof. If (x) is a constant double sequence then each subsequence of it is constant as well. Therefore both properties (i1) and (i2) are satisfied and so $((x),x) \in \mathcal{L}''$. Let $((x_{mn}),x) \in \mathcal{L}''$ and let $(z_{rs}) < (x_{mn})$. If $(y_{ij}) < (z_{rs})$ then $(y_{ij}) < (x_{mn})$ and so (i) is true. It follows $((z_{rs}),x) \in \mathcal{L}''$ and hence $(\mathcal{L}_2)$ is satisfied. Let (x_{mn}) be a sequence of points and x a point of X. Let (y_{ij}) be a subsequence of (x_{mn}). Suppose that there is a subsequence $(z_{rs}) < (y_{ij})$ such that $((z_{rs}),x) \in \mathcal{L}''$. Since each cross-subsequence in (z_{rs}) is a cross-subsequence in (y_{ij}) and each row-subsequence in (z_{rs}) is a row-subsequence in (y_{ij}) it follows that (i) is true. Hence $((x_{mn}),x) \in \mathcal{L}''$, by (i), and ($\mathcal{L}_3$) holds.

Let $(X,\mathcal{L},\lambda)$ be a convergence space. It is an $\mathcal{L}$-space $(X,\mathcal{L})$ and also a closure space (X,λ). Hence, there are a collection $\mathcal{L}''$ for $(X,\mathcal{L})$ and a collection $\mathcal{M}''$ for (X,λ). A question arises as follows: What is the relation between $\mathcal{L}''$ and $\mathcal{M}''$? This question is answered by

<u>Theorem 3</u>. Let $(X,\mathcal{L},\lambda)$ be a convergence space. Let $\mathcal{L}''$ be a collection for $(X,\mathcal{L})$ and $\mathcal{M}''$ a collection for (X,λ). Then $\mathcal{L}'' = \mathcal{M}''$.

<u>Proof.</u> Let $((x_{mn}),x) \in \mathcal{L}''$ and $(y_{ij}) \prec (x_{mn})$; then, by (i), we have $x \in \lambda \bigcup\bigcup y_{ij}$. Consequently, $((x_{mn}),x) \in \mathcal{M}''$ according to Definition 3. Now, let $((x_{mn}),x) \in \mathcal{M}''$. Suppose that (indirect proof) $((x_{mn}),x) \notin \mathcal{L}''$. By (i) there is $(y_{ij}) \prec (x_{mn})$ such that neither (i1) nor (i2) is true. Since (i2) is not true there is a subsequence $(z_{rs}) \prec (y_{ij})$ no row-subsequence in which $\mathcal{L}$-converges to x. Since (i1) is not true there is a double sequence $(t_{pq}) \prec (z_{rs})$ such that $x \notin \bigcup\bigcup t_{pq}$. On the other hand, $x \in \lambda \bigcup\bigcup t_{pq}$, by Definition 3. Hence there is a one-to-one sequence (a_k) of points $a_k \in \bigcup\bigcup t_{pq}$ such that $\mathcal{L}\text{-}\lim a_k = x$. Because no row-subsequence in (t_{pq}) $\mathcal{L}$-converges to x there is a subsequence $(a_{k_i}) \prec (a_k)$ which is a cross-subsequence in (t_{pq}). Notice that $(t_{pq}) \prec (y_{ij})$. Hence the sequence (a_{k_i}) is a cross-subsequence in (y_{ij}). Since $\mathcal{L}\text{-}\lim a_{k_i} = x$ and (i1) is not true we have a contradiction.

<u>Theorem 4</u>. Let $(X,\mathcal{L},\lambda)$ be a convergence space. Then $((x_{mn}),x) \in \mathcal{L}''$ if and only if the condition (1) is true:

(1) If $W(x)$ is a neighborhood of x in X then there is a double subsequence $(y_{ij}) \prec (x_{mn})$ of the first kind such that $\bigcup\bigcup y_{ij} \subset W(x)$.

<u>Proof.</u> Let $((x_{mn}),x) \in \mathcal{L}''$. Suppose that (1) is not true. Then there is a neighborhood $W'(x)$ of x such that if (z_{rs}) is a subsequence of (x_{mn}) of the first kind then $\bigcup\bigcup z_{rs} - W'(x) \neq \emptyset$. From this assumption we deduce that there is m_1 and a sequence $(n_j^1)_j$ such that no point $x_{m_1 n_j^1}$ belongs to $W'(x)$. (Otherwise, for each $m \in N$ there would be $n_m \in N$ such that $n > n_m$ implies $x_{mn} \in W'(x)$; hence we would have a subsequence $(z_{rs}) \prec (x_{mn})$ of the first kind such that $\bigcup\bigcup z_{rs} \subset W'(x)$.) Now, we use mathematical induction to construct a subsequence $(x_{m_i n_j^i}) \prec (x_{mn})$ such that $W'(x) \cap \bigcup\bigcup x_{m_i n_j^i} = \emptyset$. Since $((x_{mn}),x) \in \mathcal{L}''$ we have $((x_{m_i n_j^i}),x) \in \mathcal{L}''$ and hence $x \in \lambda \bigcup\bigcup x_{m_i n_j^i}$ and $W'(x) \cap \bigcup\bigcup x_{m_i n_j^i} \neq \emptyset$. This is a contradiction.

Let (x_{mn}) be a double sequence of points and x a point in X satisfying (1). Then $((x_{mn}),x) \in \mathcal{L}''$. Otherwise there would be, by Theorem

3 and Definition 3, a subsequence $(z_{rs}) \prec (x_{mn})$ and a neighborhood $W(x)$ of x such that $W(x) \cap \bigcup\bigcup z_{rs} = \emptyset$. According to (1) there is a double subsequence $(y_{ij}) \prec (x_{mn})$ of the first kind such that $\bigcup\bigcup y_{ij} \subset W(x)$. Since (y_{ij}) is of the first kind we have $\bigcup\bigcup z_{rs} \cap \bigcup\bigcup y_{ij} \neq \emptyset$ and this is a contradiction.

The following theorem instantly follows from Theorem 4.

<u>Theorem 5.</u> Let $(X, \mathcal{L}, \lambda)$ be a T_2 convergence space. Then the collection $\mathcal{L}''$ for $(X, \mathcal{L})$ is a double convergence.

We give an example of a convergence space $(X, \mathcal{L}, \lambda)$ such that the collection $\mathcal{L}''$ for X does not satisfy the axiom $(\mathcal{L}_0)$.

<u>Example 1</u>. Let X be a set of points a, b, x_{mn}. Each point x_{mn} is isolated, each cross-subsequence in (x_{mn}) converges to a, each row--subsequence converges to b. Then we have a topological convergence space $(X, \mathcal{L}, \lambda)$ which is not a T_2 space. Notice that $[(x_{mn}), a] \in \mathcal{L}''$, $[(x_{mn}), b] \in \mathcal{L}''$, $a \neq b$. Hence $\mathcal{L}''$ does not fulfil the axiom $(\mathcal{L}_0)$.

Now, we give an example of a topological space (P, w) which is not a convergence space and, in spite of it, the collection $\mathfrak{M}''$ for P is a double convergence for P.

<u>Example 2.</u> Let P be a set of points a, x_{mn}. Each point x_{mn} is isolated. A set $W \subset P$ is a neighborhood of a if $a \in W$ and there is a double subsequence $(y_{mn}) \prec (x_{mn})$ of the first kind such that $\bigcup\bigcup y_{ij} \subset \subset W$. Hence we have a topological space (P, w) such that the collection $\mathfrak{M}''$ for P is a double convergence for P. It is well known that (P, w) is not a convergence space (i.e. not a Fréchet space).

Let us formulate, in terms of double convergence, a well-known statement on Fréchet spaces.

<u>Proposition 1.</u> A convergence T_2 space $(X, \mathcal{L}, \lambda)$ is Fréchet if and only if the following statement (2) is true.

(2) If (x_{mn}) is a double sequence of points and x a point of X such that $x = \lim_m \lim_n x_{mn}$ then $x = \operatorname{Lim} x_{mn}$.

<u>Proof</u>. Let $(X, \mathcal{L}, \lambda)$ be a T_2 Fréchet space. Let (x_{mn}) be a double sequence, x a point and (x_m) a sequence of points $x_m \in X$ such that $\lim_m x_m = x$, $\lim_n x_{mn} = x_m$, $m \in N$. Let (y_{ij}) be a subsequence of (x_{mn}). Then $\lim_i \lim_j y_{ij} = x$. Because X is Fréchet, it follows that $x \in \lambda \bigcup\bigcup y_{ij}$. Since X is T_2 convergence space from Theorems 1 and 3 we have $\operatorname{Lim} x_{mn} = x$.

Now, suppose $(X, \mathcal{L}, \lambda)$ is not a Fréchet space. Then there is a one--to-one double sequence (x_{mn}), a one-to-one simple sequence (x_m) and a point x such that $x = \lim_m \lim_n x_{mn}$ and $x \notin \lambda \bigcup\bigcup x_{mn}$.

Consequently, if $(y_{ij}) < (x_{mn})$ then $x \notin \lambda \bigcup\bigcup y_{ij}$. Hence (x_{mn}) does not converge to x , by Definition 2, and so (2) does not hold.

Proposition 2. Let $(X, \mathcal{L}, \lambda)$ be a T_3 convergence space. Let $x \in X$, (x_{mn}) be a double sequence and (x_m) a simple sequence such that $\operatorname{Lim} x_{mn} = x$. Then

(3) $$\lim_n x_{mn} = x_m , \quad m \in N , \text{ implies that } \mathcal{L}^* - \lim x_m = x .$$

Proof. Let $W(x)$ be a neighborhood of x . Then there is a neighborhood $V(x)$ of x such that $\lambda V(x) \subset W(x)$, by (T_3) . By Theorem 4 there is a subsequence $(y_{ij}) < (x_{mn})$ of the first kind such that $\bigcup\bigcup y_{ij} \subset V(x)$. It follows that there is $m_0 \in N$ such that $m \geq m_0$ implies that $x_m \in \lambda \bigcup\bigcup y_{ij}$. Hence $x_m \in W(x)$, $m \geq m_0$. Therefore $((x_m), x) \in \mathcal{L}^*$.

We have proved that (3) is a necessary condition and now we show that it is not a sufficient condition for a convergence space to be a T_3 space.

Example 3. Let X be a set the points of which are pairs $[\xi, n]$, $\xi < \omega_1$, $n \in N$, and a pair $[\omega_1, 1]$. Each point $[\xi, n]$, $n \neq 1$, is isolated. We define $\lim_n [\xi, n] = [\xi, 1]$, $\xi < \omega_1$, and $\lim_k [\xi_k, n] = [\omega_1, 1]$ if $n \neq 1$ and $(\xi_k)_k$ is one-to-one. In such a way we get a convergence space $(X, \mathcal{L}, \lambda)$ where λ is a T_2 topology. The space (X, λ) is not regular. It is easy to see that the point $[\omega_1, 1]$ and the closed set $\bigcup [\xi, 1]$, $\xi < \omega_1$, cannot be separated by open sets.

It remains to prove that the condition (3) is satisfied. Notice that no one-to-one sequence $([\xi_k, 1])_k$ is convergent. If (x_m) is constant then, by $(\mathcal{L}_0)$, $x_m = x$, $m \in N$. Suppose that (x_m) is one-to-one. Then either $x = [\omega_1, 1]$ or $x \in \bigcup [\xi, 1]$, $\xi < \omega_1$. In both cases the condition (3) is satisfied.

Notice that there is no countable base of neighborhoods at the point $[\omega_1, 1]$ in the space X in Example 3. This is a substantial property of the space X as it is shown by the following

Proposition 3. Let $(X, \mathcal{L}, \lambda)$ be a first countable convergence space. Then it is a T_3 convergence space if and only if (3) is true.

Proof. The proof that (3) is a necessary condition follows instantly from Proposition 2. Now, suppose that X does not satisfy T_3 (indirect proof). Then there is a point $x \in X$ and a neighborhood $W(x)$ of x such that if V is a neighborhood of x , then $\lambda V \not\subset W(x)$. Let $\{V_k\}$ be a countable base of neighborhoods of x such that $W(x) \supset V_1 \supset V_2 \supset$

$\supset \ldots \supset V_k \supset \ldots$. Choose points $x_m \in \lambda V_m - W(x)$, $m \in N$. Then there is a simple sequence of points $(x_{mn})_n$, such that $x_{mn} \in V_m$ and $\mathcal{L}$-$\lim x_{mn} = x_m$, $m \in N$. Hence (x_{mn}) is a double sequence of points $x_{mn} \in W(x)$ such that $\operatorname{Lim} x_{mn} = x$. (Observe that X is T_2 .) As a matter of fact, if $(y_{ij}) \prec (x_{mn})$, then $V_k \cap \bigcup\bigcup y_{ij} \neq \emptyset$ for each $k \in N$ and so $x \in \lambda \bigcup\bigcup y_{ij}$. Notice that $\bigcup x_m \cap W(x) = \emptyset$. It follows that no subsequence of (x_m) $\mathcal{L}$-converges to x . We have proved that (3) does not hold.

2. Double convergence in $\mathcal{L}$-spaces

Now, we define special subcollections $\mathcal{L}_1''$, $\mathcal{L}_2''$, $\mathcal{L}_3''$ of a double convergence $\mathcal{L}''$ [1].

<u>Definition 5</u>. Let $(X,\mathcal{L})$ be an $\mathcal{L}$-space and $\mathcal{L}''$ the corresponding double convergence for $(X,\mathcal{L})$. Define a subcollection $\mathcal{L}_i'' \subset \mathcal{L}''$ to be a collection of elements $((x_{mn}),x) \in \mathcal{L}''$ such that

$i = 1$ each cross-sequence in (x_{mn}) $\mathcal{L}$-converges to x ,

$i = 2$ no cross-subsequence in (x_{mn}) $\mathcal{L}$-converges to x ,

$i = 3$ in each subsequence $(y_{ij}) \prec (x_{mn})$ there is a cross-subsequence which $\mathcal{L}$-converges to x and another cross-subsequence no subsequence of which $\mathcal{L}$-converges to x .

If $((x_{mn}),x) \in \mathcal{L}_1''$ ($\mathcal{L}_2''$, $\mathcal{L}_3''$) then $((x_{mn}),x)$ is called a π- (ρ- , σ-) *element* of the convergence $\mathcal{L}''$. A point x is called a π- (ρ- , σ-) *point* provided that there is a double sequence (x_{mn}) such that $((x_{mn}),x) \in \mathcal{L}_1''$ ($\mathcal{L}_2''$, $\mathcal{L}_3''$) .

These points were introduced in [6]. Notice that a convergence space containing a σ-point was, in fact, constructed under Continuum Hypothesis by P. Kratochvíl in [2].

From the definition it instantly follows that collections $\mathcal{L}_i''$ are disjoint and if $((x_{mn}),x) \in \mathcal{L}_i''$, $(y_{ij}) \prec (x_{mn})$ then $((y_{ij}),x) \in \mathcal{L}_i''$, $i = 1,2,3$.

<u>Proposition 4</u>. Let $\mathcal{L} = \mathcal{L}^*$ and let $((x_{mn}),x) \in \mathcal{L}''$. Then there is a subsequence $(y_{ij}) \prec (x_{mn})$ such that $((y_{ij}),x) \in \bigcup \mathcal{L}_i''$, $i = 1, 2,3$.

<u>Proof</u>. There are only three possible cases: Either there is $(y_{ij}) \prec (x_{mn})$ such that each cross-subsequence in (y_{ij}) $\mathcal{L}$-converges to x . Then $((y_{ij}),x) \in \mathcal{L}_1''$. Or there is a subsequence $(y_{ij}) \prec (x_{mn})$ such that no cross-subsequence in (y_{ij}) $\mathcal{L}$-converges to x . Then $((y_{ij}),x) \in \mathcal{L}_2''$. Or, finally, in each double subsequence $(y_{ij}) \prec (x_{mn})$

---

[1] In the sequel we always assume that the collection $\mathcal{L}''$ satisfies $(\mathcal{L}_0)$.

there is a cross-subsequence which $\mathcal{L}$-converges to x and another cross-subsequence no subsequence of which $\mathcal{L}$-converges to x. In this case $((x_{mn}),x) \in \mathcal{L}''_3$.

Proposition 5. Let $((x_{mn}),x) \in \mathcal{L}''_2 \cup \mathcal{L}''_3$. Then there is a one-to-one subsequence $(y_{ij}) \prec (x_{mn})$.

Proof. Let $((x_{mn}),x) \in \mathcal{L}''_2 \cup \mathcal{L}''_3$. First prove that there is a subsequence $(z_{ij}) \prec (x_{mn})$ each row-sequence $(z_{ij})_j$, $i \in N$, in which is one-to-one. Suppose that, on the contrary, no such subsequence (z_{ij}) exists. Then there is a subsequence $(t_{rs}) \prec (x_{mn})$ such that each row-sequence $(t_{rs})_s$ is constant, say $t_{rs} = a_r$, $s \in N$. Distinguish two cases: 1) $((x_{mn}),x) \in \mathcal{L}''_2$. Then, by (i2), there is r_0 such that $r \geq r_0$ implies that $a_r = x$. Hence $(t_{r1})_r$ is a cross-subsequence in (x_{mn}) which $\mathcal{L}$-converges to x, because $t_{r1} = x$, $r \geq r_0$. This is a contradiction. 2) $((x_{mn}),x) \in \mathcal{L}''_3$. From Definition 5 it follows that there is a subsequence $(a'_r) \prec (a_r)$ which $\mathcal{L}$-converges to x and, by the same definition, a subsequence $(a''_r)_r \prec (a'_r)_r$ no subsequence of which $\mathcal{L}$-converges to x. This contradicts the Fréchet axiom $(\mathcal{L}_2)$.

Now, let $(z_{ij}) \prec (x_{mn})$ have properties mentioned above. We construct, by mathematical induction, disjoint sets A_p, $p \in N$, and then a one-to-one subsequence $(y_{ij}) \prec (z_{ij})$. Let $A_1 = \{z_{1s_1^1}\}$ where $s_1^1 = 1$. Let $A_2 = \{z_{1s_2^1}, z_{2s_1^2}\}$ where $s_1^1 < s_2^1$, $s_1^2 \in N$, are such that the points $z_{1s_1^1}, z_{1s_2^1}, z_{2s_1^2}$ are distinct. Suppose we have just constructed sets $A_k = \{z_{is_j^i};\quad i+j = k+1\}$, $k \leq p$, such that

$$\begin{array}{l}
s_1^1 < s_2^1 < \dots < s_p^1 \\
s_1^2 < s_2^2 < \dots < s_{p-1}^2 \\
s_1^3 < s_2^3 < \dots < s_{p-2}^3 \\
\vdots \\
s_1^{p-1} < s_2^{p-1} \\
s_1^p \in N
\end{array}$$

where the natural numbers $s_j^i \in N$, $2 \leq i+j \leq p+1$ were chosen such that the points $z_{is_j^i}$, $2 \leq i+j \leq p+1$ are distinct. Since each row-sequence $(z_{ij})_j$ is one-to-one we can choose points $z_{1s_{p+1}^1}, z_{2s_p^2}, \dots, z_{p+1,s_1^{p+1}}$ such that $s_p^1 < s_{p+1}^1$, $s_{p-1}^2 < s_p^2$, ..., $s_1^p < s_2^p$, $s_1^{p+1} \in N$ such that the points $z_{is_j^i}$, $2 \leq i+j \leq p+2$ are distinct. We put $A_{p+1} = \bigcup z_{is_j^i}$, $i+j = p+2$. In such a way we have sets A_p,

$p \in N$, such that $\bigcup A_p = \bigcup z_{is^i_j}$, $2 \leq i + j$. Since $s^i_1 < s^i_2 < \ldots$ $\ldots < s^i_k < \ldots$ we have a subsequence $(z_{is^i_j})_j \prec (z_{ij})_j$, $i \in N$. Hence, we can put $(y_{ij}) = (z_{is^i_j})$.

Corollary 1. A point x is a ρ-point (σ-point) if and only if there is a one-to-one sequence (x_{mn}) such that $((x_{mn}),x) \in \mathcal{L}''_2$ ($((x_{mn}),x) \in \mathcal{L}''_3$).

Proof follows immediately from the definition of ρ-point (σ-point) and by Proposition 5.

Remark. In [3] a definition of a ρ-point was given by means of one--to-one double sequences of points and it was proved that a convergence commutative group is a topological space if and only if the zero-element is not a ρ-point. From Corollary 1 it follows that the definition in [3] is equivalent to the definition of a ρ-point given in this paper.

Definition 6. Let $(X, \mathcal{L})$ be an $\mathcal{L}$-space. Let $\mathcal{L}''$ be the corresponding double convergence for $(X, \mathcal{L})$. We say that a double sequence (x_{mn}) *strongly converges* to x or that x is a *strong limit* of (x_{mn}) and denote it $\mathcal{S}\,\mathrm{Lim}\, x_{mn} = x$ if the following implication is true:

If $(y_{ij}) \prec (x_{mn})$ then there is $(z_{rs}) \prec (y_{ij})$ such that $((z_{rs}),x) \in \mathcal{L}''_1$. The collection of all pairs $((x_{mn}),x)$ such that $\mathcal{S}\,\mathrm{Lim}\, x_{mn} = x$ is said to be a *strong double convergence* for $(X, \mathcal{L})$.

Observe that a strong double convergence for $(X, \mathcal{L})$ satisfies $(\mathcal{L}_0)$, $(\mathcal{L}_1)$, $(\mathcal{L}_2)$ and it is indeed a double convergence. Notice further that from Definitions 4 and 5 it instantly follows that if $\mathcal{S}\,\mathrm{Lim}\, x_{mn} = x$ then $\mathrm{Lim}\, x_{mn} = x$.

Proposition 6. Let $(X, \mathcal{L}, \lambda)$, $\mathcal{L} = \mathcal{L}^*$, be a T_2 convergence space. Let x be a point of X which has a countable base of neighborhoods at x . Then $\mathrm{Lim}\, x_{mn} = x$ if and only if $\mathcal{S}\,\mathrm{Lim}\, x_{mn} = x$.

Proof. If x is isolated then the assertion in the proposition is clearly true. Suppose that x is not isolated. Then there is a strictly decreasing base of neighborhoods $\{W_k\}$ of the point x . Let $\mathrm{Lim}\, x_{mn} = x$ and let (y_{ij}) be a subsequence of (x_{mn}) . From Theorem 4 it follows that there is $i_1 \in N$ and a subsequence $(z_{1s})_s \prec (y_{1j})_j$ such that $\bigcup z_{1s} \subset W_1$. We choose, by means of mathematical induction, subsequences $(z_{rs})_s \prec (y_{i_r j})_j$ such that $\bigcup_s z_{rs} \subset W_r$, $r \in N$, and such that $i_1 < i_2 < \ldots < i_r < \ldots$. Hence we have a double subsequence $(z_{rs}) \prec (y_{ij})$. We now prove that $((z_{rs}),x) \in \mathcal{L}''_1$. Let $(z_n)_n$ be a cross-sequence in (z_{rs}) . Let W be a neighborhood of x . Then there is $W_p \subset W$. Since $W_p \supset W_{p+1} \supset \ldots$ we have $z_n \in W_p$ and so $z_n \in W$ for $n \in N$ except for a finite number of n. It follows, since $\mathcal{L} = \mathcal{L}^*$, that $\mathcal{L}\text{-}\lim_n z_n = x$. Hence $\mathcal{S}\,\mathrm{Lim}\, x_{mn} = x$, by Definition 6.

From Proposition 6 we have the following

Corollary 2. Let $(X, \mathcal{L}, \lambda)$, $\mathcal{L} = \mathcal{L}^*$, be a first countable convergence space. Then the corresponding double convergence $\mathcal{L}''$ is a strong convergence.

3. Double convergence in convergence commutative groups

In this section we apply the concept of double convergence to commutative groups. We assume that all convergences satisfy ($\mathcal{L}_3$). First recall the definition of a convergence commutative group ([5]). It is an $\mathcal{L}$-space $(G, \mathcal{L})$ and a commutative group $(G, +)$ such that $\lim_m x_m = x$, $\lim_m y_m = y$ implies that $\lim (x_m - y_m) = a - b$.
(Notation: $(G, \mathcal{L}, +)$, abbreviation: cc group.)

Definition 7. Let $(G, +)$ be a commutative group, $(G, \mathcal{L})$ an $\mathcal{L}$-space and $\mathcal{L}''$ the corresponding double convergence for $(G, \mathcal{L})$. We say that $(G, \mathcal{L}'', +)$ is a *double convergence commutative group* (abbr. dcc group) if

$$\mathrm{Lim}\, x_{mn} = x\,,\quad \mathrm{Lim}\, y_{mn} = y \text{ implies that } \mathrm{Lim}\,(x_{mn} - y_{mn}) = x - y\,.$$

Proposition 7. Let $(G, +)$ be a commutative group and $(G, \mathcal{L})$ an $\mathcal{L}$-space. Let $(G, \mathcal{L}'', +)$ be a double convergence commutative group. Then $(G, \mathcal{L}, +)$ is a convergence commutative group.

Proof. Let $\lim a_m = a$, $\lim b_m = b$. Put $x_{mn} = a_m$, $y_{mn} = b_m$, $n \in N$. Then $\mathrm{Lim}\, x_{mn} = a$, $\mathrm{Lim}\, y_{mn} = b$ and, by the assumption, $\mathrm{Lim}\,(x_{mn} - y_{mn}) = a - b$. Denote $a_m - b_m = c_m$, $x_{mn} - y_{mn} = z_{mn}$. Then $z_{mn} = c_m$, $n \in N$. Let $(c_{m_i}) \prec (c_m)$. It suffices to prove that there is a subsequence of (c_{m_i}) which $\mathcal{L}$-converges to $a - b$. Since $\mathrm{Lim}\, z_{mn} = a - b$, then, either, by (i1), there is in the corresponding subsequence $(z_{m_i n}) \prec (z_{mn})$ a cross-subsequence which $\mathcal{L}$-converges to $a - b$. This cross-subsequence is a subsequence of (c_{m_i}) because $z_{m_i n} = c_{m_i}$, $n \in N$. Or there is, by (i2), i_0 such that $i \geq i_0$ implies that $\lim_n z_{m_i n} = a - b$, i.e. $z_{m_i} = a - b$, $i \geq i_0$. Hence $c_{m_i} = a - b$.

Proposition 8. Let $(G, \mathcal{L}, +)$ be a convergence commutative group. Assume that the zero element $0 \in G$ is not a ρ-point. Then $\mathcal{S}\,\mathrm{Lim}\, x_{mn} = a$, $\mathrm{Lim}\, y_{mn} = b$ implies that $\mathrm{Lim}\,(x_{mn} - y_{mn}) = a - b$.

Proof. Denote $z_{mn} = x_{mn} - y_{mn}$. Let $(z'_{ij}) \prec (z_{mn})$. It suffices to prove that there is a cross-subsequence in (z'_{ij}) which $\mathcal{L}$-converges to $a - b$. Denote $z'_{ij} = x'_{ij} - y'_{ij}$ where $(x'_{ij}) \prec (x_{mn})$, $(y'_{ij}) \prec (y_{mn})$. Since $\mathcal{S}$ Lim $x_{mn} = a$ there is a subsequence $(x''_{rs}) \prec (x'_{ij})$ each cross--sequence in which $\mathcal{L}$-converges to a. Denote (y''_{rs}) the corresponding subsequence of (y'_{ij}). Because b is not a ρ-point and Lim $y''_{rs} = b$ there is a cross-subsequence $y''_{r_i f(i)}$ in (y''_{rs}) such that $\lim_i y''_{r_i f(i)} = b$, by Definition 5. The corresponding sequence $(x''_{r_i f(i)})_i$ is a cross--subsequence in (x''_{rs}) and so $\lim_i x''_{r_i f(i)} = a$. Because $(G, \mathcal{L}, +)$ is a cc group we conclude that the sequence $(x''_{r_i f(i)} - y''_{r_i f(i)})_i$ $\mathcal{L}$-converges to $a - b$. This subsequence is a cross-subsequence in (z'_{ij}).

Theorem 6. Let $(G, \mathcal{L}, +)$ be a convergence commutative group. Let the double convergence $\mathcal{L}''$ for $(G, \mathcal{L})$ be a strong convergence. Then $(G, \mathcal{L}'', +)$ is a double convergence commutative group.

Proof. Because $\mathcal{L}''$ is a strong convergence the zero element of G is not a ρ-point. Hence Theorem 6 follows from Proposition 8.

Corollary 3. Let $(G, \mathcal{L}, +)$ be a convergence commutative group. Let $(G, \mathcal{L}, \lambda)$ be a first countable convergence space. Then $(G, \mathcal{L}'', +)$ is a double convergence commutative group.

Proof follows from Corollary 2 and Theorem 6.

The question arises whether every convergence commutative group is a double convergence commutative group. Under Continuum Hypothesis the answer is negative and the proof will appear in a forthcoming paper devoted to further study of double convergence spaces and groups.

References

[1] Frič, R., and V. Koutník: Sequential structures. Convergence Structures and Applications to Analysis, Abh. Akad. Wiss. DDR, Abt. Math.--Naturwiss - Technik, 1979, Nr. 4N. Akademie-Verlag; Berlin 1980, 37-56.

[2] Kratochvíl, P.: Sequential convergence generated by coneighborhoods and an example of D not Y space. Czechoslovak Math. J. (to appear).

[3] Novák, J., and L. Mišík: On L-spaces of continuous functions (Slovak). Matematicko-fyzikálny sborník 1 (1951), 1-17.

[4] Novák, J.: On convergence spaces and their sequential envelopes. Czechoslovak Math. J. 15 (1965), 74-100.

[5] Novák, J.: On convergence groups. Czechoslovak Math. J. 20 (1970), 357-384.

[6] Novák, J.: Concerning the topological product of two Fréchet spaces. General Topology and its Relations to Modern Analysis and Algebra, IV (Proc. Fourth Prague Topological Sympos., 1976). Part B Contributed Papers. Society of Czechoslovak Mathematicians and Physicists, Praha 1977, 342-343.

FUNCTIONAL ANALYSIS WITH K-CONVERGENCE

Endre Pap [1)]

1. Introduction

The members of the Katowice Branch of the Mathematical Institute of the Polish Academy of Sciences have developed the theory of K-convergence spaces. It turns out that the K-convergence spaces are an appropriate replacement for the complete spaces, even in the case of the pre-Hilbert spaces - [5]. The notion of K-convergence turns out to be quite useful in the study of various topics in functional analysis - - [2],[3],[4],[6],[7],[8]. This is particularly true for the uniform boundedness theorems - [2],[3]. In the paper [6] the space of linear continuous operators and in [8] the notion of K-boundedness were investigated.

In this paper we continue the study of the functional analysis with K-convergence.

In the first part we examine quotient spaces and we shall prove that the quotient is a K-space if the initial normed linear space is a K-space.

In the second part of the paper we shall prove the continuity of the adjoint operator of an arbitrary operator which is defined on a pre-Hilbert K-space. As a consequence we obtain the classical Closed Graph Theorem for Hilbert spaces - [9]. We remark that we have not used in the proof the Baire Category Theorem. Instead of it the main tool is an elementary diagonal type theorem.

First we define the notion of K-convergence. Let X be a vector topological space. We say that X is a K-space or that the topology induces in X a K-convergence if for each sequence (x_n) in X such that $x_n \to 0$ there exists a subsequence (y_n) of (x_n) such that $(\sum_{j=1}^{n} y_j)$ converges to an element y of X.

2. Quotient spaces

Let X be a vector topological space and V its vector subspace. Then we define the quotient of X by V as a vector space Y such that $Y = \{ x + V \mid x \in X \}$ with the operations $x + V + y + V = x + y + V$ and $t(x + V) = tx + V$ for t scalar and endowed with the finest vector topology which makes the function $q: X \to Y$, $q(x) = x + V$ $(x \in X)$

1) Institute of Mathematics,21 000 Novi Sad,I.Djuričića 4,Yugoslavia

continuous. q is said to be a quotient map. It can be proved that the quotient map exists for an arbitrary vector topological space.

It is well known that a quotient of a Banach space is again a Banach space. We shall prove that an analogous result for the K-property holds.

Theorem 1'. Let X be a linear seminormed space which is a K-space. Then its quotient is also a linear seminormed K-space.

Proof. Let Y be the quotient of X by its vector subspace V. Since X is a seminormed space with a seminorm $\| \|_X$, Y is also a seminormed space with a seminorm $\| \|_Y$ defined by

$$\| y\|_Y = \inf \{ \| x\|_X \mid y = q(x) \} .$$

where q is the quotient map.

Let (y_n) be a sequence in Y such that $y_n \to 0$. Then for each y_n there exists $x_n \in X$ such that $y_n = q(x_n)$ and

$$\| x_n\|_X < \| y_n\|_Y + \varepsilon_n$$

where $\varepsilon_n > 0$ and $\varepsilon_n \to 0$ for $n \to \infty$. Hence $x_n \to 0$. Then since X is a K-space, there exists a subsequence (z_n) of (x_n) such that $\sum_n z_n = z$ for some $z \in X$. Then by the continuity and linearity of the quotient map q the series $\sum_n y'_n$ converges to $q(\sum_n z_n)$, where $y'_n = q(z_n)$.

Remark 1. The preceding theorem is also true if X is a quasi--normed linear space.

Remark 2. In the general case, for an arbitrary vector topological space there are difficulties. For example, let X' be the space of all linear functionals on a Banach space X endowed with the topology $\sigma(X',X)$. Then the subspace X* of all continuous linear functionals on X is dense and sequentially closed. If $x \in X' \setminus X^*$ then $y_n = x + X^*$ tends to 0 in X'/X^* but for any choice of $x_n \in q^{-1}(y_n)$ the sequence (x_n) does not converge to 0 in X.

3. The continuity of the adjoint operator

Let X and Y be pre-Hilbert spaces. We denote by $(\mid)$ the scalar product. Let T be a map from X into Y. We define its adjoint map (operator) T* in the following way: the domain D(T*) of T* consists of those $y \in Y$ for which there exists $z \in X$ such that

$$(1) \qquad (y|T(x))_Y = (z|x)_X$$

for each $x \in X$. We then put $z = T^*(y)$.

The preceding definition is correct. Namely, T^* is a lineaar operator from $D(T^*)$ into X with the non-empty domain $D(T^*)$ since $0 \in D(T^*)$. Further, it for some $y \in Y$ there exists z such that (1) holds then this element is unique. Assume that for some $y \in D(T^*)$ there exist z_1 and z_2 such that (1) holds. Then we obtain

$$(z_1 - z_2|x) = 0 \quad (x \in X).$$

Putting $x = z_1 - z_2$ we obtain $z_1 - z_2 = 0$, i.e. $z_1 = z_2$.

If X is a Hilbert space then the preceding definition reduces to the usual definition of the adjoint operator. It is well known that if X is a Hilbert space and T a bounded linear operator on X, then T^* is also a bounded linear operator on Y. We can weaken the assumptions on X and T. Namely, we shall prove the following

Theorem 2. Let X and Y be pre-Hilbert spaces. Let $T: X \to Y$ be a map. If X is a K-space, then T^* is a linear bounded operator on $D(T^*)$.

Proof. If $D(T^*) = \{0\}$, then the theorem is trivially true. Suppose that $D(T^*) \neq \{0\}$. Let (y_n) be an arbitrary sequence from $D(T^*)$, (x_n) an arbitrary sequence in X such that $\|x_n\| \leq 1$ and (α_n) an arbitrary sequence of scalars such that $\alpha_n \to 0$. It is obvious that there exist two sequences (t_n) and (u_n) such that $\alpha_n = t_n u_n$, $t_n \geq 0$ and $t_n \to 0$ and $u_n \to 0$ for $n \to \infty$.

We define

$$x_{ij} = t_i \frac{|(u_j x_j | T^*(y_i))|}{\|y_i\|}$$

for $i \neq j$ $(i,j \in N)$ and $y_i \neq 0$, and $x_{ij} = 0$ for $i \neq j$ and $y_i = 0$. We put $x_{ii} = 0$ $(i \in N)$. Since $u_j x_j \to 0$ for $j \to \infty$ we obtain by the continuity of the scalar product $x_{ij} \to 0$ for $j \to \infty$ $(i \in N)$. We shall prove that $x_{ij} \to 0$ for $i \to \infty$ $(j \in N)$. Namely, we have

$$x_{ij} = t_i \frac{|(u_j x_j | T^*(y_i))|}{\|y_i\|} = t_i \frac{|(T(u_j x_j) | y_i)|}{\|y_i\|} \leq$$

$$\leq t_i \, \|T(u_j x_j)\| .$$

If we let in the preceding inequality $i \to \infty$ we obtain $x_{ij} \to 0$ for each fixed $j \in N$. Then since the conditions of P.Antosik's diagonal theorem [1] for the matrix $[x_{ij}]$ $(i,j \in N)$ are satisfied, we obtain an increasing sequence (p_n) of natural numbers such that

$$(2) \qquad \lim_{i\to\infty} \sum_{j=1}^{\infty} x_{p_i p_j} = 0 .$$

Since $u_j x_j \to 0$ for $j \to \infty$ and X is a K-space there exists a subsequence (s_j) of (p_j) and an element $x \in X$ such that $\sum_{j=1}^{\infty} u_{s_j} x_{s_j} = x$.

We have for each $p \in N$ and each $y_{s_i} \neq 0$

$$t_{s_i} \frac{|(u_{s_i} x_{s_i} | T^*(y_{s_i}))|}{\| y_{s_i} \|} \leq \sum_{\substack{j=1 \\ j\neq i}}^{i+p} t_{s_i} \frac{|(u_{s_j} x_{s_j} | T^*(y_{s_i}))|}{\| y_{s_i} \|} +$$

$$+ \; t_{s_i} \frac{|(\sum_{j=1}^{i+p} u_{s_j} x_{s_j} | T^*(y_{s_i}))|}{\| y_{s_i} \|} \qquad (i \in N).$$

Let $p \to \infty$. Then we obtain from the preceding inequality by the continuity of the scalar product

$$t_{s_i} \frac{|(u_{s_i} x_{s_i} | T^*(y_{s_i}))|}{\| y_{s_i} \|} \leq \sum_{j=1}^{\infty} x_{s_i s_j} + t_{s_i} \frac{|(x | T^*(y_{s_i}))|}{\| y_{s_i} \|} \leq$$

$$\leq \sum_{j=1}^{\infty} x_{s_i s_j} + t_{s_i} \| T(x) \| \qquad (i \in N).$$

If we let in the preceding inequality $i \to \infty$, we obtain by (2)

$$\alpha_{s_i} \frac{(x_{s_i} | T^*(y_{s_i}))}{\| y_{s_i} \|} \to 0.$$

Now using the Urysohn property of scalars: if each subsequence (z_n)

of a sequence of scalars (r_n) has a subsequence (v_n) such that $v_n \to 0$, then $z_n \to 0$, we conclude that

$$\alpha_n \frac{(x_n | T^*(y_n))}{\| y_n \|} \to 0 \quad \text{for } n \to \infty .$$

Since all the sequences (α_n), (x_n) and (y_n) were arbitrary there exists $M > 0$ such that

$$\sup_{\| x \| \leq 1} \frac{| (x | T^*(y)) |}{\| y \|} < M \qquad (y \in D(T^*),\ y \neq 0).$$

Then we can write for an arbitrary element $x' \in X$ and $y \neq 0$

$$\frac{| (x' | T^*(y)) |}{\| y \|} = \frac{\left| \left(\frac{x'}{\| x' \|} \,\middle|\, T^*(y) \right) \right|}{\| y \|} \| x' \| \leq M \| x' \| .$$

If we put $x' = T^*(y)$ we obtain

$$\frac{\| T^*(y) \|}{\| y \|} \leq M, \text{ i.e.}$$

$$\| T^*(y) \| \leq M \| y \| \qquad (y \in D(T^*)).$$

Corollary 1. If the same conditions as in Theorem 2 hold and T is linear and $D(T^*) = Y$ is a Hilbert space, then T is a bounded operator on X.

Proof. By Theorem 2 T^* is bounded on Y. Then $T^{**} = (T^*)^*$ is bounded on X and since it is an extension of the operator T we obtain $T = T^{**}$, i.e. T is bounded on X.

Corollary 2 (Classical Closed Graph Theorem). Let X and Y be Hilbert spaces. If $T: X \to Y$ is a linear closed operator, then T is bounded.

Proof. By Corollary 1 it is enough to prove $D(T^*) = Y$. Since T is closed operator, $D(T^*)$ is dense in Y. Hence, since X is a complete space T^* has a bounded extension to all Y.

References

[1] P.Antosik,A diagonal theorem for nonnegative matrices and equicontinuous sequences of mappings,Bull.Acad.Polon.Sci.Ser.Math.Astronom. Phys. 24 (1976),855-860.

[2] P.Antosik,On uniform boundedness of families of mappings,Proc. of the Conf.on Conv.,Szczyrk (1979),1-16.

[3] P.Antosik, A uniform convergence principle with applications, Dis.Mat. (to appear) .

[4] J.Burzyk,C.Kliś,Z.Lipecki,On metrizable Abelian Groups with a completeness-type property,Colloq.Math. (to appear).

[5] Cz.Kliś,An example of non-complete normed (K)-space,Bull.Acad. Polon.Sci.Ser.Math.Astronom.Phys. 26 (1978),415-420.

[6] E.Pap,Contributions to functional analysis on convergence spaces, Proc.of the Fifth Prague Topological Sympos. 1981, (1983),538-543.

[7] E.Pap,Funkcionalna analiza,Institute of Mathematics,Novi Sad,1982.

[8] E.Pap,C.Swartz, On summable boundedness in the convergence linear spaces (to appear) .

[9] K.Yosida,Functional Analysis,Springer-Verlag,1965.

SOME REMARKS ON d-CONVERGENCE

Caroll V. Riecke 1)

The concepts of d-compactness and ultracompactness for topological spaces were introduced by Bernstein [1] and, in a slightly different form, by Frolik [2]. d-compactness provided the essential key whereby Ginsburg and Saks [3] were able to characterize the class of Hausdorff topological spaces all of whose powers are countably compact. Both concepts may be naturally extended to convergence spaces and results obtained which are similar to those of Ginsburg and Saks. Several modifications of convergence spaces are described using the concept of d-bounded or ultrabounded sets.

Kent convergence structures will be the setting of the subsequent remarks. For d a free ultrafilter on N($d \in \beta^*N$), a d-filter on a non-empty set X is an ultrafilter on X which is a sequential image of d. An ultrafilter which contains a countable set will be called an $\aleph_0$ - ultrafilter.

1. d-compact convergence spaces

Definition 1.1. A convergence space (X, q) is d-compact for some $d \in \beta^*N$ if every d-filter on X is q-convergent and ultracompact if it is d-compact for every $d \in \beta^*N$.

Ultracompact and sequentially compact spaces are distinct classes, shown by an example in [1]. Ultracompactness for convergence spaces is equivalent to every countable subset being bounded. Hence from [5], in K-regular convergence spaces, ultracompactness is equivalent to every countable subset having bounded closure. This result contrasts with the equivalence of ultracompactness and $\aleph_0$-boundedness for regular Hausdorff topological spaces [7].

Definition 1.2. For $d \in \beta^*N$ and A a subset of a convergence space (X, q), $x \in X$ is a d-limit (ultralimit) of A if x is a limit of some d-filter ($\aleph_0$-ultrafilter) containing A. The d-closure (ultraclosure) of A is the union of A and its d-limits (ultralimits). A is d-closed (ultraclosed) if A contains all of its d-limits and d-open (ultra-open) if X - A is d-closed.

In [7], conditions were established that are equivalent to d-closure being a closure operator. Although d-closure does not enjoy idempotentcy, it defines a convergence structure finer than the original structure. For $d \in \beta^*N$, the d-modification (ultramodification) of a convergence structure is the structure defined by: $\mathfrak{F}$ dq (μq) - converges to x if $\mathfrak{F}$ q converges to x and every ultrafilter finer than $\mathfrak{F}$ is a d-filter ($\aleph_0$ - ultrafilter). This modification defines an idempotent lattice homomorphism on the lattice of convergence structures.

1) Department of Mathematics and Astronomy, Franklin and Marshall College Lancaster, PA 17604 P.O. Box #3003 USA

Proposition 1.3. For a topological space (X,q), the following are equivalent:

i) (X,q) is d-compact (ultracompact).

ii) Every countable d-open (ultraopen) cover contains a finite subcover.

iii) Every countable family of d-closed (ultraclosed) sets having the finite intersection property has nonempty intersection.

If $\rho q(\lambda q, \omega q)$ represents the pseudotopological (pretopological, topological) modification of a convergence structure q, one has:

Proposition 1.4. The d-modification (ultramodification) operator commutes with the pseudotopological operator and with the pretopological (topological) operator if $d\lambda q(d\omega q)$ is a pretopology (topology). In particular, $\omega(dq)[\omega(\mu q)]$ is the topology generated by the d-open (ultraopen) sets of q.

Proposition 1.5. d-compactness (ultracompactness) is closed hereditary and productive.

If a convergence space is countably compact when every infinite set is contained in a free, convergent ultrafilter, a primary result of [3] can be extended to convergence spaces.

Proposition 1.6. A convergence space (X,q) is d-compact for some $d \in \beta^{*}N$ if and only if every power of (X,q) is countably compact.

A mapping between convergence spaces is d-continuous (ultracontinuous) if it preserves d-limits (ultralimits). d-continuity (ultracontinuity) is equivalent to continuity if the map is considered relative to the d-modification (ultramodification).

Proposition 1.7. For any T_2-convergence space (X,q), there exists a d-compact convergence space X^d containing X as a dense subspace. If f is a continuous mapping of X into any regular, d-compact space (Y, r), then f has a continuous extension to X^d

Proof. For a T_2 convergence space (X,q), let $(\hat{X},\hat{q})$ be the Richardson compactification. In $\hat{X}$, construct a transfinite sequence (X_α, q_α) for $\alpha < \omega_1$ of subspaces containing X as follows:

Let $X_0 = X$ and assume that, for $\alpha < \beta$, X_α has been constructed with

i) $\alpha \leq \gamma < \beta$ implies $X_\alpha \subseteq X_\gamma \subseteq \hat{X}$

ii) $\alpha < \gamma < \beta$ implies X_γ contains the d-closure of X_α.

Then X_β is the d-closure of $\cup X_\alpha (\alpha < \beta)$. A d-compactification of X is $\cup X_\alpha (\alpha < \omega_1)$. Each continuous map on X to a d-compact space (Y,r) can be readily shown to have a unique continuous extension to X^d.

An ultracompactification may be constructed similarly.

A convergence space is essentially $\aleph_0$-bounded if every countable subset is contained in at most finitely many non-convergent ultrafilters.

Proposition 1.8. If (X,q) is a T_2, essentially $\aleph_0$-bounded space, the ultraclosure of X in the compactification $\hat{X}$ is ultracompact and the sequence of Proposition 1.7 has length ≤ 1.

2. Modifications by d-bounded sets

The relationships between boundedness and compactifications have been thoroughly investigated by many authors. The locally bounded modification of a convergence space was introduced in [5], while in [6] a modification was constructed which generalizes the Kuratowski localization for topological spaces. Both of these modifications may be extended to modifications by d-bounded or ultrabounded sets.

Definition 2.1. A subset A of a convergence space (X,q) is:

i) d-bounded (ultrabounded) if every d-filter ($\aleph_0^-$-filter) containing A is q-convergent.

ii) d-unbounded at x (ultra-unbounded at x) if no free d-filter (free $\aleph_0$ - ultrafilter) containing A q - converges to x.

Definition 2.2. A convergence space (X,q) is locally d-bounded (locally ultrabounded) if every convergent ultrafilter contains a d-bounded (ultrabounded) set.

The d-bounded (ultrabounded) subsets of a convergence space (X,q) form an ideal on X and the family of locally d-bounded (locally ultrabounded) convergence structures form a meet-complete subset of the lattice of convergence structures. Utilizing the d-boundedness concept, one can describe several modifications.

1. The locally d-bounded (locally ultrabounded) modification $q_d(q_\mu)$: $\mathcal{F}$ $q_d(q_\mu)$ - converges to x if $\mathcal{F}$ q-converges to x and contains a d-bounded (ultrabounded) set.

2. The d-unbounded (ultra-unbounded) modification $q_d{}^*(q_\mu{}^*)$.

$\mathcal{F}$ $q_d{}^*(q_\mu{}^*)$ - converges to x if $\mathcal{F} = \dot{x}$ or $\mathcal{F}$ contains a set which is d-unbounded (ultra-unbounded) at x with respect to q.

3. The d-localization (ultralocalization): $\mathcal{F}$ $q_d^K(q_\mu{}^K)$ - converges to x if $\mathcal{F}$ q converges to x and every ultrafilter $\mathcal{K}$ finer than $\mathcal{F}$ with $\mathcal{K} \neq \dot{x}$ contains no d-bounded (ultrabounded) set.

Each of the modifications describe lattice operators on C(X) similar to those of [5] and [6]. The structures defined by the modifications are related by:

Proposition 2.3. In the lattice C(X) of convergence structures:

i) $q_d(q_\mu)$ is the coarsest locally d-bounded (locally ultrabounded) convergence structure finer than q.

ii) The join of d_q and $q_d{}^*$ (μq and q_μ^*) is the discrete topology.

iii) For q a pseudotopology, $q_d^K(q_\mu^K)$ is the relative pseudocomplement of $q_d(q_\mu)$ with respect to q in the lattice of pseudotopologies $P_S(X)$.

Many results in the lattice of convergence structures related to modifications by bounded sets and compactification theory may be extended to d-bounded or ultrabounded sets. Also, lattice properties of modification operators defined by bounded sets may be extended to d-bounded or ultrabounded modifications.

REFERENCES

1. A. R. Bernstein, A new kind of compactness for topological spaces Fund. Math. 66(1970), 185-193.
2. Z. Frolik, Sums of ultrafilters, Bull. Amer. Math. Soc. 73(1967), 87-91.
3. J. Ginsburg and V. Saks, Some applications of ultrafilters in topology, Pac. Jrnl. Math. 57(1975),401-417.
4. G. Richardson, A Stone-Cech compactification for limit spaces, Proc. Amer. Math. Soc. 25(1970), 403-404.
5. C. Riecke, On boundedness for convergence spaces, Bolletino UMI 15-B (1978), 49-59.
6. __________, Locally - P convergence structures, Glasnik Math. 16(1981),327-332.
7. V. Saks, Ultrafilter invariants in topological spaces, Trans. Amer. Math. Soc. 241(1978), 79-97.

Band 1

Inverse and Improperly Posed Problems in Differential Equations

Proceedings of the Conference on Mathematical and Numerical Methods, held in Halle/Saale (GDR) from May 29 to June 2, 1979

Edited by Gottfried Anger
296 Seiten, 1979 (vergriffen)

Band 2

Fundamentals of Computation Theory FCT '79

Proceedings of the Conference on Algebraic, Arithmetic and Categorial Methods in Computation Theory, held in Berlin/Wendisch-Rietz (GDR) September 17—21, 1979

Edited by Lothar Budach
576 Seiten, 1979

Band 3

Decidability and Generalized Quantifiers

By Andreas Baudisch, Detlef Seese, Hans-Peter Tuschik, and Martin Weese
XII, 235 Seiten, 1980

Band 4

Zur Elastizitäts- und Thermoelastizitätstheorie

Potentialtheoretische Methoden in Anwendung auf verschiedene Modelle

Von Lothar Jentsch und Johannes Maul
157 Seiten, 1980

Band 5

Systems Analysis and Simulation 1980

Proceedings of the International Symposium held at Berlin (GDR) September 1—5, 1980

Edited by Achim Sydow
496 Seiten, 1980 (vergriffen)

Band 6

Nichtlineare diskrete Optimierung

Von Eberhard Girlich und Michael M. Kowaljow
218 Seiten, 1981

Band 7

Completeness Problems in the Structural Theory of Automata

By Jürgen Dassow
148 Seiten, 1981

Band 8

Darstellung von Lösungen linearer elliptischer Differentialgleichungen

Von Günther Wildenhain
92 Seiten, 1981

Band 9

Introduction to the Spectral Theory of Operators in Spaces with an Indefinite Metric

By I. S. Iohvidov, M. G. Krein, and H. Langer
120 Seiten, 1982 (vergriffen)

Band 10

Nonlinear Time-discrete Systems

A General Approach by Nonlinear Superposition

By Michael Gössel
115 Seiten, 1982 (vergriffen)

Band 11

Preprimal Algebras

By Klaus Denecke
162 Seiten, 1982

Band 12

Ergodic Theory and Related Topics

Proceedings of the Conference held in Vitte/Hiddensee (GDR) October 19—23, 1981

Edited by Horst Michel
232 Seiten, 1982

Band 13

Singularly Perturbed Differential Equations

By H. Goering, A. Felgenhauer, G. Lube, H.-G. Roos, and L. Tobiska
176 Seiten, 1983

Band 14

Leben wir in einer Volterra-Welt?

Ein ökologischer Zugang zur angewandten Systemanalyse

Von Manfed Peschel und Werner Mende
233 Seiten, 1983

Band 15

Markov Decision Problems with Countable State Spaces Optimality Criteria — Algorithms — Clustering

By Hans Michael Dietz and Volker Nollau
174 Seiten, 1983

Band 16

Stochastische Verfahren zur Suche nach dem Minimum einer Funktion

Von Ryszard Zieliński und Peter Neumann
133 Seiten, 1983

Band 17

Invertibility and Asymptotics of Toeplitz Matrices

By Albrecht Böttcher and Bernd Silbermann
200 Seiten, 1983

Band 18

Structural Induction on Partial Algebras

Introduction to Theory and Application of Partial Algebras — Part II

By Horst Reichel

205 pages, 1984

Band 19

Algebraic Methods for Toeplitz-like Matrices and Operators

By Georg Heinig and Karla Rost

212 pages, 1984

Band 20

Frege Conference 1984

Proceedings of the International Conference held at Schwerin (GDR), September 10—14, 1984

Edited by Gerd Wechsung

408 pages, 1984

Band 21

Numerical Analysis of Selected Semilinear Differential Equations

Edited by Christian Großmann, Thomas Riedrich, and Hartmut Schönheinz

203 pages, 1984

Band 22

Singular Ordinary Differential Operators and Pseudodifferential Equations

By Johannes Elschner

200 pages, 1985

Band 23

Lotka-Volterra-Approach to Cooperation and Competition in Dynamic Systems

Proceedings of the 5th Meeting of UNESCO's Working Group on Systems Theory held on the Wartburg, Eisenach (GDR) at March 5—9, 1984

Edited by Werner Ebeling and Manfred Peschel

271 pages, 1985

Band 24

Convergence Structures 1984

Proceedings of the International Conference on Convergence Structures and Applications held in Bechyné (Czechoslovakia), September 17—21, 1984

Edited by Josef Novák, Werner Gähler, Horst Herrlich, and Jan Mikusiński

250 pages, 1985

Band 25

Parcella '84

Proceedings of the International Workshop on Parallel Processing by Cellular Automata and Arrays held in Berlin (GDR), September 25—27, 1984

Edited by Wolfgang Händler, Tamás Legendi, and Gottfried Wolf

239 pages, 1985

Band 26

Multiobjective and Stochastic Optimization Based on Parametric Optimization

By Jürgen Guddat, Francisco Guerra Vasquez, Klaus Tammer, and Klaus Wendler

171 pages, 1985

Band 27

Systems Analysis and Simulation 1985 I: Theory and Foundations

Proceedings of the 2nd International Symposium on Systems Analysis and Simulation held in Berlin (GDR), August 26—31, 1985

Edited by Achim Sydow, Manfred Thoma, and Robert Vichnevetsky

496 pages, 1985

Band 28

Systems Analysis and Simulation 1985 II: Applications

Proceedings of the 2nd International Symposium on Systems Analysis and Simulation held in Berlin (GDR), August 26—31, 1985

Edited by Achim Sydow, Manfred Thoma, and Robert Vichnevetsky

496 pages, 1985

Band 29

Pseudo-differential Operators and Mellin Operators in Spaces of Functions with Conormal Singularity

By Stephan Rempel and Bert-Wolfgang Schulze

about 480 pages, 1986

This volume contains the lectures and some additional papers contributed to the Conference on Convergence held at Bechyně (Czechoslovakia) in Septemer 1984. The papers presented in this work represent new research results from leading specialists in the field and provide a survey of the state of art in the field of convergence.
The scope is broad and topics discussed include both filter and sequential convergence spaces and groups, completions of convergence structures, topological and uniform spaces determined by sequences such as Fréchet and sequential spaces, categorical aspects of convergence as well as applications of convergence theory to analysis, functional analysis and numerical analysis.

Dieser Band enthält die Vorträge der im September 1984 in Bechyně (ČSSR) durchgeführten Konferenz über Konvergenz sowie einige zusätzliche Beiträge. Die im Band vorgestellten Arbeiten bringen neue Forschungsergebnisse führender Fachvertreter, und sie geben einen Überblick über den Entwicklungsstand des Gebietes der Konvergenzstrukturen.
Der Themenkreis ist breit, und die behandelten Gebiete schließen die Filterkonvergenz- und Folgenkonvergenzräume und -gruppen ein, ferner Vervollständigungen von Konvergenzstrukturen, topologische und uniforme durch Folgen bestimmte Räume wie Fréchet- und sequentielle Räume, kategorielle Aspekte der Konvergenz sowie Anwendungen der Theorie der Konvergenzstrukturen in der Analysis, Funktionalanalysis und numerischen Analysis.

ISSN 0138-3019

Bei Fragen zur Produktsicherheit wenden Sie sich bitte an:
If you have any questions regarding product safety,
please contact:

Walter de Gruyter GmbH
Genthiner Straße 13
10785 Berlin
productsafety@degruyterbrill.com